Oxidative Stress Induced by Air Pollution

Oxidative Stress Induced by Air Pollution

Guest Editor

Yasuhiro Yoshida

MDPI

Basel • Beijing • Wuhan • Barcelona • Belgrade • Novi Sad • Cluj • Manchester

Guest Editor
Yasuhiro Yoshida
Department of Immunology
and Parasitology
University of Occupational
and Environmental Health
Kitakyushu
Japan

Editorial Office
MDPI AG
Grosspeteranlage 5
4052 Basel, Switzerland

This is a reprint of the Special Issue, published open access by the journal *Antioxidants* (ISSN 2076-3921), freely accessible at: www.mdpi.com/journal/antioxidants/special_issues/air_pollution_oxidative_stress.

For citation purposes, cite each article independently as indicated on the article page online and using the guide below:

Lastname, A.A.; Lastname, B.B. Article Title. *Journal Name* **Year**, *Volume Number*, Page Range.

ISBN 978-3-7258-2828-9 (Hbk)
ISBN 978-3-7258-2827-2 (PDF)
https://doi.org/10.3390/books978-3-7258-2827-2

Contents

antioxidants

Editorial

Oxidative Stress Induced by Air Pollution

Yasuhiro Yoshida

Department of Immunology and Parasitology, School of Medicine, University of Occupational and Environmental Health, 1-1 Iseigaoka, Yahatanishi-ku, Kitakyushu 807-8555, Japan; freude@med.uoeh-u.ac.jp

1. Introduction

In 2021, the World Health Organization issued new guidelines on particulate matter (PM), including PM2.5, and highlighted associated risks (WHO global air quality guidelines). Among the various types of particles, fine-particle substances (PM2.5), as well as industrial/pharmaceutical nanoparticles, differ in terms of properties and pose serious health risks. Epidemiological studies indicate that exposure to particulate matter is associated with various adverse health effects [1]. PM2.5 exposure is linked to cardiovascular diseases, respiratory disorders, and lung cancer, especially in large cities. This exposure induces inflammation and produces reactive oxygen species (ROS), leading to oxidative stress, which affects various cells and can cause cell death. Both acute and chronic inflammation contribute to these biological effects.

Many studies have explored the relationship between environmental particulates and health effects through cell and animal experiments, but many questions remain unanswered. This Special Issue will delve into the oxidative stress caused by these pollutants and cover topics related to urgent issues in this field.

2. Overview of Published Articles

PM2.5 sources include anthropogenic sources and natural sources such as volcanoes. Anthropogenic sources include facilities that generate soot, such as incinerators; facilities that generate dust; and automobiles. Exposure to PM2.5, especially in urban areas, is of great concern in terms of health effects. Sanchez-Rodriguez, L. et al. studied the relationship between traffic-related air pollution (TRAP) and oxidative stress (Contribution 1). This study revealed the effect of traffic density on plasma metabolites and urinary oxidative stress biomarkers, indicating that oxidative stress could possibly play a role as an intermediate factor in metabolic changes. PM2.5 exposure is not limited to the living environment, and exposure in the workplace is a topic that has also been attracting attention. Bellisario, V. et al. studied occupational exposure to metal nanomaterials and suggested a possible relationship between chemical composition and oxidative stress biomarkers (Contribution 2). They showed that occupational exposure to nanomaterials (NM) in particular was associated with urinary silica (Si) and titanium (Ti) concentrations and increased oxidative stress.

The skin and eyes may act as the first defense mechanism against PM2.5. Two reports on substances that could have a protective effect against skin damage are included in this Special Issue. Liyanage, N.M. et al. demonstrated that the cryonasterol-rich hexane fraction (CRHF2) from *Caulerpa racemosa* has a protective effect at the cell level and in zebrafish (Contribution 3). CRHF2 reduced intracellular and mitochondrial ROS levels, inhibited PM-induced apoptosis, downregulated the expression of apoptosis signaling pathway proteins, and reduced the accumulation of sub-G1 cells in a dose-dependent manner. Zhen, A.X. et al. also showed that the marine algae-derived compound 3-bromo-4,5-dihydroxybenzaldehyde (3-BDB) suppresses skin damage (Contribution 4). They demonstrated that 3-BDB suppresses ROS generation, mitochondrial dysfunction, DNA damage, and cellular senescence in vitro and in vivo. Both of these compounds possess the ability

Citation: Yoshida, Y. Oxidative Stress Induced by Air Pollution. *Antioxidants* **2024**, *13*, 1393. https://doi.org/10.3390/antiox13111393

Received: 11 November 2024
Accepted: 12 November 2024
Published: 15 November 2024

to suppress ROS produced by mitochondria, a discovery that raises hopes for drug development. On the other hand, there is a strong demand for in vitro research methods to develop such drugs and inhibitors, which is a global trend in line with the 3R principle (refinement, reduction, and replacement) [2]. Zeng, Z. et al. demonstrated the construction of an in vitro system to investigate the mechanism of inflammation caused by particles in the eyes. This study demonstrates that PM induces inflammation in a co-culture system of corneal epithelial cells and neutrophils and that ROS and NF-κB are involved in the mechanism (Contribution 5). The authors of this study therefore propose that NF-κB is a candidate molecular target for ROS-dependent inflammation inhibitors. From this antioxidant perspective, the search for therapeutic drugs that suppress the inflammation, damage, and diseases caused by PM may provide a guideline for the future.

Once particles enter the body, their impact on the respiratory system, especially the lungs, is of concern. Hammond, J. et al. investigated the cytotoxicity of PM2.5 collected from three different cities (Lancaster and Birmingham in the UK and Mexico City, Mexico) using lung epithelial cells (Calu-3) (Contribution 6). Samples from all cities induced the generation of ROS and the production of inflammatory cytokines. It is interesting to note that the authors pointed out that mass-based PM (particulate matter) restrictions do not fully reflect the composition and health effects of PM specific to each region. Therefore, it is important to develop and introduce new biologically relevant indicators and regionally appropriate regulations. In addition, Kim, T.Y. et al. demonstrated in experiments using mice that the aqueous extract of *Codium fragile* could potentially be a functional food or pharmaceutical ingredient (Contribution 7). *Codium fragile* suppresses pulmonary mitochondrial dysfunction by regulating ROS content and mitochondrial membrane potential levels. They pointed out the involvement of the TLR/TGF-β pathway as a mechanism. It is interesting to note that this is consistent with the involvement of NF-κB pointed out by Zeng et al. (Contribution 5).

Another important question is "Does PM affect systems other than the respiratory and cardiovascular systems?" From this perspective, the study by Park, J. et al., focusing on the effects of PM on skeletal muscle and exercise, is intriguing (Contribution 8). In a mouse study, these authors demonstrated that PM exacerbated oxidative stress and inflammation in skeletal muscle and mitochondria both at rest and during exercise, and a significant increase in the level of in vivo mitophagy was observed in the PM group.

The effects of PM on our working generation are a matter of great concern, as are the effects that the next generation could face. In a retrospective cohort study, Park, S. et al. demonstrate that indoor PM2.5 concentrations in their low birth weight (LBW) group were significantly higher than those in the normal birth weight (NBW) group (Contribution 9). In a prospective study, 8-hydroxy-2-deoxyguanosine was significantly higher in the high PM2.5 group, and these results suggest a relationship between PM2.5, oxidative stress, and fetal development.

The effects of these particles may be linked to our everyday lifestyle. Ossoli, A. et al. present a study evaluating the effect of short-term exposure to PM on high-density lipoprotein (HDL) function and the modifying effect of body mass index (BMI) (Contribution 10). A positive association between PM10 exposure and NO production was observed in participants with normal BMI but not in participants with high BMI. This suggests that an increase in BMI may cause the loss of HDL function's compensatory response to PM exposure and that PM-induced endothelial dysfunction may have a greater impact on obese individuals, raising a warning about the increased environmental risks to obese individuals.

3. Conclusions

This Special Issue highlights the diverse health risks associated with PM exposure, showing that PM2.5 and related pollutants contribute to oxidative stress and inflammation across multiple biological systems. The studies included within reveal PM's impact on metabolic and occupational health, respiratory function, and fetal development, with notable effects in urban and workplace settings. Promising protective strategies, including the use of marine algae-derived compounds, have shown potential in reducing PM-induced cellular damage.

In this Special Issue, the need for targeted regulatory measures and biologically relevant health indicators is emphasized, as is the importance of mitigating environmental risks for vulnerable groups, including those with higher BMI. Together, these findings underscore the importance of reducing PM exposure to protect public health.

Funding: This research received no external funding.

Conflicts of Interest: The author declares no conflicts of interest.

List of Contributions:

1. Sanchez-Rodriguez, L.; Galvez-Fernandez, M.; Rojas-Benedicto, A.; Domingo-Relloso, A.; Amigo, N.; Redon, J.; Monleon, D.; Saez, G.; Tellez-Plaza, M.; Martin-Escudero, J.C.; et al. Traffic Density Exposure, Oxidative Stress Biomarkers and Plasma Metabolomics in a Population-Based Sample: The Hortega Study. *Antioxidants* **2023**, *12*, 2122. https://doi.org/10.3390/antiox12122122.
2. Bellisario, V.; Garzaro, G.; Squillacioti, G.; Panizzolo, M.; Ghelli, F.; Mariella, G.; Bono, R.; Guseva Canu, I.; Bergamaschi, E. Occupational Exposure to Metal-Based Nanomaterials: A Possible Relationship between Chemical Composition and Oxidative Stress Biomarkers. *Antioxidants* **2024**, *13*, 676. https://doi.org/10.3390/antiox13060676.
3. Liyanage, N.M.; Nagahawatta, D.P.; Jayawardena, T.U.; Jayawardhana, H.; Lee, H.G.; Kim, Y.S.; Jeon, Y.J. Clionasterol-Rich Fraction of Caulerpa racemosa against Particulate Matter-Induced Skin Damage via Inhibition of Oxidative Stress and Apoptosis-Related Signaling Pathway. *Antioxidants* **2022**, *11*, 1941. https://doi.org/10.3390/antiox11101941.
4. Zhen, A.X.; Piao, M.J.; Kang, K.A.; Fernando, P.D.; Herath, H.M.; Cho, S.J.; Hyun, J.W. 3-Bromo-4,5-dihydroxybenzaldehyde Protects Keratinocytes from Particulate Matter 2.5-Induced Damages. *Antioxidants* **2023**, *12*, 1307. https://doi.org/10.3390/antiox12061307.
5. Zeng, Z.; Yoshida, Y.; Wang, D.; Fujii, Y.; Shen, M.; Mimura, T.; Tanaka, Y. Inflammatory Cytokines and Chemokines Are Synergistically Induced in a ROS-Dependent Manner by a Co-Culture of Corneal Epithelial Cells and Neutrophil-like Cells in the Presence of Particulate Matter. *Antioxidants* **2024**, *13*, 467. https://doi.org/10.3390/antiox13040467.
6. Hammond, J.; Maher, B.A.; Gonet, T.; Bautista, F.; Allsop, D. Oxidative Stress, Cytotoxic and Inflammatory Effects of Urban Ultrafine Road-Deposited Dust from the UK and Mexico in Human Epithelial Lung (Calu-3) Cells. *Antioxidants* **2022**, *11*, 1814. https://doi.org/10.3390/antiox11091814.
7. Kim, T.Y.; Kim, J.M.; Lee, H.L.; Go, M.J.; Joo, S.G.; Kim, J.H.; Lee, H.S.; Jeong, W.M.; Lee, D.Y.; Kim, H.J.; et al. Codium fragile Suppressed Chronic PM(2.5)-Exposed Pulmonary Dysfunction via TLR/TGF-beta Pathway in BALB/c Mice. *Antioxidants* **2023**, *12*, 1743. https://doi.org/10.3390/antiox12091743.
8. Park, J.; Jang, J.; So, B.; Lee, K.; Yeom, D.; Zhang, Z.; Shin, W.S.; Kang, C. Effects of Particulate Matter Inhalation during Exercise on Oxidative Stress and Mitochondrial Function in Mouse Skeletal Muscle. *Antioxidants* **2024**, *13*, 113. https://doi.org/10.3390/antiox13010113.
9. Park, S.; Kwon, E.; Lee, G.; You, Y.A.; Kim, S.M.; Hur, Y.M.; Jung, S.; Jee, Y.; Park, M.H.; Na, S.H.; et al. Effect of Particulate Matter 2.5 on Fetal Growth in Male and Preterm Infants through Oxidative Stress. *Antioxidants* **2023**, *12*, 1916. https://doi.org/10.3390/antiox12111916.
10. Ossoli, A.; Favero, C.; Vigna, L.; Pesatori, A.C.; Bollati, V.; Gomaraschi, M. Body Mass Index Modulates the Impact of Short-Term Exposure to Air Particulate Matter on High-Density Lipoprotein Function. *Antioxidants* **2022**, *11*, 1938. https://doi.org/10.3390/antiox11101938.

References

1. Atkinson, R.W.; Kang, S.; Anderson, H.R.; Mills, I.C.; Walton, H.A. Epidemiological time series studies of PM2.5 and daily mortality and hospital admissions: A systematic review and meta-analysis. *Thorax* **2014**, *69*, 660–665. [CrossRef] [PubMed]
2. Bert, B.; Dörendahl, A.; Leich, N.; Vietze, J.; Steinfath, M.; Chmielewska, J.; Hensel, A.; Grune, B.; Schönfelder, G. Rethinking 3R strategies: Digging deeper into AnimalTestInfo promotes transparency in in vivo biomedical research. *PLoS Biol.* **2017**, *15*, e2003217. [CrossRef] [PubMed]

antioxidants

Article

Traffic Density Exposure, Oxidative Stress Biomarkers and Plasma Metabolomics in a Population-Based Sample: The Hortega Study

Laura Sanchez-Rodriguez [1,2], Marta Galvez-Fernandez [1], Ayelén Rojas-Benedicto [2,3,4], Arce Domingo-Relloso [1,5], Nuria Amigo [6,7], Josep Redon [8], Daniel Monleon [8], Guillermo Saez [9], Maria Tellez-Plaza [1,*], Juan Carlos Martin-Escudero [10,†] and Rebeca Ramis [1,4,†]

1 Integrative Epidemiology Group, Department of Chronic Diseases Epidemiology, National Center for Epidemiology, Instituto de Salud Carlos III, 28029 Madrid, Spain; laura.sanchez@isciii.es (L.S.-R.); ad3531@cumc.columbia.edu (A.D.-R.); rramis@isciii.es (R.R.)
2 Joint Research Institute-National School of Health (IMIENS), National Distance Education University, 28029 Madrid, Spain
3 Department of Communicable Diseases, National Center for Epidemiology, Instituto de Salud Carlos III, 28029 Madrid, Spain
4 CIBER on Epidemiology and Public Health, Instituto de Salud Carlos III, 28029 Madrid, Spain
5 Department of Biostatistics, Mailman School of Public Health, Columbia University, New York, NY 10032, USA
6 Biosfer Teslab, 43201 Reus, Spain; namigo@biosferteslab.com
7 Department of Basic Medical Sciences, Universidad de Rovira i Virgili, 43007 Tarragona, Spain
8 Institute for Biomedical Research, Hospital Clinic de Valencia (INCLIVA), 46010 Valencia, Spain
9 Department of Biochemistry and Molecular Biology, Faculty of Medicine and Dentistry, Clinical Analysis Service, Hospital Universitario Dr. Peset-FISABIO, Universitat de Valencia, 46020 Valencia, Spain; guillermo.saez@uv.es
10 Department of Internal Medicine, Hospital Universitario Rio Hortega, University of Valladolid, 47012 Valladolid, Spain; juancarlos.martinescudero@uva.es
* Correspondence: m.tellez@isciii.es
† These authors contributed equally to this work.

check for updates

Citation: Sanchez-Rodriguez, L.; Galvez-Fernandez, M.; Rojas-Benedicto, A.; Domingo-Relloso, A.; Amigo, N.; Redon, J.; Monleon, D.; Saez, G.; Tellez-Plaza, M.; Martin-Escudero, J.C.; et al. Traffic Density Exposure, Oxidative Stress Biomarkers and Plasma Metabolomics in a Population-Based Sample: The Hortega Study. *Antioxidants* **2023**, *12*, 2122. https://doi.org/10.3390/antiox12122122

Academic Editor: Yasuhiro Yoshida

Received: 17 November 2023
Revised: 8 December 2023
Accepted: 11 December 2023
Published: 15 December 2023

Abstract: Exposure to traffic-related air pollution (TRAP) generates oxidative stress, with downstream effects at the metabolic level. Human studies of traffic density and metabolomic markers, however, are rare. The main objective of this study was to evaluate the cross-sectional association between traffic density in the street of residence with oxidative stress and metabolomic profiles measured in a population-based sample from Spain. We also explored in silico the potential biological implications of the findings. Secondarily, we assessed the contribution of oxidative stress to the association between exposure to traffic density and variation in plasma metabolite levels. Traffic density was defined as the average daily traffic volume over an entire year within a buffer of 50 m around the participants' residence. Plasma metabolomic profiles and urine oxidative stress biomarkers were measured in samples from 1181 Hortega Study participants by nuclear magnetic resonance spectroscopy and high-performance liquid chromatography, respectively. Traffic density was associated with 7 (out of 49) plasma metabolites, including amino acids, fatty acids, products of bacterial and energy metabolism and fluid balance metabolites. Regarding urine oxidative stress biomarkers, traffic associations were positive for GSSG/GSH% and negative for MDA. A total of 12 KEGG pathways were linked to traffic-related metabolites. In a protein network from genes included in over-represented pathways and 63 redox-related candidate genes, we observed relevant proteins from the glutathione cycle. GSSG/GSH% and MDA accounted for 14.6% and 12.2% of changes in isobutyrate and the CH_2CH_2CO fatty acid moiety, respectively, which is attributable to traffic exposure. At the population level, exposure to traffic density was associated with specific urine oxidative stress and plasma metabolites. Although our results support a role of oxidative stress as a biological intermediary of traffic-related metabolic alterations, with potential implications for the co-bacterial and lipid metabolism, additional mechanistic and prospective studies are needed to confirm our findings.

Keywords: traffic density; metabolomics; air pollution; population-based

1. Introduction

Traffic-related air pollution (TRAP) from motorized vehicles (passenger cars, motorbikes, heavy-duty vehicles) is a major source of air pollutant constituents, such as nitrogen oxides (NO_X) and primary particulate matter ($PM_{2.5}$), including black carbon [1–3]. TRAP has been associated with several detrimental health outcomes, including asthma onset and mortality endpoints (circulatory, ischemic heart disease and lung cancer) [4]. Some studies have reported a potential link of exposure to air pollution with alterations in specific metabolic pathways, including amino acids, purines, lipids and redox-related pathways [5,6]. In mechanistic studies, TRAP exposure consistently causes damage at the molecular level as well, such as generating reactive oxygen species (ROS) and directly altering the levels of metabolites such as fatty acids, amino acids and others, including glycine, serine, alanine and threonine and metabolites from the glycolysis and gluconeogenesis cycles [7,8]. Moreover, the presence of ROS in an organism can independently lead to additional metabolic alterations [9–11].

However, the negative health impacts of road traffic are not only attributable to exposure to air pollutants. Road-related traffic noise and the absence of green and blue spaces have been associated with metabolically unhealthy lifestyles (less physical activity, obesity) and an increased risk of certain diseases, such as type 2 diabetes [12–14]. Thus, population-based mechanistic studies on potential metabolic and redox effects from integrative traffic intensity measures are needed.

The main objective of this study was to evaluate the cross-sectional association between traffic density on the street of residence and urine oxidative stress biomarkers and plasma metabolomic profiles. We were also interested in exploring the potential biological implications of the findings through an in silico bioinformatic analysis (over-representation and network analysis of relevant metabolite pathways and redox-related candidate genes). Secondarily, we assessed whether oxidative stress could explain the association between exposure to traffic and plasma metabolite levels (i.e., the amount of change in traffic-associated metabolite levels that can be attributed to oxidative stress).

2. Methodology

2.1. Study Population

The Hortega Study cohort is a representative sample of the general population of Valladolid, Spain, obtained through a multi-stage complex sampling study. The study population consisted of beneficiaries from the universal public health system corresponding to the catchment area of the University Hospital Rio Hortega (Valladolid, Spain). In 2001–2003, the Hortega Study participants were examined and interviewed and provided biological samples. Details of the study design and data collection have been previously reported [15]. Among the 1502 recruited participants, we excluded 310 participants with insufficient plasma sample for metabolomic determinations, 40 participants missing BMI, 11 participants missing urine cotinine, 6 participants missing tobacco smoking variables, 3 participants missing education level and 2 participants missing oxidative stress markers, leaving 1181 subjects for the final analysis. The Ethics Committee of the Rio Hortega University Hospital approved the research protocol, and every participant provided informed consent.

2.2. Traffic Density

To assess the effect of traffic exposure on oxidative stress and metabolite levels, we used the traffic density of the closest roads to the home address. For this, we first merged the Navteq cartography with the official cartography from the Ministry of Transport, Mobility and Urban Agenda [16], which provides data on the total volume of vehicles circulating every Spanish road or street. We subsequently estimated individual exposure to traffic in the residence road by creating a 50 m buffer around the geographic coordinates of each participant's home address. Finally, we computed the traffic density within the buffer as the total number of vehicles passing through the buffer over a year divided into 365.25 days (continuous variable expressed in vehicles per day) [17], which is a proxy reflecting how active a road is. For the descriptive analysis, we categorized traffic density into three groups using tertiles (<20.71 (low), 20.71–45.71 (moderate) and >45.71 (high)) cars/day.

2.3. Plasma Metabolite Levels

Metabolomic profile was determined by nuclear magnetic resonance (NMR) spectroscopy in non-fasting plasma. An amount of 82 μL of D_2O was added to 418 μL of blood plasma and placed in a 5 mm NMR tube. NMR spectra were recorded using a Bruker Avance DRX 600 spectrometer (Bruker GmbH, Berlin, Germany). A single-pulse pre-saturation experiment was conducted in all samples, which were kept at 37 °C. To reference the spectra, the doublet of alanine at 1478 ppm was used. To eliminate differences in metabolite total concentration, the spectra were binned into 0.01 buckets and normalized to total aliphatic spectral area. Signals belonging to selected metabolites were quantified using semi-automated in-house MATLAB 6.5 (The MathWorks Inc., Natick, MA, USA) integration and peak-fitting routines. Chenomx NMR Suite V.4.5 software and two-dimensional (2D) NMR methods including homonuclear correlation spectroscopy and heteronuclear single-quantum correlation spectroscopy were used to identify and subsequently confirm the results [18].

In addition, an extended lipoprotein profile was assessed using the Liposcale® methods for NMR spectra analysis. An amount of 500 μL of blood plasma samples was shipped on dry ice to the Biosfer Teslab (Reus, Spain) to determine lipoprotein lipid composition, size and the particle concentration of their respective subclasses (large, medium and small). Particle concentrations and lipoprotein subtypes were determined using the distinctive signals of the lipid methyl group. The size of a given subtype was evaluated by its diffusion coefficient. Common conversion factors were used to convert concentration units into volume units. The particle numbers of each lipoprotein subtype were estimated dividing the lipid volume by the particle volume of a given class.

All metabolites were adjusted by fasting time (hours) using linear regression. We then recalibrated the distribution of resulting metabolite residuals to metabolite-specific mean levels observed in the subset of individuals reporting fasting condition.

2.4. Oxidative Stress Biomarkers

The percentage ratio of oxidized (GSSG) to reduced (GSH) glutathione (GSSG/GSH%) and malondialdehyde (MDA) and the presence of the damaged base 8-oxo-7,8-dihydro-2′-deoxy-guanine (8-oxo-dG) were measured in urine. Analysis of GSSG and GSH levels was performed using high-performance liquid chromatography (HPLC) [19,20]. Additionally, MDA was quantified through spectrophotometric measurement at 532 nm following the MDA-thiobarbituric acid method [21]. Detection of 8-oxo-dG was achieved using high-performance liquid chromatography with electrochemical detection (HPLC-EC) [22,23]. To account for urine dilution, oxidative stress biomarker data were divided by urine creatinine levels and reported in nanomoles per millimole of creatinine. The measurement of urine creatinine was carried out using the modified kinetic Jaffé method. The coefficients of variation for GSSG, GSH, MDA and 8-oxo-dG were recorded at 11.4%, 4.7%, 5.5% and 11.9%, respectively.

2.5. Other Relevant Variables

Participants were interviewed by qualified staff to collect information on sociodemographic data, lifestyle habits and cardiovascular risk factors. Alcohol intake and smoking were classified as never, former and current status based on self-report. Physical activity was estimated in metabolic equivalents (METs) per minute/week based on standardized intensity scores [24] using reported type of activity and amount of time dedicated to each activity per week. Body mass index (BMI) was calculated using measured weight (kilograms) by height (meters) squared. Obesity was defined as a BMI equal or higher than 30 kg/m^2. Urine cotinine was measured with an enzyme-linked immunosorbent assay (ELISA) ("Analysis DRI® Cotinine" Kit, Ref. 0395 Microgenics laboratories). Concentrations below the lower limit (34 ng/mL) were detected in 77% of participants. Urine and serum creatinine were measured by the modified kinetic Jaffé method by isotope dilution mass spectrometry on a Hitachi 917 analyzer (Roche Boehringer). Urine albumin was measured by automated nephelometric immunochemistry (Behring, Germany). Renal function was assessed by the glomerular filtration rate, as estimated using the CKD-EPI equation [25].

2.6. Statistical Analysis

Descriptive and association analysis. Statistical analyses were conducted with the "survey" package of the R software (version 4.1.14) to account for the complex sampling. We reported participant characteristics and the median and interquartile range (IQR) of plasma metabolites overall and by tertiles of traffic density to compare moderate and high-to-low exposure levels. We conducted linear regression models to evaluate the association between traffic density exposure (continuous independent variable) and oxidative stress biomarkers and plasma metabolite level (dependent variables in separate models). The resultant regression coefficients were re-scaled to compare the 80th and the 20th percentiles of traffic density. All models were adjusted for sex (men, women), age (years), BMI (kg/m^2), high education (no/yes), smoking status (never, former, current), cumulative smoking (number of pack-years), drinking status (never, former, current), glomerular filtration rate (mL/min/1.73 m^2), physical activity (METs min/week) and urine cotinine level (mg/dL). For all the association analyses, we established a p-value threshold (α) of 0.05 as the statistical significance level.

Bioinformatic exploration of potential biological implications of the findings. For statistically significant metabolites from the association analysis, we ran Metabolite Set Enrichment Analysis (MSEA) to explore over-represented metabolites within pre-specified metabolite sets from the KEGG database [26] with MetaboAnalyst 5.0 [27]. MetaboAnalyst conducts a hypergeometric test to yield p-values that are interpreted as the probability of having a particular metabolite represented within a given set more than expected by chance. Subsequently, we constructed a protein interaction network. For this, we first extracted genes from the KEGG pathways with suggestively over-represented traffic-related metabolites (nominal p-value < 0.10). We further extended the in silico characterization of interconnections with oxidative stress by selecting additional candidate genes associated with redox balance from a published review [28]. Finally, we obtained protein–protein and protein–compound interaction networks (i.e., the interactions of proteins encoded by these genes and metabolites compounds from the IntAct database Release 243 [29] built-in feature of Cytoscape v3.9 [30]). The resulting network was filtered by removing self-loops, networks with fewer than three nodes and selecting only interactions with a Mutual Information score (MI score) of at least 0.5. Only nodes corresponding to human proteins or molecules associated with chemical functions were kept.

Mediation analysis. In secondary analysis, we formally tested the potential mediating role of oxidative stress as an intermediary variable in the association between traffic density with relevant metabolomic markers. Our conceptual mediation model can be found in the Supplementary File S1, Supplementary Methods. To assess natural indirect (i.e., mediated) effects, we used a counterfactual mediation framework as implemented by the multimediate R package [31]. The multimediate algorithm is able to conduct mediation analysis using the counterfactuals method. In this setting, our mediator models were separate linear models in which relevant oxidative stress biomarkers were entered as the dependent variable and traffic density (exposure) was entered as the independent variable. The outcome linear model included relevant metabolites (i.e., statistically significant in previous analysis) as the dependent variable in separate models, traffic density as the exposure and oxidative stress biomarkers as mediators. Both outcome and mediator models were adjusted for age, sex, education, BMI, smoking status, accumulated smoking (packs-year), alcohol intake, urine cotinine levels, physical activity (METs/week), triglycerides and lipid-lowering medication.

As result, absolute mediated effects (i.e., natural indirect effects) were reported as the mean difference in changes in traffic-related metabolite levels attributed to differences in oxidative stress. The direct effect was reported as the mean difference in changes in traffic-related metabolite levels not attributable to differences in oxidative stress. The total effect corresponds to the sum of the direct and the indirect effect. The relative mediated effect was calculated as the ratio between the indirect and the total effect. Confidence intervals were calculated using a resampling method based on simulations from a multivariate normal distribution [31].

3. Results

Descriptive and association analysis. Table 1 shows the crude (unadjusted) characteristics of our study population according to traffic density levels (low, moderate, high). The mean age was 52.74 years and 49.78% were women. The group exposed to the highest traffic density had higher accumulated smoking (pack-years) and was more physically active compared to less exposed subjects. Participants with higher exposure to traffic density showed lower levels of amino acids (cysteine, proline, tryptophan), products of bacterial co-metabolism (phenylpropionate) and energy metabolism (acetate) metabolite subclasses, and higher levels of cholesterol, triglycerides (LDL and IDL triglycerides), lipoprotein particle subclasses (large and medium LDL and HDL) and the oxidative stress marker GSSG/GSH% (Supplementary File S1, Supplementary Table S1). The association of traffic density (per 80th to 20th percentiles of traffic distribution comparison) with NMR metabolites (unitless) was positive for some of the measured fatty acid moieties, including CH_2CH_2CO and CH_2N (MD [95% CI] was 0.155 [0.033, 0.276] and 1.744 [0.277, 3.211], respectively). Alternatively, the association of traffic density was inverse for the amino acid cysteine (MD [95% CI] −0.010 [−0.020, −0.001]); the fatty acid isobutyrate (MD [95% CI] −0.042 [−0.077, −0.007]); some products of bacterial co-metabolism, such as trimethylamines (MD [95% CI] −0.065 [−0.113, −0.018]); acetate, a product of energy metabolism (MD [95% CI] −0.030 [−0.058, −0.002]); and the fluid-balance-associated metabolite albumin (MD [95% CI] −0.073 [−0.135, −0.011]). Regarding the association with oxidative stress biomarkers, traffic exposure was positively associated with urine GSSG/GSH (%) (3.142 [0.049, 6.236]) but inversely associated with MDA levels (nmol/mmol creatinine) (−0.097 [−0.170, −0.023]) (Table 2). All the traffic-related oxidative stress biomarkers (independent variable) were also associated with the traffic-related metabolites (dependent variables), except for the association of isobutyrate with GSSG/GSH% and that of CH_2N and acetate, and possibly cysteine, trimethylamine and albumin, with MDA (Supplementary File S1, Supplementary Table S2).

Table 1. Age- and gender-adjusted baseline characteristics based on traffic density (N = 1181). Mean, overall and by tertiles of traffic density.

		Traffic Density at Home Address (Cars per Day)		
	Overall	**Low Traffic Density**	**Moderate Traffic Density**	**High Traffic Density**
Age, years; mean	52.74	51.90	51.66	54.59
Women; %	49.78	49.21	46.23	53.72
BMI, kg/m^2; mean	26.36	26.67	26.06	26.36

Table 1. Cont.

	Overall	Traffic Density at Home Address (Cars per Day)		
		Low Traffic Density	Moderate Traffic Density	High Traffic Density
Smoking status				
Never; %	47.23	47.40	44.15	49.75
Former; %	29.53	28.39	32.71	28.50
Current; %	23.22	24.19	22.34	21.75
Cumulative smoking, pack-year; mean	9.08	8.05	9.59	9.59
Urine cotinine, mg/dL				
<12 mg/dL	77.3	95.52	77.95	78.40
12–500 mg/dL	4.73	6.05	5.1	3.08
>500 mg/dL	17.97	18.42	16.93	18.51
Alcohol intake status				
Never; %	39.26	44.73	32.79	40.10
Former; %	8.41	8.42	8.06	8.74
Current; %	52.32	46.84	56.55	51.15
eGFR, mL/min/1.73 m^2	90.84	91.7	92.13	88.77
High education; %	72.92	69.21	78.19	72.00
Physical activity, METs min/week; mean	3135.94	3110.10	3050.08	3242.95

Table 2. Mean difference (95% CI) * of NMR-metabolites and oxidative stress markers when comparing the 80th and 20th percentiles of traffic density distribution in the Hortega Study (N = 1181).

Group	Metabolite	MD (95% CI)	*p*-Value
Lipoprotein profile	Cholesterol, mg/dL	1.430 (−2.340, 5.201)	0.457
	VLDL cholesterol, mg/dL	0.638 (−0.130, 1.407)	0.104
	LDL cholesterol, mg/dL	1.833 (−2.232, 5.898)	0.377
	HDL cholesterol, mg/dL	−0.832 (−2.664, 1.001)	0.374
	IDL cholesterol, mg/dL	0.487 (−0.067, 1.042)	0.085
	Total VLDL, nmol/L	0.796 (−1.492, 3.083)	0.496
	Large VLDL, nmol/L	−0.015 (−0.084, 0.054)	0.496
	Medium VLDL, nmol/L	0.002 (−0.459, 0.463)	0.994
	Small VLDL, nmol/L	0.808 (−1.252, 2.869)	0.442
	Total LDL, nmol/L	17.08 (−22.022, 56.182)	0.392
	Large LDL, nmol/L	2.497 (−3.041, 8.035)	0.377
	Medium LDL, nmol/L	11.853 (−6.439, 30.146)	0.204
	Small LDL, nmol/L	2.830 (−17.158, 22.819)	0.781
	Total HDL, nmol/L	−0.311 (−1.207, 0.524)	0.440
	Large HDL, µmol/L	0.002 (−0.006, 0.010)	0.695
	Medium HDL, µmol/L	0.004 (−0.301, 0.309)	0.979
	Small HDL, µmol/L	−0.347 (−0.966, 0.271)	0.271

Table 2. *Cont.*

Group	Metabolite	MD (95% CI)	p-Value
Amino acids	Alanine	0.013 (−0.058, 0.084)	0.718
	Creatine phosphate	−0.008 (−0.021, 0.004)	0.187
	Creatine	−0.010 (−0.021, 0.002)	0.090
	Cysteine	−0.010 (−0.020, −0.001)	0.038
	Glutamine	−0.008 (−0.063, 0.047)	0.772
	Proline	0.017 (−0.044, 0.078)	0.595
	Tryptophan	0.017 (−0.026, 0.060)	0.432
	Tyrosine	−0.014 (−0.036, 0.007)	0.198
	Isoleucine	0.004 (−0.042, 0.049)	0.875
	Leucine	−0.002 (−0.040, 0.036)	0.909
	Valine	−0.038 (−0.083, 0.006)	0.093
Inflamation marker	N-acetylglutamine	−0.031 (−0.068, 0.007)	0.109
Fatty acids	CH_2CH_2CO	0.155 (0.033, 0.276)	0.013
	CH_2CH_3	0.121 (−0.033, 0.274)	0.123
	CH_2N	1.744 (0.277, 3.211)	0.020
	CH_3	0.063 (−0.358, 0.483)	0.770
	$CHCH_2CH$	0.005 (−0.095, 0.105)	0.919
	Isobutyrate	−0.042 (−0.077, −0.007)	0.020
Products of bacterial co-metabolism	Ethanol	−0.004 (−0.196, 0.187)	0.964
	Isopropanol	−0.041 (−0.087, 0.006)	0.087
	Methanol	−0.012 (−0.024, 0.001)	0.065
	Trimethylamines	−0.065 (−0.113, −0.018)	0.008
	Phenylpropionate	0.030 (−0.039, 0.098)	0.395
	O-phosphoethanolamine	−0.036 (−0.079, 0.008)	0.111
Energy metabolism	*Glycolisis*		
	Citrate	−0.024 (−0.060, 0.013)	0.208
	Lactate	0.299 (−0.065, 0.663)	0.108
	Pyruvate	0.001 (−0.013, 0.014)	0.905
	Ketone bodies		
	Acetate	−0.030 (−0.058, −0.002)	0.034
	Acetone	0.062 (−0.002, 0.127)	0.057
	3-Hydroxybutyrate	−0.006 (−0.063, 0.050)	0.822
Fluid balance	Albumin	−0.073 (−0.135, −0.011)	0.022
	Creatinine	−0.006 (−0.027, 0.015)	0.572
Oxidative stress markers	GSSG/GSH,%	3.142 (0.049, 6.236)	0.047
	Malondialdehyde (MDA), nmol/mmol creatinine	−0.097 (−0.170, −0.023)	0.010
	8-oxo-7,8-dihydroguanine (8-oxo-dG), nmol/mmol creatinine	0.116 (−0.068, 0.300)	0.216

Models were adjusted for age, sex, education, BMI, smoking status, cigarette packages per year, alcohol intake, urine cotinine levels, physical activity per week, triglycerides and lipid-lowering medication. * We normalized the spectral vector to the total spectral area, excluding residual water signals to minimize the effects of variable dilution of the sample. The metabolic content is therefore expressed in relative metabolic content (unitless), unless other units are indicated.

Bioinformatic exploration of potential biological implications of the findings. Table 3 shows the results for the Metabolite Set Enrichment Analysis (MSEA). Valine and cysteine were the most over-represented metabolites, followed by acetate. At the p-value threshold of 0.10, out of the 84 KEGG-based metabolite sets included in the MSEA, 12 included over-represented statistically significant metabolites from our association analysis. These pathways were mainly associated with amino acids, carbohydrates and co-factors metabolism. The three most enriched pathways were "Pantothenate and CoA biosynthesis" (hsa00770), "Glycine, serine and threonine metabolism" (hsa00260) and "Aminoacyl-tRNA biosynthesis" (hsa00970) (Supplementary File S1, Supplementary Figure S1). Figure 1 shows the protein–protein interaction network resulting from displaying IntAct-based interactions from the list of proteins encoded by our redox-related candidate genes with proteins encoded by genes within the metabolite sets with over-represented metabolites from our association analysis (Supplementary File S2, Supplementary Sheet S1). The initial network had 1330 nodes and 4503 interactions (Supplementary File S2, Supplementary Sheet S2). We excluded 143 nodes not corresponding to human proteins or molecules associated with chemical molecules, 628 nodes with an MI score below 0.5 and 17 nodes with self-loops without identifiable ID (Supplementary File S2, Supplementary Sheet S3). Finally, the resulting protein network after filtering (Figure 1) retained a total of 468 unique proteins and 493 interactions (Supplementary File S2, Supplementary Sheet S4). Among the enriched pathways and the oxidative-stress-related proteins, 10 common proteins were found, most of which are associated with the glutathione metabolism (*GPX1* to *7*, *GSR* and *TXNDC12*), while *CAT* encodes another key enzyme involved in redox balance. Nevertheless, after filtering the interaction network, only the glutathione peroxidase *GPX7*, an endoplasmic catalase involved in the cellular response to oxidative stress, remained (Figure 1). Furthermore, only one direct interaction between a protein in a KEGG pathway with over-represented metabolites, *MAT2A* and an oxidative stress-related protein, *MT2A*, prevailed. The proteins *FARS2*, *GLYCTK*, *ALPP* and *ALAS1* are some of the largest nodes as they have the greatest number of interactions.

Table 3. Metabolite Enrichment Analysis results for the significant metabolites associated with traffic density.

KEGG-Based Pathways	Total	Expected	Hits	Enrich. Ratio	Raw p-Value	Holm p-Value	FDR p-Value	Metabolites
Pantothenate and CoA biosynthesis	19	0.050	2	40.404	0.001	0.072	0.072	Cysteine, Valine
Glycine, serine and threonine metabolism	33	0.086	2	23.283	0.003	0.217	0.110	Creatine, Cysteine
Aminoacyl-tRNA biosynthesis	48	0.125	2	16.000	0.006	0.452	0.154	Cysteine, Valine
Thiamine metabolism	7	0.018	1	54.945	0.018	1	0.290	Cysteine
Valine, leucine and isoleucine biosynthesis	8	0.021	1	48.077	0.021	1	0.290	Valine
Taurine and hypotaurine metabolism	8	0.021	1	48.077	0.021	1	0.290	Cysteine
Pyruvate metabolism	22	0.057	1	17.452	0.056	1	0.636	Acetate
Glycolysis/gluconeogenesis	26	0.068	1	14.771	0.066	1	0.636	Acetate
Glutathione metabolism	28	0.073	1	13.717	0.071	1	0.636	Cysteine
Glyoxylate and dicarboxylate metabolism	32	0.083	1	12.005	0.081	1	0.636	Acetate
Cysteine and methionine metabolism	33	0.086	1	11.641	0.083	1	0.636	Cysteine
Arginine and proline metabolism	38	0.099	1	10.101	0.095	1	0.648	Creatine

Total is the total number of compounds in the pathway; Expected is the number of matched compounds expected by chance given the pathway size; Hits is the number of matched compounds from the data; Enrichment Ratio is the number of hits divided by the expected number of hits. Raw p-value calculated from the enrichment analysis; Holm p-value was adjusted using the Holm–Bonferroni method; FDR p-value was adjusted using False Discovery Rate (FDR); metabolites showed over-represented statistically significant metabolites from the association analysis.

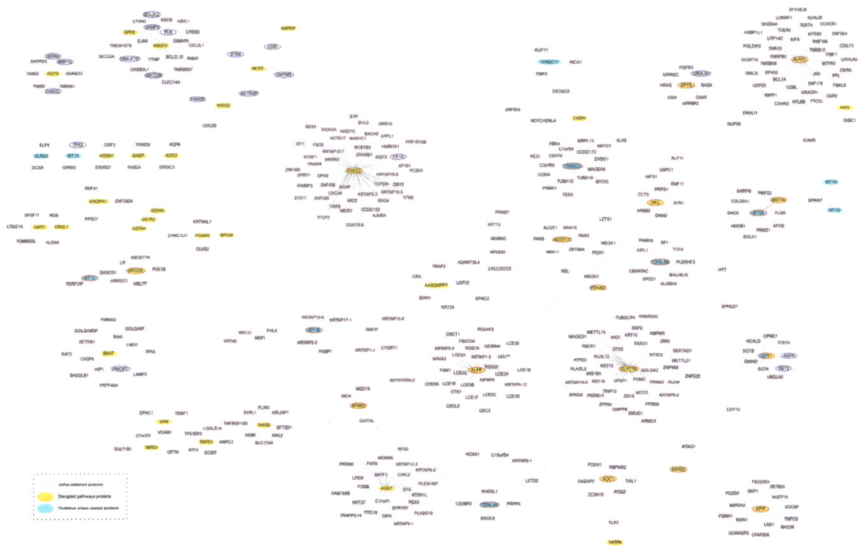

Figure 1. Protein interaction network generated using IntAct by redox-related proteins and pathways with over-represented traffic-exposure-associated metabolites (468 proteins connected by 493 edges). Nodes are colored according to protein data origin: oxidative-stress-related proteins are blue, proteins identified by the enrichment analysis are yellow, and IntAct-identified proteins are purple. Proteins circled in red are, at least, second-degree neighbors between the proteins of interest. Proteins circled in blue interact with a GPX protein (see Supplemental File S2, Supplemental Sheet S5).

Contribution of oxidative stress to traffic-density-related NMR-metabolites. MDA accounted for 21.88 (0.19, 59.24) and 20.48 (2.26, 83.97) % of the traffic-related isobutyrate and CH_2CH_2CO fatty acid moiety variation, respectively. In absolute terms, of the 0.139 (0.018, 0.256) CH_2CH_2CO fatty acid moiety units and −0.038 (−0.0727, −0.005) isobutyrate units, according to an 80th versus the 20th percentile comparison in traffic density, 0.014 (0.003, 0.032) and −0.006 (−0.012, −0.002) were attributable to variations in MDA, respectively (Supplementary File S1, Supplementary Table S3).

4. Discussion

In our cross-sectional study, traffic density was positively associated with fatty acid moieties (CH_2N, CH_2CH_2CO) and with markers of oxidative stress, such as GSSG/GSH in urine. On the other hand, traffic exposure was inversely associated with the amino acid cysteine and fatty acids such as isobutyrate, with products of bacterial co-metabolism (trimethylamines), energy metabolism (acetate), fluid balance metabolites (albumin) and the oxidative stress biomarker malondialdehyde (MDA). The association of traffic density with cysteine, acetate and albumin was partly explained by oxidative stress biomarkers, supporting the idea that oxidative stress is a biological intermediary in traffic-related disease. The statistical models accounted for known oxidative stress determinants, such as alcohol consumption and smoking status.

Traffic density and metabolomics. Evidence has shown over the years that traffic-associated air pollution (TRAP) is a known risk factor for the development of health conditions such as cardiovascular disease [32], metabolic disorders [33] and respiratory diseases, including lung cancer and asthma [34,35]. A study conducted by the University of Bari showed that exposure to TRAP produces neutrophilic airway inflammation [36]. However, the mechanisms by which TRAP causes adverse health effects still remain poorly understood, and little is known regarding the potential role of traffic exposure as a determinant of metabolite levels [37]. In our study, a broad panel of metabolites were used to assess potential metabolic traffic effects. Next, we reviewed the consistency of our results with those of other available epidemiological studies.

Amino acids. In our study, exposure to traffic density was inversely associated with amino acids levels, which are components of key enzymes in metabolic pathways involved in cell homeostasis, nutrition and the regulation of the immune system [38,39]. Dysfunctional levels of essential amino acids have been associated with several pathologies, such as cardiovascular and neurological disorders and certain types of cancer [40,41]. Consistently with our results, an intervention study conducted in healthy subjects (N = 43) found that exposure to PM2.5 was associated with a decrease in glutamate, aspartate and taurine levels [42]. A longitudinal study on healthy adults (N = 73) found that short-term air pollution exposure was associated with significant reductions in the levels of plasma alanine, threonine and glutamic acid [43]. A cross-sectional study (N = 54) observed an inverse association between histidine and outdoor PM2.5 level [44].

Bacterial co-metabolism. The gut microbiota play a vital role in human homeostasis [45], stimulate the immune system and contribute to metabolism [46]. Microbiome imbalance caused by exposure to air pollutants could have a role in cardiometabolic, infectious and inflammatory disease [47,48]. Some studies have pointed to a potential role of trimethylamines (TMAO), a component of bacterial co-metabolism involved in human physiological processes [49] and in the pathogenesis of numerous diseases, including kidney and cardiovascular diseases [47,50]. A study conducted in China (N = 114) found that oropharyngeal microbiota of healthy volunteers differed within regions of high, medium and low TRAP [51]. In addition, a pilot study conducted with obese adolescents (N = 43) demonstrated that TRAP exposure can alter the composition and abundance of the gut microbiota [52]. Consistently, in our study, traffic exposure was inversely associated with all products of bacterial co-metabolism evaluated (ethanol, methanol, isopropanol, trimethylamines and phenylpropionate).

Fatty acids and lipoprotein subclasses. The association of traffic density with fatty acids and most of the lipoprotein subclasses, except for HDL cholesterol, was positive, although the uncertainty of these associations was substantial. While there are no studies specifically reporting the association between traffic exposure and lipids, air pollution studies carried out in large cohorts (MESA-Air Study, META-Air Study) identified that pollutant components such as PM2.5 and NOx were positively associated with total and LDL cholesterol levels [53,54]. The fatty acid isobutyrate is also considered a by-product of bacterial co-metabolism [55–57].

The role of oxidative stress in traffic-related metabolomics. There is growing evidence in support of a role of exposure to environmental pollution in altering redox balance [58,59]. In our data, higher traffic density exposure was positively associated with GSSG/GSH levels and negatively associated with MDA levels. An increase in GSSG/GSH is indicative of increased oxidative stress at the cellular level [60–62], consistently with our hypotheses. Malondialdehyde is a product of lipid peroxidation, especially at the membrane level; it is derived from polyunsaturated fatty acids and increases in situations of oxidative stress [63]. Thus, identifying lower levels in subjects more exposed to traffic with respect to those less exposed was unexpected. In a small study from Cracow (N = 40), exposure to carbon monoxide (CO), a major pollutant from road traffic, was positively associated with MDA and GSSG [64]. Inconsistently with our data, short-term exposure to traffic-related black carbon concentrations in the air was positively associated with 8-isoprostane, a marker of oxidative stress in lipids [65], and 8-oxo-OhdG, a marker of oxidative stress in the cellular nucleus [66]. In our study, MDA substantially explained the association of traffic density with metabolites within the fatty acids and subproducts of bacterial co-metabolism groups. Studies on experimental models show that MDA levels were dependent on fatty acid unsaturation and correlated to carbonyls in fatty acids (CH_2CH_2CO) [67]. In addition, alteration in gut and lung microbiota has been observed in situations of TRAP exposure, with oxidative stress playing an important role in these alterations [68,69].

Overall, our results are consistent with the possibility that redox imbalance is a biological mediator of trafficking-related metabolic alterations, especially at the cell membrane level, and related to the microbiota [37,70].

Bioinformatic exploration of traffic- and redox-related metabolic pathways. The most enriched pathway was Pantothenate and CoA biosynthesis, which has been associated with mitochondrial function and energy metabolism (the most over-represented metabolites were cysteine and valine, which were decreased in the subjects most exposed to traffic density). An experimental study conducted in human epidermal keratinocytes showed that a pantothenate derivative reduced cell damage by stimulating the intracellular defense system against ROS [71]. Little is known, however, about the effect that exposure to traffic density may have on energy metabolism. In our study, higher traffic density was inversely related to acetate. Studies conducted in mice and rats also concluded that certain air pollutants, such as lead and other PM, function as deregulators of energy metabolism [55–57].

The molecular interactions (edges) in our protein network reflect the accumulated evidence linking proteins involved in oxidative stress and proteins in pathways with over-represented metabolites and provide an overview of the potential downstream biological implications of the most interesting findings. For instance, *MT2A* encodes for metallothionein proteins, which play an important antioxidant role and have been associated with the progression of various chronic diseases [72,73]. *MAT2A* encodes for an essential enzyme that synthesizes S-adenosylmethionine (SAM), a precursor of glutathione (GSH) [74]. Glutathione peroxidase isoform 7 (GPx7) is activated upon redox situations, mainly at the endoplasmic reticulum level, and participates in the oxidative folding of proteins [75,76]. It is known that the GPx7 structure contains cysteine [70], an amino acid that is associated with high traffic density in our data. FARS2, GLYCTK, ALAS1 and ALPP intervene in various metabolic pathways whose alteration has been associated with exposure to traffic and various pollutants [77–79]. For a deeper review of other interesting potential mechanisms identified in bioinformatic analysis, see Supplementary File S1, Supplementary Discussion. Given the connection of most relevant metabolites with the glutathione cycle, as a post hoc analysis, we descriptively report the interaction network of the *GPX* family of proteins (isoforms 1 to 7) with other proteins and metabolites also obtained from IntAct (Figure 1 and Supplementary File S2, Supplementary Sheet S5).

Limitations and strengths. Our study has several limitations. For instance, the interpretation of our findings requires some caution because the traffic exposure assessment in our study was based on the participants home address, and it does not take into account movements made throughout the day. To address this potential source of heterogeneity in the traffic exposure measurement, we used average annual traffic density as a measure of traffic exposure to select streets with a similar density and then created tertiles of traffic density to compare between low, moderate and high exposure levels, as previously done in other epidemiological studies [16]. Some other studies have employed other variables, such as nitrogen dioxide, benzene [80–82] and volatile organic compounds [79], or have assessed personal exposure to air pollution using rechargeable devices that can be carried by the subjects themselves or via satellite-derived data [83]. Further, it should be taken into account that humans are not only exposed to outdoor environmental pollution but also indoor pollutants (aromatic hydrocarbons, aldehydes and others) [84]. An additional limitation is related to the targeted metabolomic approach, which quantified a predefined set of metabolites. Consequently, some relevant metabolites may have been missed. Also, our findings must be interpreted with caution because diet and microbiota can affect metabolomic profiles [85]. Several sensitivity analyses have been performed including adjustment for total energy, fat, carbohydrate and protein intake from 24 h recall questionnaires, yielding essentially similar findings, suggesting that possible confounding by dietary factors is probably not relevant in our data. Last, the assumptions for mediation analysis include no unmeasured confounding by the relationship between the exposure, the outcome and the mediators, an assumption that is impossible to verify in observational studies. Another important limitation is the cross-sectional nature of our data.

Our study also has several strengths. To our knowledge, this is the first study to analyze the association between traffic density exposure and metabolomic determinations while exploring the potential role of oxidative stress as an intermediary in this association. In addition, our approach did not provide individual compound measurements but summarized the total exposure to traffic pollution, which is preferable for the purpose of our study, which included a large number of metabolites with unknown relation to traffic exposure as a whole. The strengths of this study also include the complex survey design and sample size, which allows our results to be representative for the general population of a region in Spain. Furthermore, the unique availability of a considerable panel of metabolites measured with high-quality procedures is an additional strength of the study.

5. Conclusions

In our study, we observed a clear association of exposure to traffic density with differences in certain metabolic patterns that have traditionally been linked to the development of chronic conditions in the general population. Our results supported the idea that oxidative stress might be a relevant mechanism of traffic-related health effects, especially for lipidic membranes and bacterial co-metabolism. Our findings need to be confirmed by prospective studies with longitudinal metabolite measurements but suggest that reinforcing public health interventions to reduce exposure to traffic in the population is needed.

Supplementary Materials: The following supporting information can be downloaded at: https://www.mdpi.com/article/10.3390/antiox12122122/s1. References [86–122] are cited in the Supplementary Materials.

Author Contributions: Conceptualization, M.T.-P. and R.R.; Methodology, N.A., D.M., G.S., M.T.-P. and R.R.; Software, A.D.-R.; Formal Analysis, M.G.-F., A.R.-B., R.R. and M.T.-P.; Resources, N.A., D.M., G.S., J.R., M.T.-P. and J.C.M.-E.; Data Curation, M.G.-F., M.T.-P., J.C.M.-E. and R.R.; Writing—Original Draft Preparation, L.S.-R. and A.R.-B.; Writing—Review and Editing, M.G.-F., A.R.-B., A.D.-R., N.A., J.R., D.M., G.S., M.T.-P., J.C.M.-E. and R.R.; Supervision, R.R. and M.T.-P.; Project Administration, M.T.-P. and J.C.M.-E.; Funding Acquisition, J.R. and J.C.M.-E. All authors have read and agreed to the published version of the manuscript.

Funding: This research was funded by the State Agency for Research (PID2019-108973RB-C21 and C22), by Strategic Action for Research in Health Sciences (PI15/00071 and PI22CIII/00029) from the Spanish Ministry of Economy and Competitiveness and co-funded with European Funds for Regional Development (FEDER), and IDIFEDER/2021/072, CIAICO/2022/181 and INVEST/2023/180 from the Generalitat Valenciana of Spain.

Institutional Review Board Statement: The study was conducted in accordance with the Declaration of Helsinki, and approved by by the Ethics Committee of Clinical Research of the University Hospital Rio Hortega (protocol code 01/2003 and date of approval 12 January 2003).

Informed Consent Statement: Written informed consent was obtained from all subjects involved in the study.

Data Availability Statement: The data presented in this study are available on request to the corresponding author, upon a reasonable request. The data are not publicly available because unrestricted data sharing is not possible due to privacy or ethical restrictions.

Conflicts of Interest: N.A. owns stocks in the company Biosfer Teslab and has a patent for a "Method for lipoprotein characterization". M.G.-F. was affiliated with AstraZeneca at the time of manuscript submission. The authors declare no other conflict of interest.

References

1. Jereb, B.; Gajšek, B.; Šipek, G.; Kovše, Š.; Obrecht, M. Traffic Density-Related Black Carbon Distribution: Impact of Wind in a Basin Town. *Int. J. Environ. Res. Public Health* **2021**, *18*, 6490. [CrossRef] [PubMed]
2. Long, E.; Carlsten, C. Controlled Human Exposure to Diesel Exhaust: Results Illuminate Health Effects of Traffic-Related Air Pollution and Inform Future Directions. *Part. Fibre Toxicol.* **2022**, *19*, 11. [CrossRef] [PubMed]
3. Mak, H.W.L.; Ng, D.C.Y. Spatial and Socio-Classification of Traffic Pollutant Emissions and Associated Mortality Rates in High-Density Hong Kong via Improved Data Analytic Approaches. *Int. J. Environ. Res. Public Health* **2021**, *18*, 6532. [CrossRef] [PubMed]
4. Boogaard, H.; Patton, A.P.; Atkinson, R.W.; Brook, J.R.; Chang, H.H.; Crouse, D.L.; Fussell, J.C.; Hoek, G.; Hoffmann, B.; Kappeler, R.; et al. Long-Term Exposure to Traffic-Related Air Pollution and Selected Health Outcomes: A Systematic Review and Meta-Analysis. *Environ. Int.* **2022**, *164*, 107262. [CrossRef] [PubMed]
5. Arayasiri, M.; Mahidol, C.; Navasumrit, P.; Autrup, H.; Ruchirawat, M. Biomonitoring of Benzene and 1,3-Butadiene Exposure and Early Biological Effects in Traffic Policemen. *Sci. Total Environ.* **2010**, *408*, 4855–4862. [CrossRef] [PubMed]
6. Brower, J.B.; Doyle-Eisele, M.; Moeller, B.; Stirdivant, S.; McDonald, J.D.; Campen, M.J. Metabolomic Changes in Murine Serum Following Inhalation Exposure to Gasoline and Diesel Engine Emissions. *Inhal. Toxicol.* **2016**, *28*, 241–250. [CrossRef] [PubMed]
7. Ritz, B.; Yan, Q.; He, D.; Wu, J.; Walker, D.I.; Uppal, K.; Jones, D.P.; Heck, J.E. Child Serum Metabolome and Traffic-Related Air Pollution Exposure in Pregnancy. *Environ. Res.* **2022**, *203*, 111907. [CrossRef]
8. Yan, Q.; Liew, Z.; Uppal, K.; Cui, X.; Ling, C.; Heck, J.E.; von Ehrenstein, O.S.; Wu, J.; Walker, D.I.; Jones, D.P.; et al. Maternal Serum Metabolome and Traffic-Related Air Pollution Exposure in Pregnancy. *Environ. Int.* **2019**, *130*, 104872. [CrossRef]
9. Bhatia, S.S.; Pillai, S.D. A Comparative Analysis of the Metabolomic Response of Electron Beam Inactivated *E. coli* O26:H11 and Salmonella Typhimurium ATCC 13311. *Front. Microbiol.* **2019**, *10*, 694. [CrossRef]
10. Bie, P.; Yang, X.; Zhang, C.; Wu, Q. Identification of Small-Molecule Inhibitors of Brucella Diaminopimelate Decarboxylase by Using a High-Throughput Screening Assay. *Front. Microbiol.* **2020**, *10*, 2936. [CrossRef]
11. Noctor, G.; Lelarge-Trouverie, C.; Mhamdi, A. The Metabolomics of Oxidative Stress. *Phytochemistry* **2015**, *112*, 33–53. [CrossRef] [PubMed]
12. Sørensen, M.; Poulsen, A.H.; Hvidtfeldt, U.A.; Brandt, J.; Frohn, L.M.; Ketzel, M.; Christensen, J.H.; Im, U.; Khan, J.; Münzel, T.; et al. Air Pollution, Road Traffic Noise and Lack of Greenness and Risk of Type 2 Diabetes: A Multi-Exposure Prospective Study Covering Denmark. *Environ. Int.* **2022**, *170*, 107570. [CrossRef]
13. Teixeira, A.; Gabriel, R.; Quaresma, L.; Alencoão, A.; Martinho, J.; Moreira, H. Obesity and Natural Spaces in Adults and Older People: A Systematic Review. *J. Phys. Act. Health* **2021**, *18*, 714–727. [CrossRef] [PubMed]
14. Yuen, J.W.M.; Chang, K.K.P.; Wong, F.K.Y.; Wong, F.Y.; Siu, J.Y.M.; Ho, H.C.; Wong, M.S.; Ho, J.Y.S.; Chan, K.L.; Yang, L. Influence of Urban Green Space and Facility Accessibility on Exercise and Healthy Diet in Hong Kong. *Int. J. Environ. Res. Public Health* **2019**, *16*, 1514. [CrossRef] [PubMed]
15. Tellez-Plaza, M.; Briongos-Figuero, L.; Pichler, G.; Dominguez-Lucas, A.; Simal-Blanco, F.; Mena-Martin, F.J.; Bellido-Casado, J.; Arzua-Mouronte, D.; Chaves, F.J.; Redon, J.; et al. Cohort Profile: The Hortega Study for the Evaluation of Non-Traditional Risk Factors of Cardiometabolic and Other Chronic Diseases in a General Population from Spain. *BMJ Open* **2019**, *9*, e024073. [CrossRef]
16. Tamayo-Uria, I.; Boldo, E.; García-Pérez, J.; Gómez-Barroso, D.; Romaguera, E.P.; Cirach, M.; Ramis, R. Childhood Leukaemia Risk and Residential Proximity to Busy Roads. *Environ. Int.* **2018**, *121*, 332–339. [CrossRef]
17. Ojeda Sánchez, C.; García-Pérez, J.; Gómez-Barroso, D.; Domínguez-Castillo, A.; Pardo Romaguera, E.; Cañete, A.; Ortega-García, J.A.; Ramis, R. Exploring Blue Spaces' Effects on Childhood Leukaemia Incidence: A Population-Based Case–Control Study in Spain. *Int. J. Environ. Res. Public Health* **2022**, *19*, 5232. [CrossRef]
18. Galvez-Fernandez, M.; Sanchez-Saez, F.; Domingo-Relloso, A.; Rodriguez Hernandez, Z.; Tarazona, S.; Gonzalez-Marrachelli, V.; Grau-Perez, M.; Morales-Tatay, J.M.; Amigo, N.; Garcia-Barrera, T.; et al. Gene-Environment Interaction Analysis of Redox-Related Metals and Genetic Variants with Plasma Metabolic Patterns in a General Population from Spain: The Hortega Study. *Redox Biol.* **2022**, *52*, 102314. [CrossRef]
19. Brigelius, R.; Muckel, C.; Akerboom, T.P.M.; Sies, H. Identification and Quantitation of Glutathione in Hepatic Protein Mixed Disulfides and Its Relationship to Glutathione Disulfide. *Biochem. Pharmacol.* **1983**, *32*, 2529–2534. [CrossRef]
20. Navarro, J.; Obrador, E.; Pellicer, J.A.; Aseni, M.; Viña, J.; Estrela, J.M. Blood Glutathione as an Index of Radiation-Induced Oxidative Stress in Mice and Humans. *Free Radic. Biol. Med.* **1997**, *22*, 1203–1209. [CrossRef]
21. Moselhy, H.F.; Reid, R.G.; Yousef, S.; Boyle, S.P. A Specific, Accurate, and Sensitive Measure of Total Plasma Malondialdehyde by HPLC. *J. Lipid Res.* **2013**, *54*, 852–858. [CrossRef] [PubMed]
22. Espinosa, O.; Jiménez-Almazán, J.; Chaves, F.J.; Tormos, M.C.; Clapes, S.; Iradi, A.; Salvador, A.; Fandos, M.; Redón, J.; Sáez, G.T. Urinary 8-Oxo-7,8-Dihydro-2′-Deoxyguanosine (8-Oxo-dG), a Reliable Oxidative Stress Marker in Hypertension. *Free Radic. Res.* **2007**, *41*, 546–554. [CrossRef] [PubMed]
23. Loft, S.; Deng, X.-S.; Tuo, J.; Wellejus, A.; Sørensen, M.; Poulsen, H.E. Experimental Study of Oxidative DNA Damage. *Free Radic. Res.* **1998**, *29*, 525–539. [CrossRef] [PubMed]

24. Ainsworth, B.E.; Haskell, W.L.; Herrmann, S.D.; Meckes, N.; Bassett, D.R.; Tudor-Locke, C.; Greer, J.L.; Vezina, J.; Whitt-Glover, M.C.; Leon, A.S. 2011 Compendium of Physical Activities: A Second Update of Codes and MET Values. *Med. Sci. Sports Exerc.* **2011**, *43*, 1575–1581. [CrossRef]
25. Levey, A.S.; Stevens, L.A.; Schmid, C.H.; Zhang, Y.; Castro, A.F.; Feldman, H.I.; Kusek, J.W.; Eggers, P.; Van Lente, F.; Greene, T.; et al. A New Equation to Estimate Glomerular Filtration Rate. *Ann. Intern. Med.* **2009**, *150*, 604–612. [CrossRef]
26. Kanehisa, M.; Furumichi, M.; Tanabe, M.; Sato, Y.; Murishima, K. KEGG: New perspectives on genomes, pathways, diseases and drugs. *Nucleic Acids Res.* **2017**, *45*, 353–361. [CrossRef]
27. Pang, Z.; Chong, J.; Zhou, G.; de Lima Morais, D.A.; Chang, L.; Barrette, M.; Gauthier, C.; Jacques, P.E.; Li, S.; Xia, J. MetaboAnalyst 5.0: Narrowing the gap between raw spectra and functional insights. *Nucleics Acids Res.* **2021**, *49*, 388–396. [CrossRef]
28. Gelain, D.P. A Systematic Review of Human Antioxidant Genes. *Front. Biosci.* **2009**, *14*, 4457–4463. [CrossRef]
29. Del Toro, N.; Shrivastava, A.; Ragueneau, E.; Meldal, B.; Combe, C.; Barrera, E.; Perfetto, L.; How, K.; Ratan, P.; Shirodkar, G.; et al. The IntAct Database: Efficient Access to Fine-Grained Molecular Interaction Data. *Nucleic Acids Res.* **2022**, *50*, D648–D653. [CrossRef]
30. Shannon, P.; Markiel, A.; Ozier, O.; Baliga, N.S.; Wang, J.T.; Ramage, D.; Amin, N.; Schwikowski, B.; Ideker, T. Cytoscape: A Software Environment for Integrated Models of Biomolecular Interaction Networks. *Genome Res.* **2003**, *13*, 2498–2504. [CrossRef]
31. Jérolon, A.; Baglietto, L.; Birmelé, E.; Alarcon, F.; Perduca, V. Causal Mediation Analysis in Presence of Multiple Mediators Uncausally Related. *Int. J. Biostat.* **2021**, *17*, 191–221. [CrossRef]
32. Morishita, M.; Wang, L.; Speth, K.; Zhou, N.; Bard, R.L.; Li, F.; Brook, J.R.; Rajagopalan, S.; Brook, R.D. Acute Blood Pressure and Cardiovascular Effects of Near-Roadway Exposures With and Without N95 Respirators. *Am. J. Hypertens.* **2019**, *32*, 1054–1065. [CrossRef]
33. Rider, C.F.; Carlsten, C. Air Pollution and DNA Methylation: Effects of Exposure in Humans. *Clin. Epigenetics* **2019**, *11*, 131. [CrossRef]
34. Ribeiro, A.G.; Downward, G.S.; de Freitas, C.U.; Chiaravalloti Neto, F.; Cardoso, M.R.A.; Latorre, M. do R.D. de O.; Hystad, P.; Vermeulen, R.; Nardocci, A.C. Incidence and Mortality for Respiratory Cancer and Traffic-Related Air Pollution in São Paulo, Brazil. *Environ. Res.* **2019**, *170*, 243–251. [CrossRef]
35. Tiotiu, A.I.; Novakova, P.; Nedeva, D.; Chong-Neto, H.J.; Novakova, S.; Steiropoulos, P.; Kowal, K. Impact of Air Pollution on Asthma Outcomes. *Int. J. Environ. Res. Public Health* **2020**, *17*, 6212. [CrossRef] [PubMed]
36. Dragonieri, S.; Musti, M.; Izzo, C.; Esposito, L.M.; Foschino Barbaro, M.P.; Resta, O.; Spanevello, A. Sputum Induced Cellularity in a Group of Traffic Policemen. *Sci. Total Environ.* **2006**, *367*, 433–436. [CrossRef]
37. Zhang, Q.; Du, X.; Li, H.; Jiang, Y.; Zhu, X.; Zhang, Y.; Niu, Y.; Liu, C.; Ji, J.; Chillrud, S.N.; et al. Cardiovascular Effects of Traffic-Related Air Pollution: A Multi-Omics Analysis from a Randomized, Crossover Trial. *J. Hazard. Mater.* **2022**, *435*, 129031. [CrossRef] [PubMed]
38. Miyajima, M. Amino Acids: Key Sources for Immunometabolites and Immunotransmitters. *Int. Immunol.* **2020**, *32*, 435–446. [CrossRef]
39. Wu, G. Amino Acids: Metabolism, Functions, and Nutrition. *Amino Acids* **2009**, *37*, 1–17. [CrossRef] [PubMed]
40. Tinkov, A.A.; Skalnaya, M.G.; Skalny, A.V. Serum Trace Element and Amino Acid Profile in Children with Cerebral Palsy. *J. Trace Elem. Med. Biol. Organ Soc. Miner. Trace Elem. GMS* **2021**, *64*, 126685. [CrossRef]
41. Zhang, Z.-Y.; Monleon, D.; Verhamme, P.; Staessen, J.A. Branched-Chain Amino Acids as Critical Switches in Health and Disease. *Hypertens. Dallas Tex 1979* **2018**, *72*, 1012–1022. [CrossRef] [PubMed]
42. Hu, X.; Yan, M.; He, L.; Qiu, X.; Zhang, J.; Zhang, Y.; Mo, J.; Day, D.B.; Xiang, J.; Gong, J. Associations between Time-Weighted Personal Air Pollution Exposure and Amino Acid Metabolism in Healthy Adults. *Environ. Int.* **2021**, *156*, 106623. [CrossRef] [PubMed]
43. Feng, B.; Liu, C.; Yi, T.; Song, X.; Wang, Y.; Liu, S.; Chen, J.; Zhao, Q.; Zhang, Y.; Wang, T.; et al. Perturbation of Amino Acid Metabolism Mediates Air Pollution Associated Vascular Dysfunction in Healthy Adults. *Environ. Res.* **2021**, *201*, 111512. [CrossRef] [PubMed]
44. Yu, Z.; Mao, X.; Tang, M.; Chen, Y.; Wu, M.; Jin, M.; Wang, J.; Xu, L.; Ye, G.; Ding, J.; et al. Association between Past Exposure to Fine Particulate Matter (PM2.5) and Peptic Ulcer: A Cross-Sectional Study in Eastern China. *Chemosphere* **2021**, *265*, 128706. [CrossRef] [PubMed]
45. Flint, H.J.; Scott, K.P.; Louis, P.; Duncan, S.H. The Role of the Gut Microbiota in Nutrition and Health. *Nat. Rev. Gastroenterol. Hepatol.* **2012**, *9*, 577–589. [CrossRef]
46. Adak, A.; Khan, M.R. An Insight into Gut Microbiota and Its Functionalities. *Cell Mol. Life Sci.* **2019**, *76*, 473–493. [CrossRef] [PubMed]
47. Abbasalizad Farhangi, M.; Vajdi, M. Gut Microbiota–Associated Trimethylamine N-oxide and Increased Cardiometabolic Risk in Adults: A Systematic Review and Dose-Response Meta-Analysis. *Nutr. Rev.* **2021**, *79*, 1022–1042. [CrossRef]
48. Hayes, R.B.; Lim, C.; Zhang, Y.; Cromar, K.; Shao, Y.; Reynolds, H.R.; Silverman, D.T.; Jones, R.R.; Park, Y.; Jerrett, M.; et al. PM2.5 Air Pollution and Cause-Specific Cardiovascular Disease Mortality. *Int. J. Epidemiol.* **2020**, *49*, 25–35. [CrossRef]
49. Qiu, Z.; Li, G.; An, T. In Vitro Toxic Synergistic Effects of Exogenous Pollutants-Trimethylamine and Its Metabolites on Human Respiratory Tract Cells. *Sci. Total Environ.* **2021**, *783*, 146915. [CrossRef]

50. Tang, W.H.W.; Wang, Z.; Kennedy, D.J.; Wu, Y.; Buffa, J.A.; Agatisa-Boyle, B.; Li, X.S.; Levison, B.S.; Hazen, S.L. Gut Microbiota-Dependent Trimethylamine N-Oxide (TMAO) Pathway Contributes to Both Development of Renal Insufficiency and Mortality Risk in Chronic Kidney Disease. *Circ. Res.* **2015**, *116*, 448–455. [CrossRef]
51. Liang, D.; Ladva, C.N.; Golan, R.; Yu, T.; Walker, D.I.; Sarnat, S.E.; Greenwald, R.; Uppal, K.; Tran, V.; Jones, D.P.; et al. Perturbations of the Arginine Metabolome Following Exposures to Traffic-Related Air Pollution in a Panel of Commuters with and without Asthma. *Environ. Int.* **2019**, *127*, 503–513. [CrossRef]
52. Alderete, T.L.; Jones, R.B.; Chen, Z.; Kim, J.S.; Habre, R.; Lurmann, F.; Gilliland, F.D.; Goran, M.I. Exposure to Traffic-Related Air Pollution and the Composition of the Gut Microbiota in Overweight and Obese Adolescents. *Environ. Res.* **2018**, *161*, 472–478. [CrossRef]
53. Bell, G.; Mora, S.; Greenland, P.; Tsai, M.; Gill, E.; Kaufman, J.D. Association of Air Pollution Exposures With High-Density Lipoprotein Cholesterol and Particle Number. *Arterioscler. Thromb. Vasc. Biol.* **2017**, *37*, 976–982. [CrossRef] [PubMed]
54. Kim, J.S.; Chen, Z.; Alderete, T.L.; Toledo-Corral, C.; Lurmann, F.; Berhane, K.; Gilliland, F.D. Associations of Air Pollution, Obesity and Cardiometabolic Health in Young Adults: The Meta-AIR Study. *Environ. Int.* **2019**, *133*, 105180. [CrossRef] [PubMed]
55. Araujo, J.A. Particulate Air Pollution, Systemic Oxidative Stress, Inflammation, and Atherosclerosis. *Air Qual. Atmos. Health* **2011**, *4*, 79–93. [CrossRef] [PubMed]
56. Feng, J.; Cavallero, S.; Hsiai, T.; Li, R. Impact of Air Pollution on Intestinal Redox Lipidome and Microbiome. *Free Radic. Biol. Med.* **2020**, *151*, 99–110. [CrossRef]
57. Gao, B.; Chi, L.; Mahbub, R.; Bian, X.; Tu, P.; Ru, H.; Lu, K. Multi-Omics Reveals That Lead Exposure Disturbs Gut Microbiome Development, Key Metabolites, and Metabolic Pathways. *Chem. Res. Toxicol.* **2017**, *30*, 996–1005. [CrossRef]
58. Block, M.L.; Calderón-Garcidueñas, L. Air Pollution: Mechanisms of Neuroinflammation and CNS Disease. *Trends Neurosci.* **2009**, *32*, 506–516. [CrossRef]
59. Craig, L.; Brook, J.R.; Chiotti, Q.; Croes, B.; Gower, S.; Hedley, A.; Krewski, D.; Krupnick, A.; Krzyzanowski, M.; Moran, M.D.; et al. Air Pollution and Public Health: A Guidance Document for Risk Managers. *J. Toxicol. Environ. Health A* **2008**, *71*, 588–698. [CrossRef]
60. Delgado-Roche, L.; Mesta, F. Oxidative Stress as Key Player in Severe Acute Respiratory Syndrome Coronavirus (SARS-CoV) Infection. *Arch. Med. Res.* **2020**, *51*, 384–387. [CrossRef]
61. Larosa, V.; Remacle, C. Insights into the Respiratory Chain and Oxidative Stress. *Biosci. Rep.* **2018**, *38*, BSR20171492. [CrossRef] [PubMed]
62. Rodríguez-Moro, G.; Roldán, F.N.; Baya-Arenas, R.; Arias-Borrego, A.; Callejón-Leblic, B.; Gómez-Ariza, J.L.; García-Barrera, T. Metabolic Impairments, Metal Traffic, and Dyshomeostasis Caused by the Antagonistic Interaction of Cadmium and Selenium Using Organic and Inorganic Mass Spectrometry. *Environ. Sci. Pollut. Res.* **2020**, *27*, 1762–1775. [CrossRef] [PubMed]
63. Tsikas, D. Assessment of Lipid Peroxidation by Measuring Malondialdehyde (MDA) and Relatives in Biological Samples: Analytical and Biological Challenges. *Anal. Biochem.* **2017**, *524*, 13–30. [CrossRef] [PubMed]
64. Gregorczyk-Maga, I.; Celejewska-Wojcik, N.; Gosiewska-Pawlica, D.; Darczuk, D.; Kesek, B.; Maga, M.; Wojcik, K. Exposure to Air Pollution and Oxidative Stress Markers in Patients with Potentially Malignant Oral Disorders. *J. Physiol. Pharmacol. Off. J. Pol. Physiol. Soc.* **2019**, *70*, 115–120. [CrossRef]
65. Patel, M.M.; Chillrud, S.N.; Deepti, K.C.; Ross, J.M.; Kinney, P.L. Traffic-Related Air Pollutants and Exhaled Markers of Airway Inflammation and Oxidative Stress in New York City Adolescents. *Environ. Res.* **2013**, *121*, 71–78. [CrossRef] [PubMed]
66. Guilbert, A.; De Cremer, K.; Heene, B.; Demoury, C.; Aerts, R.; Declerck, P.; Brasseur, O.; Van Nieuwenhuyse, A. Personal Exposure to Traffic-Related Air Pollutants and Relationships with Respiratory Symptoms and Oxidative Stress: A Pilot Cross-Sectional Study among Urban Green Space Workers. *Sci. Total Environ.* **2019**, *649*, 620–628. [CrossRef] [PubMed]
67. Martín-Grau, M.; Pardo-Tendero, M.; Casanova, P.; Dromant, M.; Marrachelli, V.G.; Morales, J.M.; Borrás, C.; Pisoni, S.; Maestrini, S.; Di Blasio, A.M.; et al. Altered Lipid Moieties and Carbonyls in a Wistar Rat Dietary Model of Subclinical Fatty Liver: Potential Sex-Specific Biomarkers of Early Fatty Liver Disease? *Antioxidants* **2023**, *12*, 1808. [CrossRef] [PubMed]
68. Luca, M.; Di Mauro, M.; Di Mauro, M.; Luca, A. Gut Microbiota in Alzheimer's Disease, Depression, and Type 2 Diabetes Mellitus: The Role of Oxidative Stress. *Hindawi* **2019**, *2019*, 4730539. [CrossRef]
69. Wang, S.; Zhou, Q.; Tian, Y.; Hu, X. The Lung Microbiota Affects Pulmonary Inflammation and Oxidative Stress Induced by PM2.5 Exposure. *Environ. Sci. Technol.* **2022**, *56*, 12368–12379. [CrossRef]
70. Axelsen, P.H.; Komatsu, H.; Murray, I.V.J. Oxidative Stress and Cell Membranes in the Pathogenesis of Alzheimer's Disease. *Physiology* **2011**, *26*, 54–69. [CrossRef]
71. Yokota, M.; Yahagi, S.; Masaki, H. Ethyl 2,4-Dicarboethoxy Pantothenate, a Derivative of Pantothenic Acid, Prevents Cellular Damage Initiated by Environmental Pollutants through Nrf2 Activation. *J. Dermatol. Sci.* **2018**, *92*, 162–171. [CrossRef] [PubMed]
72. Kayaaltı, Z.; Aliyev, V.; Söylemezoğlu, T. The Potential Effect of Metallothionein 2A −5 A/G Single Nucleotide Polymorphism on Blood Cadmium, Lead, Zinc and Copper Levels. *Toxicol. Appl. Pharmacol.* **2011**, *256*, 1–7. [CrossRef] [PubMed]
73. Ling, X.-B.; Wei, H.-W.; Wang, J.; Kong, Y.-Q.; Wu, Y.-Y.; Guo, J.-L.; Li, T.-F.; Li, J.-K. Mammalian Metallothionein-2A and Oxidative Stress. *Int. J. Mol. Sci.* **2016**, *17*, 1483. [CrossRef] [PubMed]
74. Wang, K.; Fang, S.; Liu, Q.; Gao, J.; Wang, X.; Zhu, H.; Zhu, Z.; Ji, F.; Wu, J.; Ma, Y.; et al. TGF-B1/P65/MAT2A Pathway Regulates Liver Fibrogenesis via Intracellular SAM. *EBioMedicine* **2019**, *42*, 458–469. [CrossRef]

75. Chen, Y.-I.; Wei, P.-C.; Hsu, J.-L.; Su, F.-Y.; Lee, W.-H. NPGPx (GPx7): A Novel Oxidative Stress Sensor/Transmitter with Multiple Roles in Redox Homeostasis. *Am. J. Transl. Res.* **2016**, *8*, 1626–1640. [PubMed]
76. Wang, L.; Zhang, L.; Niu, Y.; Sitia, R.; Wang, C. Glutathione Peroxidase 7 Utilizes Hydrogen Peroxide Generated by Ero1α to Promote Oxidative Protein Folding. *Antioxid. Redox Signal.* **2014**, *20*, 545–556. [CrossRef]
77. Niittynen, M.; Tuomisto, J.T.; Pohjanvirta, R. Effect of 2,3,7,8-Tetrachlorodibenzo-p-Dioxin (TCDD) on Heme Oxygenase-1, Biliverdin IXalpha Reductase and Delta-Aminolevulinic Acid Synthetase 1 in Rats with Wild-Type or Variant AH Receptor. *Toxicology* **2008**, *250*, 132–142. [CrossRef]
78. Tsaprouni, L.G.; Yang, T.-P.; Bell, J.; Dick, K.J.; Kanoni, S.; Nisbet, J.; Viñuela, A.; Grundberg, E.; Nelson, C.P.; Meduri, E.; et al. Cigarette Smoking Reduces DNA Methylation Levels at Multiple Genomic Loci but the Effect Is Partially Reversible upon Cessation. *Epigenetics* **2014**, *9*, 1382–1396. [CrossRef]
79. Yang, C.; Rodionov, D.A.; Rodionova, I.A.; Li, X.; Osterman, A.L. Glycerate 2-Kinase of Thermotoga Maritima and Genomic Reconstruction of Related Metabolic Pathways. *J. Bacteriol.* **2008**, *190*, 1773–1782. [CrossRef]
80. Boothe, V.L.; Boehmer, T.K.; Wendel, A.M.; Yip, F.Y. Residential Traffic Exposure and Childhood Leukemia: A Systematic Review and Meta-Analysis. *Am. J. Prev. Med.* **2014**, *46*, 413–422. [CrossRef]
81. Carlos-Wallace, F.M.; Zhang, L.; Smith, M.T.; Rader, G.; Steinmaus, C. Parental, In Utero, and Early-Life Exposure to Benzene and the Risk of Childhood Leukemia: A Meta-Analysis. *Am. J. Epidemiol.* **2016**, *183*, 1–14. [CrossRef]
82. Filippini, T.; Heck, J.E.; Malagoli, C.; Giovane, C.D.; Vinceti, M. A Review and Meta-Analysis of Outdoor Air Pollution and Risk of Childhood Leukemia. *J. Environ. Sci. Health Part C* **2015**, *33*, 36–66. [CrossRef] [PubMed]
83. Cheng, M.; Galbally, I.E.; Molloy, S.B.; Selleck, P.W.; Keywood, M.D.; Lawson, S.J.; Powell, J.C.; Gillett, R.W.; Dunne, E. Factors Controlling Volatile Organic Compounds in Dwellings in Melbourne, Australia. *Indoor Air* **2016**, *26*, 219–230. [CrossRef] [PubMed]
84. Ilacqua, V.; Dawson, J.; Breen, M.; Singer, S.; Berg, A. Effects of Climate Change on Residential Infiltration and Air Pollution Exposure. *J. Expo. Sci. Environ. Epidemiol.* **2017**, *27*, 16–23. [CrossRef] [PubMed]
85. Deffner, V.; Küchenhoff, H.; Maier, V.; Pitz, M.; Cyrys, J.; Breitner, S.; Schneider, A.; Gu, J.; Geruschkat, U.; Peters, A. Personal Exposure to Ultrafine Particles: Two-Level Statistical Modeling of Background Exposure and Time-Activity Patterns during Three Seasons. *J. Expo. Sci. Environ. Epidemiol.* **2016**, *26*, 17–25. [CrossRef]
86. Depeint, F.; Bruce, W.R.; Shangari, N.; Mehta, R.; O'Brien, P.J. Mitochondrial Function and Toxicity: Role of the B Vitamin Family on Mitochondrial Energy Metabolism. *Chem. Biol. Interact.* **2006**, *163*, 94–112. [CrossRef] [PubMed]
87. Leonardi, R.; Jackowski, S. Biosynthesis of Pantothenic Acid and Coenzyme A. *EcoSal Plus* **2007**, *2*. [CrossRef]
88. Santambrogio, P.; Ripamonti, M.; Cozzi, A.; Raimondi, M.; Cavestro, C.; Di Meo, I.; Rubio, A.; Taverna, S.; Tiranti, V.; Levi, S. Massive Iron Accumulation in PKAN-Derived Neurons and Astrocytes: Light on the Human Pathological Phenotype. *Cell Death Dis.* **2022**, *13*, 1–12. [CrossRef]
89. Yamanaka, K.; Urano, Y.; Takabe, W.; Saito, Y.; Noguchi, N. Induction of Apoptosis and Necroptosis by 24(S)-Hydroxycholesterol Is Dependent on Activity of Acyl-CoA:Cholesterol Acyltransferase 1. *Cell Death Dis.* **2014**, *5*, e990. [CrossRef]
90. Swanson, M.A.; Miller, K.; Young, S.P.; Tong, S.; Ghaloul-Gonzalez, L.; Neira-Fresneda, J.; Schlichting, L.; Peck, C.; Gabel, L.; Friederich, M.W.; et al. Cerebrospinal Fluid Amino Acids Glycine, Serine, and Threonine in Nonketotic Hyperglycinemia. *J. Inherit. Metab. Dis.* **2022**, *45*, 734–747. [CrossRef]
91. Razak, M.A.; Begum, P.S.; Viswanath, B.; Rajagopal, S. Multifarious Beneficial Effect of Nonessential Amino Acid, Glycine: A Review. *Oxid. Med. Cell. Longev.* **2017**, *2017*, 1716701. [CrossRef]
92. Ecker, K.; Hengst, L. Skp2: Caught in the Akt. *Nat. Cell Biol.* **2009**, *11*, 377–379. [CrossRef] [PubMed]
93. Ibba, M.; Söll, D. Aminoacyl-TRNA Synthesis. *Annu. Rev. Biochem.* **2000**, *69*, 617–650. [CrossRef] [PubMed]
94. Goodarzi, H.; Nguyen, H.C.B.; Zhang, S.; Dill, B.D.; Molina, H.; Tavazoie, S.F. Modulated Expression of Specific TRNAs Drives Gene Expression and Cancer Progression. *Cell* **2016**, *165*, 1416–1427. [CrossRef] [PubMed]
95. Torres, A.G.; Batlle, E.; Ribas de Pouplana, L. Role of TRNA Modifications in Human Diseases. *Trends Mol. Med.* **2014**, *20*, 306–314. [CrossRef] [PubMed]
96. Jones, O.A.; Maguire, M.L.; Griffin, J.L. Environmental Pollution and Diabetes: A Neglected Association. *Lancet* **2008**, *371*, 287–288. [CrossRef] [PubMed]
97. Akifumi, E.; Hidenobu, M.; Chisato, M. The Effects of Early Postnatal Exposure to a Low Dose of Decabromodiphenyl Ether (BDE-209) on Serum Metabolites in Male Mice. *J. Toxicol. Sci.* **2016**, *41*, 667–675. [CrossRef]
98. Gray, L.R.; Tompkins, S.C.; Taylor, E.B. Regulation of Pyruvate Metabolism and Human Disease. *Cell. Mol. Life Sci.* **2014**, *71*, 2577–2604. [CrossRef]
99. Kunze, M.; Hartig, A. Permeability of the Peroxisomal Membrane: Lessons from the Glyoxylate Cycle. *Front. Physiol.* **2013**, *4*. [CrossRef]
100. Li, J.; Hu, Y.; Liu, L.; Wang, Q.; Zeng, J.; Chen, C. PM2.5 Exposure Perturbs Lung Microbiome and Its Metabolic Profile in Mice. *Sci. Total Environ.* **2020**, *721*, 137432. [CrossRef]
101. Almannai, M.; Wang, J.; Dai, H.; El-Hattab, A.W.; Faqeih, E.A.; Saleh, M.A.; Al Asmari, A.; Alwadei, A.H.; Aljadhai, Y.I.; AlHashem, A.; et al. FARS2 Deficiency; New Cases, Review of Clinical, Biochemical, and Molecular Spectra, and Variants Interpretation Based on Structural, Functional, and Evolutionary Significance. *Mol. Genet. Metab.* **2018**, *125*, 281–291. [CrossRef] [PubMed]

102. Cao, T.; Gao, Z.; Gu, L.; Chen, M.; Yang, B.; Cao, K.; Huang, H.; Li, M. AdipoR1/APPL1 Potentiates the Protective Effects of Globular Adiponectin on Angiotensin II-Induced Cardiac Hypertrophy and Fibrosis in Neonatal Rat Atrial Myocytes and Fibroblasts. *PLoS ONE* **2014**, *9*, e103793. [CrossRef]
103. Liu, R.; Meng, J.; Lou, D. Adiponectin Inhibits D-gal-induced Cardiomyocyte Senescence via AdipoR1/APPL1. *Mol. Med. Rep.* **2021**, *24*, 1–10. [CrossRef] [PubMed]
104. Yamano, K.; Youle, R.J. Two Different Axes CALCOCO2-RB1CC1 and OPTN-ATG9A Initiate PRKN-Mediated Mitophagy. *Autophagy* **2020**, *16*, 2105–2107. [CrossRef]
105. Zhang, C.; Nie, P.; Zhou, C.; Hu, Y.; Duan, S.; Gu, M.; Jiang, D.; Wang, Y.; Deng, Z.; Chen, J.; et al. Oxidative Stress-Induced Mitophagy Is Suppressed by the MiR-106b-93-25 Cluster in a Protective Manner. *Cell Death Dis.* **2021**, *12*, 209. [CrossRef]
106. De Freitas-Silva, L.; Rodríguez-Ruiz, M.; Houmani, H.; da Silva, L.C.; Palma, J.M.; Corpas, F.J. Glyphosate-Induced Oxidative Stress in Arabidopsis Thaliana Affecting Peroxisomal Metabolism and Triggers Activity in the Oxidative Phase of the Pentose Phosphate Pathway (OxPPP) Involved in NADPH Generation. *J. Plant Physiol.* **2017**, *218*, 196–205. [CrossRef]
107. Li, X.; Schimenti, J.C. Mouse Pachytene Checkpoint 2 (Trip13) Is Required for Completing Meiotic Recombination but Not Synapsis. *PLoS Genet.* **2007**, *3*, e130. [CrossRef]
108. Clairmont, C.S.; Sarangi, P.; Ponnienselvan, K.; Galli, L.D.; Csete, I.; Moreau, L.; Adelmant, G.; Chowdhury, D.; Marto, J.A.; D'Andrea, A.D. TRIP13 Regulates DNA Repair Pathway Choice through REV7 Conformational Change. *Nat. Cell Biol.* **2020**, *22*, 87–96. [CrossRef]
109. Yuan, Q.; Tang, B.; Zhang, C. Signaling Pathways of Chronic Kidney Diseases, Implications for Therapeutics. *Signal Transduct. Target. Ther.* **2022**, *7*, 182. [CrossRef]
110. Park, S.; Kim, S.; Kim, M.J.; Hong, Y.; Lee, A.Y.; Lee, H.; Tran, Q.; Kim, M.; Cho, H.; Park, J.; et al. GOLGA2 Loss Causes Fibrosis with Autophagy in the Mouse Lung and Liver. *Biochem. Biophys. Res. Commun.* **2018**, *495*, 594–600. [CrossRef]
111. Kubota, Y.; Nomura, K.; Katoh, Y.; Yamashita, R.; Kaneko, K.; Furuyama, K. Novel Mechanisms for Heme-Dependent Degradation of ALAS1 Protein as a Component of Negative Feedback Regulation of Heme Biosynthesis. *J. Biol. Chem.* **2016**, *291*, 20516–20529. [CrossRef]
112. Papaconstantinou, J. The Role of Signaling Pathways of Inflammation and Oxidative Stress in Development of Senescence and Aging Phenotypes in Cardiovascular Disease. *Cells* **2019**, *8*, 1383. [CrossRef]
113. Xie, X.; Deng, T.; Duan, J.; Xie, J.; Yuan, J.; Chen, M. Exposure to Polystyrene Microplastics Causes Reproductive Toxicity through Oxidative Stress and Activation of the P38 MAPK Signaling Pathway. *Ecotoxicol. Environ. Saf.* **2020**, *190*, 110133. [CrossRef] [PubMed]
114. Newey, P.J.; Bowl, M.R.; Thakker, R.V. Parafibromin--Functional Insights. *J. Intern. Med.* **2009**, *266*, 84–98. [CrossRef] [PubMed]
115. Deska, M.; Romuk, E.; Segiet, O.; Polczyk, J.; Buła, G.; Gawrychowski, J. Oxidative Stress in Proliferative Lesions of Parathyroid Gland. *Pol. Przegl. Chir.* **2018**, *91*, 29–34. [CrossRef]
116. Moulin, P.; Vaysse, F.; Bieth, E.; Mornet, E.; Gennero, I.; Dalicieux-Laurencin, S.; Baunin, C.; Tauber, M.T.; De Gauzy, J.S.; Salles, J.P. Hypophosphatasia May Lead to Bone Fragility: Don't Miss It. *Eur. J. Pediatr.* **2009**, *168*, 783–788. [CrossRef] [PubMed]
117. Wu, D.-D.; Irwin, D.M. Evolution of Trichocyte Keratin Associated Proteins. In *The Hair Fibre: Proteins, Structure and Development*; Plowman, J.E., Harland, D.P., Deb-Choudhury, S., Eds.; Advances in Experimental Medicine and Biology; Springer: Singapore, 2018; pp. 47–56. ISBN 978-981-10-8195-8.
118. Si, M.; Lang, J. The Roles of Metallothioneins in Carcinogenesis. *J. Hematol. Oncol.* **2018**, *11*, 107. [CrossRef]
119. Baszuk, P.; Janasik, B.; Pietrzak, S.; Marciniak, W.; Reszka, E.; Białkowska, K.; Jabłońska, E.; Muszyńska, M.; Lesicka, M.; Derkacz, R.; et al. Lung Cancer Occurrence-Correlation with Serum Chromium Levels and Genotypes. *Biol. Trace Elem. Res.* **2021**, *199*, 1228–1236. [CrossRef]
120. Williams, E.L.; Acquaviva, C.; Amoroso, A.; Chevalier, F.; Coulter-Mackie, M.; Monico, C.G.; Giachino, D.; Owen, T.; Robbiano, A.; Salido, E.; et al. Primary Hyperoxaluria Type 1: Update and Additional Mutation Analysis of the AGXT Gene. *Hum. Mutat.* **2009**, *30*, 910–917. [CrossRef] [PubMed]
121. Apanasets, O.; Grou, C.P.; Van Veldhoven, P.P.; Brees, C.; Wang, B.; Nordgren, M.; Dodt, G.; Azevedo, J.E.; Fransen, M. PEX5, the Shuttling Import Receptor for Peroxisomal Matrix Proteins, Is a Redox-Sensitive Protein. *Traffic* **2014**, *15*, 94–103. [CrossRef]
122. Buday, K.; Conrad, M. Emerging Roles for Non-Selenium Containing ER-Resident Glutathione Peroxidases in Cell Signaling and Disease. *Biol. Chem.* **2021**, *402*, 271–287. [CrossRef] [PubMed]

antioxidants

MDPI

Article

Occupational Exposure to Metal-Based Nanomaterials: A Possible Relationship between Chemical Composition and Oxidative Stress Biomarkers

Valeria Bellisario [1,†], Giacomo Garzaro [1,†], Giulia Squillacioti [1], Marco Panizzolo [1], Federica Ghelli [1], Giuseppe Mariella [1], Roberto Bono [1,*], Irina Guseva Canu [2,‡] and Enrico Bergamaschi [1,‡]

1 Department of Public Health and Pediatrics, University of Turin, 10126 Turin, Italy; valeria.bellisario@unito.it (V.B.); giacomo.garzaro@unito.it (G.G.); giulia.squillacioti@unito.it (G.S.); marco.panizzolo@unito.it (M.P.); federica.ghelli@unito.it (F.G.); giuseppe.mariella@unito.it (G.M.); enrico.bergamaschi@unito.it (E.B.)

2 Department of Occupational and Environmental Health, Center for Primary Care and Public Health (Unisanté), University of Lausanne, 1010 Lausanne, Switzerland; irina.guseva-canu@unisante.ch

* Correspondence: roberto.bono@unito.it

† These authors contributed equally to this work and share first authorship.

‡ These authors contributed equally to this work and share senior authorship.

Abstract: Nanomaterials (NMs) are in high demand for a wide range of practical applications; however, comprehensively understanding the toxicity of these materials is a complex challenge, due to the limited availability of epidemiological evidence on the human health effects arising from workplace exposures. The aim of this work is to assess whether and how urinary metal concentrations could be reliable and useful in NM biomonitoring. In the framework of "NanoExplore Project" [EU LIFE17 Grant ENV/GR/000285], 43 not-exposed subjects and 40 exposed workers were recruited to measure exposure to NMs (PCN and LDSA) in the proximity of the workstations and biological biomarkers (urinary metal concentrations—Aluminum (Al), Silica (Si), Titanium (Ti), and Chromium (Cr); urinary OS biomarkers—TAP, Isop, and MDA). The results showed that Si and Ti were directly associated with NM exposure (both PCN and LDSA), as well as with OS biomarkers, especially in exposed workers. Moreover, the mediation analyses showed that Si could account for about 2.8% in the relationship between LDSA and OS biomarkers, possibly by decreasing OS antioxidant defenses in exposed people. In conclusion, our study provides evidence that occupational exposure to mixtures containing NMs can represent an underestimated hazard for exposed people, increasing the body burden and the oxidative balance.

Keywords: occupational air pollution exposure; NM biomonitoring; NM occupational exposure; nanosized metal; body burden; oxidative imbalance; antioxidant defenses

Citation: Bellisario, V.; Garzaro, G.; Squillacioti, G.; Panizzolo, M.; Ghelli, F.; Mariella, G.; Bono, R.; Guseva Canu, I.; Bergamaschi, E. Occupational Exposure to Metal-Based Nanomaterials: A Possible Relationship between Chemical Composition and Oxidative Stress Biomarkers. *Antioxidants* **2024**, *13*, 676. https://doi.org/10.3390/antiox13060676

Academic Editor: Yasuhiro Yoshida

Received: 24 April 2024
Revised: 28 May 2024
Accepted: 29 May 2024
Published: 31 May 2024

1. Introduction

Nanomaterials (NMs) have emerged as an exciting class of materials that are in high demand for a wide range of practical applications, with an approximate annual production volume of NMs reaching 60,000 tons [1–3]. As an excellent example of emerging nanotechnology, NMs currently show great potential and improved performances [4] in scratch-free paints, surface coatings, electronics, cosmetics, environmental remediation, sports equipment, sensors, and energy storage devices [5]. According to the European Commission, a nanomaterial is a natural, incidental, or manufactured material consisting of solid particles characterized by at least (a) one or more external dimensions in the size range of 1 nm–100 nm; (b) an elongated shape, such as a rod, fiber, or tube, where two external dimensions are smaller than 1 nm and the other dimension is larger than 100 nm; and (c) a plate-like shape, where one external dimension is smaller than 1 nm and the other dimensions are larger than 100 nm [6]. Carbon, metal, metal oxides, or organic substances are special classes of NMs, and the release of nanoparticles originating from the handling of micron-sized materials may be substantial in some occupational scenarios [7,8].

While the number of NM types and applications continues to increase, the knowledge on the health effects of nanoparticles exposure is still limited, even if the number of efforts aimed at determining the health risks associated with NM exposure continues to grow. As these nanoparticles are intentionally engineered to interact with cells, it is fundamental to ensure that these enhancements are not causing any adverse effects, undergoing biodegradation, or causing damage in the cellular environment [9]. For example, biodegraded nanoparticles may accumulate within cells and lead to intracellular changes, such as disruption of organelle integrity or gene alterations [10]. In vitro studies, using different cell lines and protocols, of chemically different nanoparticles have revealed a potential hazard; however, these surrogate models could only estimate what could occur after the interactions with biological systems. Moreover, these studies include a wide range of nanoparticle concentrations, making it difficult to determine whether the toxicity endpoints are relevant for human beings [11]. While in vitro nanoparticle applications allow less stringent toxicological characterization, in vivo studies of NMs require thorough understanding of the kinetics and toxicology of the particles. Although the toxicology of metals is a mature science, many interactions of metal nanoparticles with biological systems are still controversial.

Metal NMs interact with complex biological systems, such as the human body, and many epidemiological studies [8,12–15] have addressed the general role of these nanoparticles in neurodegeneration, neurotoxicity, and oxidative imbalance, as well as lung injury [16,17]. In particular, some works [17–19] suggest that Al could have some redox capacity and could be linked to oxidative damage, reacting with H_2O_2, producing Al superoxide radicals ($AlO_2\bullet^-$), and promoting the generation of ROS. Owing to its high reactivity, Al is primarily found forming insoluble oxides, whose toxicity seems to be related to the displacement of other biological cations (Ca II, Fe II, or Mg Ife). Silica and Titanium may impair antioxidant/oxidant status and activate the immune system, which is indicative of inflammatory responses, but the mechanisms are not clearly understood [20–22]. Chromium (Cr III and IV) [23] could induce DNA damage, as MN or aberrations, instead of oxidative stress imbalance and inflammation.

Therefore, fully understanding the toxicity of these materials is a complex task, especially because there has been limited occupational epidemiologic evidence of human health effects from workplace exposures to NMs [24]. Possible adverse health effects can result from additional sources of metal NMs, thus increasing their body burden. This can occur especially in occupational settings, where the mass of airborne particles, instead of more appropriate metrics, such as the particle number concentration, is still considered [7,25].

In this context, we aimed to assess whether and how urinary metal concentrations, particularly Aluminum (Al), Titanium (Ti), Chromium (Cr), and Silica (Si), a metalloid, could be reliable biomarkers of exposure for NM biomonitoring. Moreover, we aimed to assess whether urinary metals are associated with oxidative stress biomarkers with a mediation and trigger role. Finally, this work assessed whether NM exposure can affect oxidative stress imbalance, in the perspectives of improving surveillance protocols and preventive strategies correlated.

2. Materials and Methods

2.1. Study Sample and Exposure to ENMs

This work is based on sampling and data gathered in the framework of the multicenter prospective cohort study "NanoExplore Project" [EU LIFE17 Grant ENV/GR/000285] [25], which aimed to investigate the association between occupational exposure to NMs and different biomarkers in workers within the European Union and Switzerland. In this work, within this main and general aim, we assessed the association between nanoparticle exposure and some urinary metals and a metalloid excreted with urine, namely Aluminum (Al), Titanium (Ti), and Chromium (Cr) and Silica (Si).

The companies were enrolled in each country based on confirmed prior knowledge of their activities related to the handling of NMs. A preliminary visit was conducted in each of the enrolled companies by the company's environmental health and safety manager and an occupational hygienist associated with the NanoExplore study. Workers who had been previously identified as potentially exposed were invited to participate in the study. We analyzed 40 workers recruited in 4 industrial plants handling mixtures of materials containing a fraction of sub-micrometric and nanosized metal oxides (e.g., Alumina—Al_2O_3, Titanium dioxide—TiO_2, Chromium ores, Siliceous sands—SiO_2) for the production of paints, coatings, or construction chemicals. Furthermore, 43 people not exposed to chemicals and dusts were enrolled as a control group. All the subjects provided written informed consent to participate. The harmonized protocol, the cohort characteristics, the recruitment procedure, the study design, and the preliminary results have been reported elsewhere [25–27].

Exposure to airborne nanoparticles was measured by portable nanoparticle counters "DiSCminiTM" (Testo, Mönchaltorf, Switzerland) placed in the proximity of the workstations for two up to four consecutive working days, as already described [25,26]. The DiSCmini allowed measurements to be obtained based on two different metrics: (1) the particle number concentration (PCN, expressed as number of particles/cm^3) and (2) the lung-deposited surface area (LDSA, expressed as $\mu m_2/cm^3$). Exposure was defined a priori according to the working tasks, and data provided by the DiSCmini measurements were used as independent variables.

2.2. Biological Sampling and Biomarkers

Biological samples were collected twice, the first on the first day of the campaign week, before the shift, and the second on the fourth day, after the shift [25]. Spot urine samples were collected in the morning before the beginning of shift and were immediately aliquoted and frozen at −20 °C until analysis. Two types of biomarkers were measured in the urine samples: (1) urinary metal concentrations (Aluminum (Al), Titanium (Ti), and Chromium (Cr)), and a metalloid (Silica (Si)) and (2) urinary OS biomarkers (Isoprostane (Isop), Malondialdehyde (MDA), and Total Antioxidant Power (TAP)). The technical information on the analytical methods adopted are reported in the Supplementary Materials (Table S1). The urinary metal concentrations were determined by inductively coupled plasma mass spectrometry (ICP-MS). In accordance with the manufacturer's instructions, the OS biomarkers were analyzed as follows: IsoP concentrations using a competitive enzyme-linked immunoassay (EA85, Oxford, MI, USA); MDA concentrations (Thiobarbituric Acid Reactive Substances assay FR40 Oxford, MI, USA) and TAP concentrations using a colorimetric assay kit (Cupric ion reducing antioxidant capacity assay TA02, Oxford, MI, USA), respectively. In addition, the urinary creatinine concentration was measured, using the kinetic Jaffé technique, to normalize the concentration of urinary biomarkers according to urinary volume (μg/mg of creatinine). Only TAP was chosen for the analysis here, mainly because this biomarker has previously resulted in being the most sensitive urinary biomarker of OS [25].

2.3. Statistical Analysis

The statistical approach has been deeply described in a previous work [25]. In the above analysis, each urinary metal biomarker was considered as independent of the other biomarkers. A directed acyclic graph (DAG) was again performed using Dagitty (^{TM}), to better identify the potential confounders in the association between NM exposure, urinary metal concentrations, and OS biomarkers and, also, to identify the set of variables that should be included in the advanced models. As shown in Figure 1, these variables were working age, sex, smoking and alcohol habits, and the body mass index (BMI).

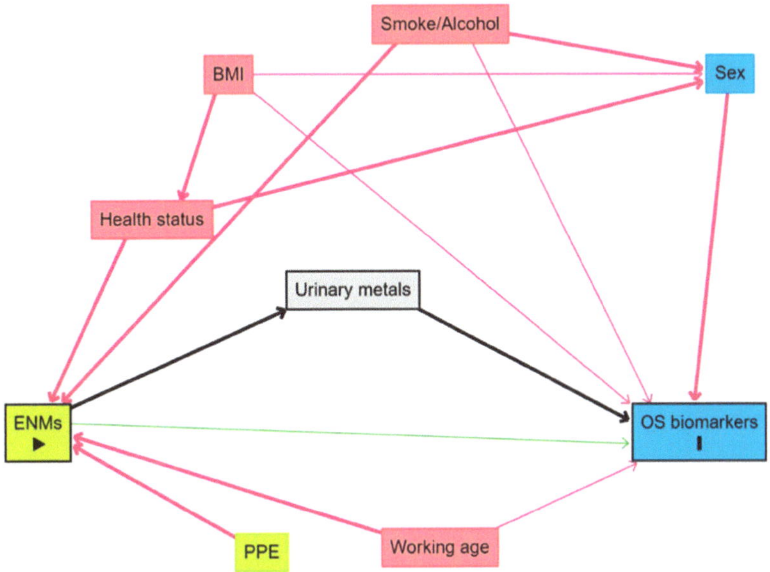

Figure 1. Directed acyclic graph (DAG) showing the assumed causal relationship between nanomaterial exposure and OS biomarkers. Abbreviations: exposure to nanomaterials (ENMs), personal protective equipment (PPE), body mass index (BMI).

For this specific work, we decided to maintain the original exposure classification of the workers, thus analyzing only the high exposures vs. the control group, excluding the low-exposure subjects from the analysis. The descriptive general analyses were reported elsewhere [25], but for the foreplay and descriptive analyses referred to in this work, the distributions of quantitative variables were summarized with means ± SD, medians, and minimum and maximum values, while categorical variables were presented as number and percentage. As the statistical distribution of some quantitative parameters was found to be non-Gaussian (Kolmogorov–Smirnov test), non-parametric tests were used to assess between-group differences (Mann–Whitney U test). A two-sided p-value < 0.05 was considered to indicate statistical significance. Finally, associations among the exposure variables were evaluated using Pearson's r coefficients.

The direct association between the urinary elements (Al, Si, Ti, and Cr), as the dependent variables, and the NM exposure metrics (PCN and LDSA), as the independent variables, was analyzed using a multilevel mixed-effects model, using $\log_{10}$-transformed variables. All the models considered urinary parameters as a single entity and were controlled for different potential confounders. The results were reported as relative risks (RR) with 95% confidence intervals (CIs). All the analyses were carried out using the STATA 16.1 software (StataCorp LLC: College Station, TX, USA).

Finally, the role of urinary metals as effect modifiers when assessing the association between urinary metals and OS biomarkers, represented by TAP concentrations, was explored using the medmod package (for Jamovi 2.3.28 and Rstudio 3.6 softwares), which showed the mediation estimates, as a percentage, and the individual path estimates with 95% confidence intervals (CIs).

3. Results

In this work, the main outcome was fixed by the analysis of the possible associations between NM exposure, represented by PCN and LDSA measurements, and some urinary biomarkers of exposure, i.e., urinary metal concentrations, and early biological effects, such as TAP concentrations. Table 1 summarizes the main characteristics of the sample extracted from the cohort described previously [25].

Table 1. Descriptive statistics of the sample extracted from the cohort.

	SAMPLE					
	Not Exposed (n = 43)		**Exposed (n = 40)**		**Total (n = 83)**	
Subjects	Male	21	Male	32	Male	53
	Female	22	Female	8	Female	30
BMI	Male	24.7 ± 3.4	Male	26.9 ± 4.1	Male	25.8 ± 5.8
	Female	22.2 ± 3.2	Female	22 ± 2.3	Female	22.1 ± 2.6
Tobacco smoke	No	39	No	28	No	67
	Yes	4	Yes	12	Yes	16
Alcohol	No	24	No	18	No	42
	Yes	19	Yes	22	Yes	41
Health score	76.6 ± 17.7		66.9 ± 21.1		70.5 ± 20	
Employment duration	<5 years	30	<5 years	20	<5 years	50
(years)	>5 years	13	>5 years	20	>5 years	33
PPE use	No	43	No	5	No	48
	Yes	-	Yes	35	Yes	35

	BIOLOGICAL BIOMARKERS			
	Not Exposed	**Exposed**	**Total**	**Non-Parametric Test**
PCN [#/cm^3]	3.7 ± 0.3	4.8 ± 0.3	4.4 ± 0.6	0.002
LDSA [μm^2/cm^3]	1.1 ± 0.3	1.9 ± 0.2	1.6 ± 0.5	0.003
Aluminum [μg/L]	44.7 ± 45.7	37 ± 37.9	44 ± 48.9	0.7
Silica [mg/L]	12.3 ± 5.1	19.6 ± 8.3	15.7 ± 8	0.02
Titanium [μg/L]	25.8 ± 10.1	33 ± 11.5	29.6 ± 11.4	0.03
Chromium [μg/L]	0.6 ± 0.04	0.3 ± 0.2	0.4 ± 0.5	0.3
MDA [μg/mg$_{CREA}$]	243 ± 196	235 ± 252	240 ± 218	0.5
Isop [μg/mg$_{CREA}$]	4.3 ± 2.8	4.1 ± 2.5	76.6 ± 17.7	0.16
TAP [μg/mg$_{CREA}$]	1.3 ± 0.4	0.9 ± 0.3	1.01 ± 0.9	0.001

Non-parametric correlation analyses were performed between the exposure data, urinary concentrations of nanosized metals, and OS biomarkers. TAP was associated with Silica (p = 0.03), Titanium (p < 0.001), Chromium (p = 0.004), PCN (p < 0.001), and LDSA (p < 0.001). Conversely, no associations were found with the other OS biomarkers (Isop and MDA).

All the models' results were presented (1) first in the whole sample and then (2) as stratified by exposure pattern (exposed vs. not-exposed subjects). Since the correlations did not show significant associations between the metals, the metals were analyzed one by one. We adopted mixed multilevel regression models with ID and center as random variables, and each model was adjusted according to the subjects' personal characteristics (BMI, smoking, health score, alcohol consumption, gender), working characteristics (exposure YES/NO, employment duration, use of PPE) and the quantitative variables measured (PCN, LDSA, urinary OS biomarkers). The results were summarized according to the type of exposure metrics (PCN and LDSA), and the significant relative risk with 95% confidence intervals (CIs) was reported.

3.1. Exposure

Table 2 and Figure 2 summarize the results of the models of urinary metal and NM exposure (PCN and LDSA). The Silica and Titanium urinary concentrations increased directly and were associated with increased PCN levels (Table 2, part A; coefficient = 4.7, p = 0.05, 95% C.I. 0.08/4.5; coefficient = 15.5, p = 0.02, 95% C.I. 2.2/23.2, respectively) in the whole-sample model. Interestingly, in the models stratified by exposure, the same associations were confirmed, albeit only in the exposed subjects and only for Silica and

Titanium (Table 2, part B; coefficient = 14.5, $p = 0.03$, 95% C.I. 1.1/29.03; coefficient = 9.3 $p = 0.04$, 95% C.I. 3.6/12.2, respectively). Silica and Titanium (Figure 2) were directly associated with LDSA in the whole-sample model (Table 2, part C; coefficient = 4.5, $p = 0.04$, 95% C.I. 0.6/16.7; coefficient = 11.4, $p = 0.03$, 95% C.I. 1.1/16.2, respectively). The same associations were confirmed in the models stratified by exposure (Figure 2), but only in the exposed subjects and only for Silica and Titanium (Table 2, part D; coefficient = 5, $p = 0.04$, 95% C.I. 0.8/8.8; coefficient = 2.5, $p = 0.02$, 95% C.I. 0.5/4.2, respectively).

Table 2. Multilevel mixed-effects models of urinary metal (dependent variables) and NM exposure metrics (PCN and LDSA) in the whole sample (parts A and C) and stratified by exposure (parts B and D).

		PCN			
Part A		**Multilevel Mixed-Effects Model (Whole Sample)**			
		Coeff.	**Std.Err.**	***p***	**[95% CI]**
Al		−5.6	6.2	0.08	−5.6/5.7
Si		4.7	4.8	0.05	0.08/4.5
Ti		15.5	7.2	0.02	2.2/23.2
Cr		−0.7	0.4	0.9	−1.5/0.11
Part B		**Multilevel Mixed-Effects Model (Exposure Stratified)**			
		Coeff.	**Std.Err.**	***p***	**[95% CI]**
Al	Exposed	−1.4	2.2	0.9	−4.1/1.3
	Not exposed	−6.3	4.8	0.3	−14.2/4.1
Si	Exposed	14.5	7.4	0.03	1.1/29.03
	Not exposed	3.7	10.9	0.7	−5.1/7.9
Ti	Exposed	9.3	11.7	0.04	3.6/12.2
	Not exposed	7.8	7.7	0.3	−6.8/12.6
Cr	Exposed	0.2	0.3	0.5	−0.3/0.7
	Not exposed	−1.9	0.9	0.4	−3.7/0.08
		LDSA			
Part C		**Multilevel Mixed-Effects Model (Whole Sample)**			
		Coeff.	**Std.Err.**	***p***	**[95% CI]**
Al		−7.3	9.3	0.8	−8.4/0.11
Si		4.5	6.2	0.04	0.6/16.7
Ti		11.4	10.1	0.03	1.1/16.2
Cr		0.05	0.5	0.9	−0.8/0.9
Part D		**Multilevel Mixed-Effects Model (Exposure Stratified)**			
		Coeff.	**Std.Err.**	***p***	**[95% CI]**
Al	Exposed	−7.8	6.7	0.2	−8.2/4.4
	Not exposed	−2.9	4.3	0.9	−4.3/6.1
Si	Exposed	5	3.6	0.04	0.8/8.8
	Not exposed	3.5	4.4	0.4	−4.7/7.1
Ti	Exposed	2.5	1.05	0.02	0.5/4.2
	Not exposed	−2.3	3.3	0.3	−5.6/2.6
Cr	Exposed	−0.6	0.4	0.1	−1.3/0.1
	Not exposed	0.7	0.8	0.4	−0.9/2.3

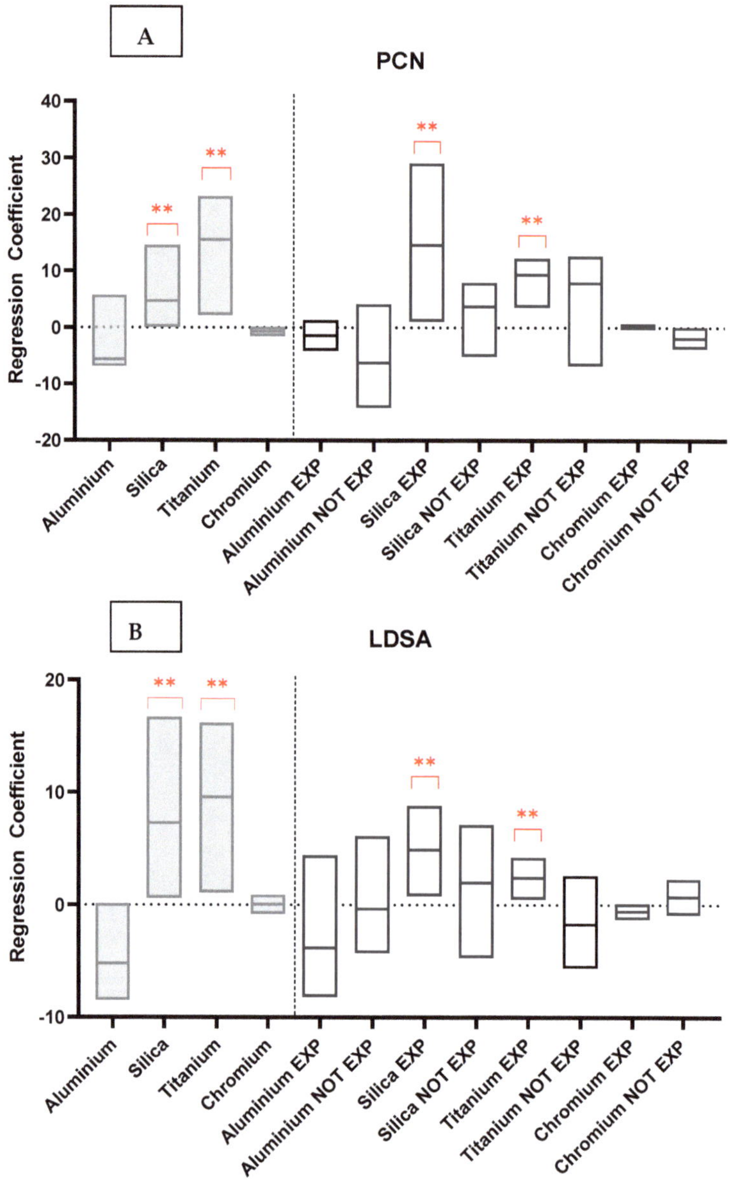

Figure 2. Multilevel mixed-effects models of urinary metal (dependent variables) and NM exposure metrics (PCN part **A** and LDSA part **B**) in the whole sample (left part in grey) and stratified by exposure (right part in white). Red stars show significant *p*-value.

3.2. Oxidative Stress

We assessed whether urinary elements were associated with OS biomarkers, and with TAP in particular; these were the only biomarkers that showed significant associations in previous studies [7,25,27].

Owing to the small size and the heterogeneity of the sample under study, the models were performed only in the whole sample (Figure 3). The models showed a significant decrease in TAP concentration with increasing urinary Silica and Titanium concentrations (coefficient = −0.7, p = 0.01, 95% C.I. −0.02/−0.002; coefficient = −0–008, p = 0.04, 95% C.I. −0.05/−0.0001, respectively).

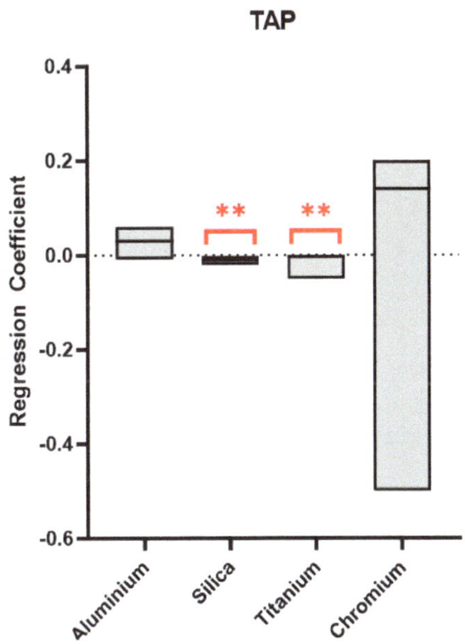

Figure 3. Multilevel mixed-effects models graph of TAP (dependent variable) and urinary element concentrations in the whole sample. Red stars show significant p-value.

3.3. Mediation Analysis

On the basis of the previous findings, we assessed whether metallic elements were associated with redox imbalance. Thus, we ran a raw mediation analysis between the exposure metrics (PCN and LDSA) and TAP to investigate the possible mediating role of urinary Silica and Titanium, as a proxy of NM exposure.

3.3.1. Silica Mediation Analysis of Particle Metrics towards TAP

This mediation analysis was conducted to examine the mediating effect of Silica on the relationship between PCN and TAP (Table 3-part A). The total effect of the model was statistically significant (b = −4.5, z = −6.5, 95% C.I. −5.9/−3.2, p < 0.001). It was also found that there was a statistically significant direct effect of PCN on TAP (b = −4.8, z = −6.9, 95% C.I. −6.7/−3.5, p < 0.001) and a slight mediated indirect effect (b = −0.26, z = −1.5, 95% C.I. −0.7/0.05, p = 0.04). These results suggest that Si could account for about 1.1% of the relationship between PCN and TAP, affecting OS antioxidant defenses.

The same model was applied to examine the mediating effect of Silica between LDSA and TAP (Table 3-part B). The total effect of the model was statistically significant (b = −6.6, z = −7.02, 95% C.I. −8.4/−4.7], p < 0.001). The direct effect of LDSA on TAP (b = −7.1, z = −7.7, 95% C.I. −8.9/−5.3], p < 0.001) and the mediated indirect effect of Si were also statistically significant (b = 0.14, z = 1.9, 95% C.I. 0.04/1.03, p = 0.05). These results suggest that Silica could account for about 2.8% in the relationship between LDSA and TAP, possibly by decreasing OS antioxidant defenses.

Table 3. Mediating analysis between exposure biomarkers (PCN and LDSA) and TAP concentrations with Si (part A and B) and Ti (part C and D) as mediating factors.

				Mediation Models: SILICA				
Part A				**PCN/Si/TAP Mediation estimates**				
Effect	**Label**	**Estimate**	**SE**	**Lower**	**Upper**	**Z**	***p***	**% Mediation**
Indirect	a × b	−0.26	0.169	−0.714	−0.059	−1.54	0.042	1.12
Direct	c	−4.826	0.691	−6.1798	−3.471	−6.98	<0.001	94.88
Total	c + a × b	−4.565	0.702	−5.9416	−3.189	−6.5	<0.001	100
Part B				**LDSA/Si/TAP Mediation estimates**				
Effect	**Label**	**Estimate**	**SE**	**Lower**	**Upper**	**Z**	***p***	**% Mediation**
Indirect	a × b	0.14	0.262	0.045	1.03	−1.96	0.05	2.78
Direct	c	−7.063	0.917	−8.86	−5.27	−7.7	<0.001	93.22
Total	c + a × b	−6.549	0.933	−8.38	−4.72	−7.02	<0.001	100
				Mediation Models: TITANIUM				
Part C				**PCN/Ti/TAP Mediation estimates**				
Effect	**Label**	**Estimate**	**SE**	**Lower**	**Upper**	**Z**	***p***	**% Mediation**
Indirect	a × b	−0.101	0.0974	−0.292	0.0897	−1.04	0.298	2.22
Direct	c	−4.464	0.7019	−5.84	−3.088	−6.36	<0.001	97.78
Total	c + a × b	−4.565	0.7023	−5.942	−3.1887	−6.5	<0.001	100
Part D				**LDSA/Ti/TAP Mediation estimates**				
Effect	**Label**	**Estimate**	**SE**	**Lower**	**Upper**	**Z**	***p***	**% Mediation**
Indirect	a × b	−0.14	0.133	−0.4	0.12	−1.06	0.29	2.14
Direct	c	−6.409	0.934	−8.239	−4.579	−6.86	<0.001	97.86
Total	c + a × b	−6.549	0.933	−8.377	−4.722	−7.02	<0.001	100

3.3.2. Titanium Mediation Analysis towards PCN–TAP Relationship

When assessing Titanium's mediating effect on the relationship between PCN and TAP (Table 3-part C), the total effect and the direct effect of the model were found to be statistically significant, (total: b = −4.6, z = −6.5, 95% C.I. −5.9/−3.2, $p < 0.001$; direct (b = −4.5, z = −6.4, 95% C.I. −5.8/−3.1, $p < 0.001$, respectively). The mediating indirect effect resulted in being not statistically significant (b = −0.1, z = −1.04, 95% C.I. −0.3/0.09, $p = 0.3$). These results suggest that Ti did not contribute to the relationship between PCN and TAP.

Finally, Ti showed a statistically significant mediating effect (Table 3-part D), for both total and direct effects, on the relationship between LDSA and TAP (total: b = −6.6, z = −7.02, 95% C.I. −8.4/−4.8, $p < 0.001$; direct: b = −6.4, z = −6.9, 95% C.I. −8.3/−4.6, $p < 0.001$). However, the mediating indirect effect resulted in being not significant (b = −0.1, z = −1.06, 95% C.I. −0.4/0.1, $p = 0.3$). These results suggest that Titanium did not affect the relationship between LDSA and TAP.

4. Discussion

The present study assessed whether the urinary concentration of metals (Al, Ti and Cr) and a metalloid (Si) released from NMs could be reliably used in the biomonitoring of exposure to NMs. The results showed that exposed workers had higher metal urinary concentrations than the not-exposed group, especially for Ti and Si. In addition, these metals, particularly Si, mediated the relationship between NM exposure and the OS biomarkers.

Although exposure assessment for NMs has dramatically improved in recent years, relying on innovative approaches that allow the sampling in the breathing zone of workers and the translation of the aerosol characteristics into relevant metrics [28], there is still the need to assess both the internal dose and the possible effects on the target organ [8,26]. There is a general consensus regarding the likelihood that particles between 0.1 and 1.0 μm in diameter reach the lower respiratory tract, but their deposition is low compared with particles with lower and higher aerodynamic diameter due to symmetric human lung morphology [29,30]. However, the available biokinetic data suggest that the translocation rates of nanoparticles from the portal of entry—the respiratory tract—to secondary organs, is size- and charge-dependent, but the number of particles reaching the systemic circulation is actually very low [30,31]). The demonstration of the translocation of NMs from the lungs into systemic circulation is theoretically possible for metallic nanoparticles, which release metal ions or dissolve in biological media. Similarly, metallic elements are measurable in blood and urine with appropriate analytical methods, giving an estimate of current or past exposure. As a result, excretion via the kidney is also presumed to be of a low quantity, although it is possible. This limited amount can be related to the absorption of the chemical constituents of nanomaterials in the body, thus supporting the usefulness of their urinary excretion as exposure biomarkers of nanomaterials [8].

Even there is few literatures describing, the in vivo and in vitro studies confirmed that urinary metals could be useful in the working and professional exposure context. In fact, urinary metal concentrations could be a possible product of the metabolism related to NM working exposure. In addition, generally, urinary metal concentrations could be related to redox oxidative stress imbalance, resulting from their high surface-to-volume ratio and from surface characteristics such as defects of crystal structure, surface atoms with free valence electrons, and adsorbed redox-active metal ions [32–34].

Our results confirmed that urinary metal concentrations can be reliable biomarkers of exposure and can correctly reflect NM levels of exposure. In fact, we found that urinary metal concentrations are higher in the high-exposure subjects, and this is even more true for Si and Ti concentrations. Mediation analysis between the exposure metrics (PCN and LDSA) and TAP, carried out for urinary Si and Ti, revealed a mild, albeit significant, contribution of Si (from SiO_2) and Ti (from TiO_2) in the relationships between one of the exposure metrics (LDSA) and the decline in the antioxidant defense pool.

This finding seems consistent with one of the main mechanisms by which nanoparticles induce adverse health effects., i.e., the generation of ROS and oxidative stress [15]. In our study, the only statistically significant finding concerning OS was the lower values of the urinary TAP in exposed workers. TAP reflects the cumulative effects of all the antioxidants from various endogenous anti-oxidative defense systems; thus, it limits the noxious effects caused by OS. However, it is also likely that the extent of the exposure of workers did not reach levels inducing overt oxidative changes. The absence of an increase in OS biomarkers in EBC [25] and the negative TAP relationship in urine have suggested that efficient antioxidant defense mechanisms could have counterbalanced the noxious effects of metal oxides, leading to the maintenance of the redox balance in exposed workers [8,25].

The relevance of oxidative markers due to exposure to particles and fibers and especially nanomaterials has been recently reviewed [34,35]. In addition, in some cases, ions released from nanoparticles such as silver, gold, and iron can be measured in urine and in blood [7]. Biomarkers of OS and inflammation have been shown to have an association with the biopersistence of particles and fibers, resulting in frustrated phagocytosis and oxidative cellular stress, especially in the lungs [25,26,35].

Regarding the limitations of this study, the most important is linked to the exposure assessment strategy adopted. As already reported in previous studies [25,26], airborne nanoparticle concentrations were measured with a stationary device (DiSCmini), without personal exposure devices, likely underestimating exposure by inhalation.

5. Conclusions

Our study provides evidence that occupational exposure to air mixtures containing NMs composed of metal oxides and other dusts can increase the body burden of these foreign chemicals.

Metals released by NMs can represent an underestimated hazard for people handling technological products, increasing their body burden and potentially affecting physiological functions, e.g., the redox balance within the body. Although the health significance of such findings needs to be further elucidated, the assessment of metals in the urine of nanotechnology workers may represent a tool for estimating the body burden following long-term exposure. As a whole, a combination of biomarkers of exposure and oxidative stress can suggest early health effects, indicating the need to carry out longitudinal studies on nanotechnology workers and predictable analytical techniques, in the highest perspectives of primary prevention and health promotion.

Supplementary Materials: The following supporting information can be downloaded at: https://www.mdpi.com/article/10.3390/antiox13060676/s1, Table S1: Technical information.

Author Contributions: Conceptualization, V.B., G.G., G.S., R.B., I.G.C., and E.B.; methodology, V.B., G.G., G.S., G.M., R.B., I.G.C., and E.B.; validation V.B., G.G., G.S., M.P., and F.G.; formal analysis and data curation V.B., G.G., and G.S.; writing—original draft preparation, V.B., G.G., G.S., and E.B.; writing—review and editing, M.P., F.G., G.M., R.B., and I.G.C. supervision, R.B., I.G.C., and E.B. All authors have read and agreed to the published version of the manuscript.

Funding: This work is supported by the European Commission LIFE program [Grant LIFE17 ENV/GR/000285].

Institutional Review Board Statement: Approvals have been obtained from the local Ethics Committees: The Swiss ethics in Switzerland (approval 2020-01098); the Bioethical Committee of the University of Torino in Italy (approval 336577 8.08.2020); and the Health and Safety Board of the Catalan Institute of Nanoscience and Nanotechnology, in Spain (approval ICN2-22-03-2022).

Informed Consent Statement: Written informed consent was obtained from all subjects involved in the study.

Data Availability Statement: The data that were used are confidential.

Acknowledgments: We acknowledge all members of NanoExplore consortium.

Conflicts of Interest: The authors declare no conflicts of interest.

References

1. Maynard, A.D.; Wilson, W.; Hamilton, L.H.; Billington, J.H.; Cartwright, C.; Cook, R.; Garcia, D.E.; Gelb, B.S.; Glazer, C.L.; Longaberger, T.; et al. *Nanotechnology: A Research Strategy for Addressing Risk*; International Center for Scholars: Washington, DC, USA, 2006.
2. Baig, N.; Kammakakam, I.; Falath, W.; Kammakakam, I. Nanomaterials: A Review of Synthesis Methods, Properties, Recent Progress, and Challenges. *Mater. Adv.* **2021**, *2*, 1821–1871. [CrossRef]
3. OECD. *Environment Directorate Chemicals and Biotechnology Committee Important Issues on Risk Assessment of Manufactured Nanomaterials Series on the Safety of Manufactured Nanomaterials*; OECD: Paris, France, 2022.
4. Marques, A.C.; Vale, M.; Vicente, D.; Schreck, M.; Tervoort, E.; Niederberger C Marques, M.A.; Vale, M.; Vicente, D.; Schreck, M.; Tervoort, E.; et al. Porous Silica Microspheres with Immobilized Titania Nanoparticles for In-Flow Solar-Driven Purification of Wastewater. *Glob. Chall.* **2021**, *5*, 2000116. [CrossRef]
5. Sharifi, S.; Behzadi, S.; Laurent, S.; Forrest, M.L.; Stroevee, P.; Mahmoudi, M. Toxicity of Nanomaterials. *Chem. Soc. Rev.* **2012**, *41*, 2323–2343. [CrossRef]
6. Audignon-Durand, S.; Gramond, C.; Ducamp, S.; Manangama, G.; Garrigou, A.; Delva, F.; Brochard, P.; Lacourt, A. Development of a Job-Exposure Matrix for Ultrafine Particle Exposure: The MatPUF JEM. *Ann. Work Expo Health* **2021**, *65*, 516–527. [CrossRef] [PubMed] [PubMed Central]
7. Ghelli, F.; Panizzolo, M.; Garzaro, G.; Squillacioti, G.; Bellisario, V.; Colombi, N.; Bergamaschi, E.; Guseva Canu, I.; Bono, R. Inflammatory Biomarkers in Exhaled Breath Condensate: A Systematic Review. *Int. J. Mol. Sci.* **2022**, *23*, 9820. [CrossRef]
8. Bergamaschi, E.; Bellisario, V.; Macrì, M.; Buglisi, M.; Garzaro, G.; Squillacioti, G.; Ghelli, F.; Bono, R.; Fenoglio, I.; Barbero, F.; et al. A Biomonitoring Pilot Study in Workers from a Paints Production Plant Exposed to Pigment-Grade Titanium Dioxide (TiO_2). *Toxics* **2022**, *10*, 171. [CrossRef]
9. Luo, X.; Xie, D.; Hu, J.; Su, J.; Xue, Z. Oxidative Stress and Inflammatory Biomarkers for Populations with Occupational Exposure to Nanomaterials: A Systematic Review and Meta-Analysis. *Antioxidants* **2022**, *11*, 2182. [CrossRef]
10. Lewinski, N.; Colvin, V.; Drezek, R. Cytotoxicity of Nanopartides. *Small* **2008**, *4*, 26–49. [CrossRef]
11. Kim, J.; Yu, I.J. National Survey of Workplaces Handling and Manufacturing Nanomaterials, Exposure to and Health Effects of Nanomaterials, and Evaluation of Nanomaterial Safety Data Sheets. *Biomed. Res. Int.* **2016**, *2016*, 8389129. [CrossRef]
12. Jusko, T.A.; Henderson, C.R.; Lanphear, B.P.; Cory-Slechta, D.A.; Parsons, P.J.; Canfield, R.L. Blood Lead Concentrations < 10 Microg/DL and Child Intelligence at 6 Years of Age. *Environ. Health Perspect.* **2008**, *116*, 243–248. [CrossRef]

13. Khan, K.; Wasserman, G.A.; Liu, X.; Ahmed, E.; Parvez, F.; Slavkovich, V.; Levy, D.; Mey, J.; van Geen, A.; Graziano, J.H.; et al. Manganese Exposure from Drinking Water and Children's Academic Achievement. *Neurotoxicology* **2012**, *33*, 91–97. [CrossRef]
14. Valko, M.; Jomova, K.; Rhodes, C.J.; Kuča, K.; Musílek, K. Redox- and Non-Redox-Metal-Induced Formation of Free Radicals and Their Role in Human Disease. *Arch. Toxicol.* **2016**, *90*, 1–37. [CrossRef]
15. Farina, M.; Avila, D.S.; Da Rocha, J.B.T.; Aschner, M. Metals, Oxidative Stress and Neurodegeneration: A Focus on Iron, Manganese and Mercury. *Neurochem. Int.* **2013**, *62*, 575–594. [CrossRef]
16. Donaldson, K.; Brown, D.; Clouter, A.; Duffin, R.; MacNee, W.; Renwick, L.; Tran, L.; Stone, V. The Pulmonary Toxicology of Ultrafine Particles. *J. Aerosol. Med.* **2002**, *15*, 213–220. [CrossRef]
17. Garza-Lombó, C.; Posadas, Y.; Quintanar, L.; Gonsebatt, M.E.; Franco, R. Neurotoxicity Linked to Dysfunctional Metal Ion Homeostasis and Xenobiotic Metal Exposure: Redox Signaling and Oxidative Stress. *Antioxid. Redox Signal.* **2018**, *28*, 1669–1703. [CrossRef]
18. McCormick, S.; Niang, M.; Dahm, M.M. Occupational Exposures to Engineered Nanomaterials: A Review of Workplace Exposure Assessment Methods. *Curr. Environ. Health Rep.* **2021**, *8*, 223–234. [CrossRef]
19. Samir, A.M.; Rashed, L.A. Effects of Occupational Exposure to Aluminium on Some Oxidative Stress and DNA Damage Parameters. *Hum. Exp. Toxicol.* **2018**, *37*, 901–908. [CrossRef]
20. Marie, C.; Ravanat, J.L.; Badouard, C.; Marques, M.; Balducci, F.; Maître, A. Urinary Levels of Oxidative DNA and RNA Damage among Workers Exposed to Polycyclic Aromatic Hydrocarbons in Silicon Production: Comparison with 1-Hydroxypyrene. *Environ. Mol. Mutagen.* **2009**, *50*, 88–95. [CrossRef]
21. Wu, W.T.; Jung, W.T.; Lee, H.L. Lipid Peroxidation Metabolites Associated with Biomarkers of Inflammation and Oxidation Stress in Workers Handling Carbon Nanotubes and Metal Oxide Nanoparticles. *Nanotoxicology* **2021**, *15*, 577–587. [CrossRef]
22. Anlar, H.G.; Bacanli, M.; İritaş, S.; Bal, C.; Kurt, T.; Tutkun, E.; Hinc Yilmaz, O.; Basaran, N. Effects of Occupational Silica Exposure on OXIDATIVE Stress and Immune System Parameters in Ceramic Workers in TURKEY. *J. Toxicol. Environ. Health A* **2017**, *80*, 688–696. [CrossRef]
23. Junaid, M.; Hashmi, M.Z.; Malik, R.N.; Pei, D.S. Toxicity and Oxidative Stress Induced by Chromium in Workers Exposed from Different Occupational Settings around the Globe: A Review. *Environ Sci. Pollut. Res. Int.* **2016**, *23*, 20151–20167. [CrossRef]
24. Schulte, P.A.; Leso, V.; Niang, M.; Iavicoli, I. Current State of Knowledge on the Health Effects of Engineered Nanomaterials in Workers: A Systematic Review of Human Studies and Epidemiological Investigations. *Scand. J. Work Environ. Health* **2019**, *45*, 217–238. [CrossRef]
25. Hemmendinger, M.; Squillacioti, G.; Charreau, T.; Garzaro, G.; Ghelli, F.; Bono, R.; Sauvain, J.J.; Suarez, G.; Hopf, N.B.; Wild, P.; et al. Occupational Exposure to Nanomaterials and Biomarkers in Exhaled Air and Urine: Insights from the NanoExplore International Cohort. *Environ. Int.* **2023**, *179*, 108157. [CrossRef]
26. Guseva Canu, I.; Plys, E.; Velarde Crézé, C.; Fito, C.; Hopf, N.B.; Progiou, A.; Riganti, C.; Sauvain, J.J.; Squillacioti, G.; Suarez, G.; et al. A Harmonized Protocol for an International Multicenter Prospective Study of Nanotechnology Workers: The NanoExplore Cohort. *Nanotoxicology* **2023**, *17*, 1–19. [CrossRef]
27. Panizzolo, M.; Barbero, F.; Ghelli, F.; Garzaro, G.; Bellisario, V.; Guseva Canu, I.; Fenoglio, I.; Bergamaschi, E.; Bono, R. Assessing the Inhaled Dose of Nanomaterials by Nanoparticle Tracking Analysis (NTA) of Exhaled Breath Condensate (EBC) and Its Relationship with Lung Inflammatory Biomarkers. *Chemosphere* **2024**, *358*, 142139. [CrossRef]
28. Iavicoli, I.; Fontana, L.; Pingue, P.; Todea, A.M.; Asbach, C. Assessment of Occupational Exposure to Engineered Nanomaterials in Research Laboratories Using Personal Monitors. *Sci. Total Environ.* **2018**, *627*, 689–702. [CrossRef]
29. Broday, D.M.; Agnon, Y. Asymmetric Human Lung Morphology Induce Particle Deposition Variation. *J. Aerosol. Sci.* **2007**, *38*, 701–718. [CrossRef]
30. Kreyling, W.G.; Semmler, M.; Erbe, F.; Mayer, P.; Takenaka, S.; Schulz, H.; Oberdörster, G.; Ziesenis, A. Translocation of Ultrafine Insoluble Iridium Particles from Lung Epithelium to Extrapulmonary Organs Is Size Dependent but Very Low. *J. Toxicol. Environ. Health A* **2002**, *65*, 1513–1530. [CrossRef]
31. Kreyling, W.G.; Semmler-Behnke, M.; Seitz, J.; Scymczak, W.; Wenk, A.; Mayer, P.; Takenaka, S.; Oberdrster, G. Size Dependence of the Translocation of Inhaled Iridium and Carbon Nanoparticle Aggregates from the Lung of Rats to the Blood and Secondary Target Organs. *Inhal. Toxicol.* **2009**, *21* (Suppl. 1), 55–60. [CrossRef]
32. Papp, A.; Horváth, T.; Igaz, N.; Gopisetty, M.K.; Kiricsi, M.; Berkesi, D.S.; Kozma, G.; Kónya, Z.; Wilhelm, I.; Patai, R.; et al. Presence of Titanium and Toxic Effects Observed in Rat Lungs, Kidneys, and Central Nervous System in Vivo and in Cultured Astrocytes in Vitro on Exposure by Titanium Dioxide Nanorods. *Int. J. Nanomed.* **2020**, *15*, 9939–9960. [CrossRef]
33. Nel, A.; Xia, T.; Mädler, L.; Li, N. Toxic Potential of Materials at the Nanolevel. *Science* **2006**, *311*, 622–627. [CrossRef]
34. Buzea, C.; Pacheco, I.I.; Robbie, K. Nanomaterials and Nanoparticles: Sources and Toxicity. *Biointerphases* **2007**, *2*, MR17–MR71. [CrossRef]
35. Iavicoli, I.; Leso, V.; Manno, M.; Schulte, P.A. Biomarkers of Nanomaterial Exposure and Effect: Current Status. *J. Nanoparticle Res.* **2014**, *16*, 2302. [CrossRef]

antioxidants

Article

Clionasterol-Rich Fraction of *Caulerpa racemosa* against Particulate Matter-Induced Skin Damage via Inhibition of Oxidative Stress and Apoptosis-Related Signaling Pathway

N. M. Liyanage [1,†], D. P. Nagahawatta [1,†], Thilina U. Jayawardena [1,2], H. H. A. C. K. Jayawardhana [1], Hyo-Geun Lee [1], Young-Sang Kim [1] and You-Jin Jeon [1,*]

1 Department of Marine Life Sciences, Jeju National University, Jeju 63243, Korea
2 Department of Chemistry, Biochemistry, and Physics, Université du Québec à Trois-Rivières, Trois-Rivières, QC G8Z 4M3, Canada
* Correspondence: youjinj@jejunu.ac.kr; Tel.: +82 64-754-3475
† These authors contributed equally to this work.

Abstract: The increasing airborne particulate matter (PM) consisting of environmental contaminants such as dust, aerosols, and fibers has become a global concern by causing oxidative stress that leads to apoptosis and skin damage. The current study evaluated the protective effect of *Caulerpa racemosa* (CR) against PM-induced skin damage using human keratinocytes and a zebrafish model. The clionasterol-rich hexane fraction (CRHF2) of CR exhibited superior protective activity through downregulating intracellular reactive oxygen species levels and mitochondrial ROS levels, as well as the PM-induced increase in apoptotic body formation and upregulation of apoptotic signaling pathway proteins, along with sub-G1 cell accumulation dose-dependently. Furthermore, in vivo results showed that CRHF2 potentially downregulates PM-induced cell death, ROS, and NO production in the zebrafish model. Hence, the results evidenced that the protective effect of CRHF2 is caused by inhibiting oxidative stress and mitochondrial-mediated apoptosis in cells. Therefore, *C. racemosa* has the potential to be used in the development of pharmaceuticals to attenuate PM-induced skin diseases.

Keywords: *Caulerpa racemose*; particulate matter; mitochondria; apoptosis; oxidative stress; clionasterol; zebrafish model

check for updates

Citation: Liyanage, N.M.; Nagahawatta, D.P.; Jayawardena, T.U.; Jayawardhana, H.H.A.C.K.; Lee, H.-G.; Kim, Y.-S.; Jeon, Y.-J. Clionasterol-Rich Fraction of *Caulerpa racemosa* against Particulate Matter-Induced Skin Damage via Inhibition of Oxidative Stress and Apoptosis-Related Signaling Pathway. *Antioxidants* **2022**, *11*, 1941. https://doi.org/10.3390/antiox11101941

Academic Editor: Stanley Omaye

Received: 5 September 2022
Accepted: 26 September 2022
Published: 28 September 2022

1. Introduction

Air pollution by environmental pollutants such as particulate matter (PM) is a matter of interest worldwide, particularly in China, Korea, and Japan. It has caused several deleterious effects on human health as well as regional and global climatic changes. Any toxic gas or airborne particles with a diameter less than 10 μm are included in air pollutants. They can be composed of organic or inorganic compounds or of both. The size, composition, and the origin of these particles are based on their microenvironment [1]. PM contains organic compounds that can readily penetrate the skin [2]. The skin is the largest organ in the human body and provides the largest interface between the body and external environment. Therefore, the emergence of skin diseases caused by external factors is common. Recently, the deleterious effects of PM on skin have attracted the attention of dermatologists and scientists worldwide [3,4]. PM results in oxidative stress injuries such as DNA damage and repair, cell apoptosis, and inflammation in skin keratinocytes [5–7].

Natural antioxidants present in the cells protect against injuries caused by reactive oxygen species (ROS). Superoxide dismutase, catalase, and glutathione transferase are primary enzymatic antioxidants, whereas low-molecular-weight ascorbic acids, α-tocopherol (vitamin E), β-carotene, and glutathione are important non-enzymatic antioxidants [8]. In addition, supplementation with commercially synthesized antioxidants such as butylated hydroxyanisole, butylated hydroxytoluene, and propyl gallate is being applied to prevent the adverse effects caused by excess ROS accumulation [9]. However, considering their synthetic nature, PM has been reported to have toxic and carcinogenic effects on humans. This has led to an increased interest in developing natural treatments that would be safe for humans.

Marine seaweeds are a valuable resource for important bioactive compounds, particularly phytosterols. These are lipid-rich compounds produced by plants and are the main lipid constituents of the biological plant cell membranes. Phytosterols isolated from algae possess antioxidant activity [10]. Clionasterol is a member of the class of phytosterols having a role as a marine metabolite. It is proven to possess antibacterial and antifungal activities [11]. It has also been tested for its influence on the classical and alternative pathways of the activation of the human complement system [12]. The activity of marine phytosterols in preventing skin damage was not studied extensively in the past.

Recently, marine seaweeds have gained attention as natural antioxidants and skin protectants for the production of pharmaceuticals and cosmeceuticals, owing to their availability and safety in long-term use. *Caulerpa racemosa* is a green marine alga distributed in warm waters in tropical regions. It is popular as a raw material for salads in most South Asian countries, including Sri Lanka. It is also used as an ingredient in folk medicine to treat hypertension and rheumatism. Due to their harsh environmental growth conditions, these seaweeds are exposed to extreme stress, leading to the production of free radicals and other oxidants [13]. In a previous study, it was shown that *C. racemosa* contains higher antioxidant and phenolic components than those in red seaweeds [14]. Although extensive studies have been conducted on various seaweed extracts, to the best of our knowledge, limited data are available on the antioxidant and antiapoptotic potential of *C. racemosa* and its application in pharmaceutical and cosmeceutical production. Hence, the present study aimed to explore the application of *C. racemosa* as a potential skin protective agent.

2. Materials and Methods

2.1. Materials

Human keratinocytes (HaCaT) were purchased from the Korean Cell Line Bank (KCLB, Seoul, Korea). Cell line was maintained in Dulbecco's modified essential medium (DMEM, Gibco) with L-glutamine, 1% (*v*/*v*) antibiotics (penicillin, streptomycin), and 5% (*v*/*v*) fetal bovine serum (FBS) purchased from Gibco-BRL (Grand Island, NY, USA). The certified reference material (CRM) for PM (CRM No.28 Urban Aerosols) was purchased from the Center for Environmental Measurements and Analysis, National Institute for Environmental Studies (NIES), Ibaraki, Japan. Chemicals such as 3-(4,5-dimethylthiazol-2-yl)-2,5-diphenyltetrazolium bromide (MTT), 2′,7′-dichlorodihydrofluorescein diacetate (DCFH2-DA), acridine orange, Hoechst 33342, and Ethidium bromide were purchased from Sigma-Aldrich (St. Louis, MO, USA). Antibodies used in this study were purchased from Santa Cruz Biotechnology (Santa Cruz, CA, USA). Analytical-grade solvents such as hexane, ethanol, and methanol were obtained from Sigma-Aldrich (St. Louis, MO, USA).

2.2. Collection of Seaweeds and Extraction

The green seaweed *C. racemosa* was collected from coastal areas of Maldives and Sri Lanka. Samples were immediately washed to remove impurities. The washed seaweeds were dried using a hybrid hot water Goodle dryer [15] and lyophilized. Extraction was performed as previously described [16]. Briefly, 50 g of the ground sample was extracted with 70% ethanol to acquire *C. racemosa* ethanolic extract (CRE). It was suspended in deionized water and fractionated using hexane, chloroform, and ethyl acetate. The hexane fraction was further resolved into five fractions, CRHF1, CRHF2, CRHF3, CRHF4, and CRHF5, using a silica open column due to its potential bioactivities. Gas chromatography–mass spectrometry (GC/MS) analysis was performed to characterize the five fractions, from which CRHF2 was selected for further experiments due to the presence of a key compound, clionasterol. The bioassay-guided purification process is illustrated in Figure 1a. Fractions were evaporated using a rotary evaporator followed by freeze-drying.

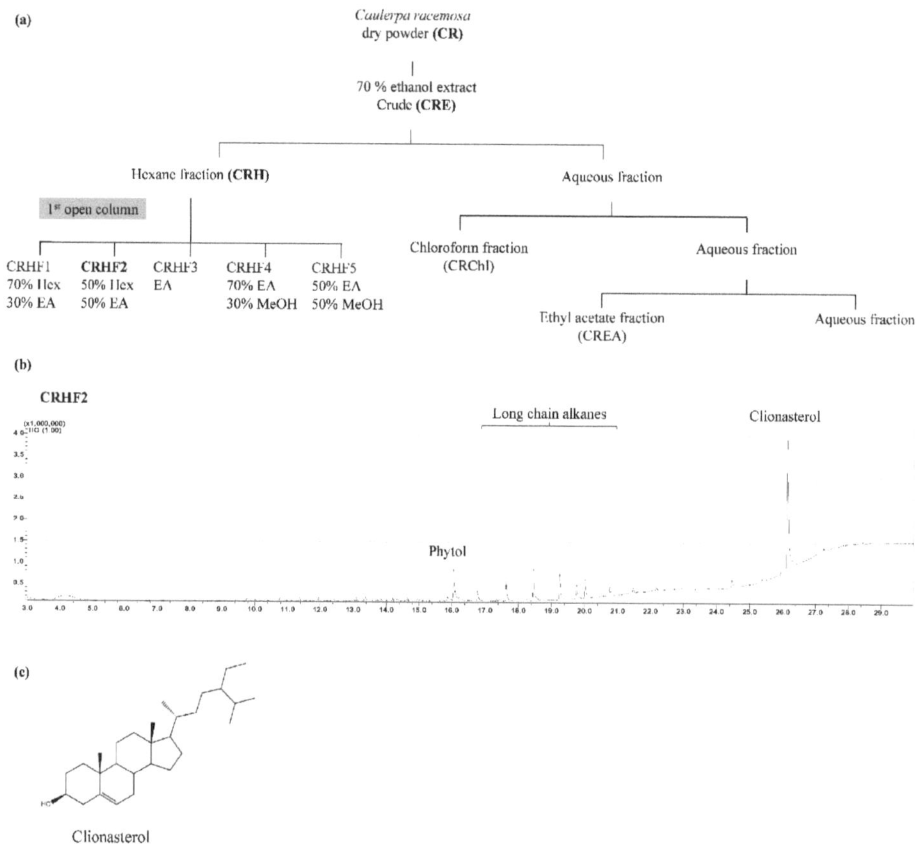

Figure 1. (**a**) Separation procedure implementing solvent/solvent extraction method leading to active fraction CRHF2, (**b**) GCMS chromatogram of CRHF2, (**c**) chemical structure of clionasterol.

2.3. CRM No: 28 for Particulate Matter

CRM was developed by the NIES to identify the elements in PM. Detailed characterization of PM was conducted in previous studies [17,18]. Their analyses were based on detailed spectroscopic analyses of atomic absorption spectroscopy, instrumental neutron activation analysis, particle-induced X-ray emission, and X-ray fluorescence. The size distribution analysis revealed that majority of the PM had a diameter of approximately 2 µm. Furthermore, it consisted of eight polycyclic aromatic hydrocarbons, and among them, the highest mass fraction was benzo[b]fluoranthene. The earth metals detected were magnesium, calcium, strontium, barium, and inorganic materials, while the transition metals detected were manganese and lead, which contributed to the higher mass fraction.

2.4. Cell Viability

HaCaT cells were cultured in DMEM supplemented with 10% FBS and 1% penicillin/streptomycin mixture at 37 °C, under 5% CO_2 humidified atmosphere. PM stock solution was prepared by suspending the PM in DMEM and diluting when necessary. Cells were seeded (1×10^5 cells/mL) in 96-well plates and incubated for 24 h. Column fractions (CRHF1-CRHF5) were added to cells at final concentrations of 25, 50, 100, and 200 µg/mL. After 2 h of incubation, cells were stimulated with 200 µg/mL PM. The working concentration of the PM to be treated was established based on preliminary studies. After 24 h of incubation with PM, cell viability was measured using MTT assay [19].

2.5. Intracellular ROS Scavenging Activity of PM-Induced Cells

DCFH2-DA assay was used to measure the intracellular ROS scavenging potential of the samples. A 96-well plate was seeded with HaCaT cells (1×10^5 cells/mL) and incubated for 24 h. Cells were pretreated with CRHF2 for 2 h before stimulating with PM. The wells were treated with DCFH2-DA (25 μg/mL) for 10 min and fluorescence measurements (excitation—485 nm, emission—530 nm) were performed using a microplate reader (Bio-Tech, Winooski, VT, USA).

The superior antioxidant and protective activities of the CRHF2 fraction were further confirmed by flow cytometry using DCFH2-DA staining of cells [20]. After CRHF2 pretreatment and PM stimulation, the cells were washed and treated with DCFH2-DA for 30 min at 37 °C in dark. Flow cytometry (CytoFLEX, Beckman Coulter, PA, USA) was used to analyze the stained cells.

2.6. PM-Induced Cell Apoptosis

To determine cell apoptosis due to PM, the nuclear morphology of cells was evaluated using DNA dye Hoechst 33342 and nuclear double-staining method. HaCaT cells were seeded in 24-well plates and pretreated with CRHF2 (25 and 50 μg/mL) prior to PM stimulation for 24 h. Cells were treated with Hoechst 33342 at a final concentration of 10 μg/mL and further incubated for 10 min. Stained cells were observed under a fluorescence microscope (Olympus, Tokyo, Japan) and apoptosis levels were measured using ImageJ software [21]. A mixture of acridine orange and ethidium bromide was used for double staining of PM-induced HaCaT cells. After incubation for 10 min, cellular morphology was assessed using a fluorescence microscope equipped with CoolSNAP-Pro color digital camera [22].

2.7. Analysis of Cell Cycle

To identify apoptotic sub-G1 cells, flow cytometry analysis was performed according to a previously reported method [23]. Briefly, cells were seeded in 6-well plates, treated with CRHF2, and incubated for 12 h. Following this, the cells were harvested and fixed with 1 mL of 70% ethanol at 4 °C overnight. Fixed cells were then washed twice with cold phosphate saline buffer (PBS) and the supernatant was separated. The cells were further incubated in PBS containing propidium iodide, 100 μg RNase A, and ethylenediaminetetraacetic acid EDTA at 37 °C for 30 min. Cells were analyzed using a fluorescence-activated cell sorting (FACS) Calibur flow cytometer (Becton Dickinson, San Jose, CA, USA).

2.8. Mitochondrial ROS Generation Measurement

Mitochondrial membrane permeability is related to apoptosis through caspase-associated protein activation, and membrane permeability depends on mitochondrial oxidative stress [24]. Mitochondrial ROS generation in PM-induced HaCaT cells was analyzed using dihydrorhodamine-123 (DHR123) staining. Seeded cells were pretreated with CRHF2 and stimulated with PM for 24 h. the stimulated cells were incubated with 25 μM DHR 123 (Sigma-Aldrich, Poole, UK) for 20 min in dark conditions. Excess DHR123 was removed and cells were observed under a fluorescence microscope.

2.9. Western Blot Analysis

Western blotting was performed for the evaluation of intracellular expression levels of selected key molecular mediators related to apoptosis, including Bax, Bcl-xL, Caspase-3, P53, cleaved PARP, and cleaved caspase-9. Cells were seeded and treated with CRHF2 sample for 24 h, following incubation with PM. Finally, cells were harvested for protein extraction and proteins were analyzed using a BCA protein assay kit. Electrophoresis was carried out using 12% sodium dodecyl sulphate-polyacrylamide gels. Resolved bands were blotted onto nitrocellulose membranes and blocked with 5% skimmed milk in tris-buffered saline with 0.1% Tween 20. Membranes were consecutively incubated with selective primary and secondary antibodies and signals were developed using chemiluminescent substrate (Cyanagen Srl, Bologna, Italy). Fluorescence images were obtained using the FUSION SOLO Vilber Lourment system. ImageJ program was used for densitometric analysis of proteins.

2.10. *In Vivo Antioxidant Activity of CRHF2 Using Zebrafish Model*

Animal experiments were conducted in accordance with the experimental animal guidelines provided by Jeju National University Animal center and were authorized by the Animal Care Use Committee (IACUC) of Jeju National University (protocol 2020-0049).

2.10.1. Maintenance of Zebrafish

Adult zebrafish were purchased from the Seoul Aquarium, Korea and maintained in acrylic tanks under controlled conditions (28 °C, 14/10 h light/dark cycle). Embryos were obtained by natural spawning and were collected within 30 min.

2.10.2. Application of CRHF2 to Zebrafish Embryos

Zebrafish embryos were transferred to a 12-well plate (15 embryos/well) containing embryo medium within 7–9 h postfertilization. They were treated with CRHF2 (25, 50, and 100 μg/mL) and incubated for 1 h. Stimulation with PM (400 μg/mL) was performed and incubated for 24 h. The survival rate of PM-stimulated zebrafish was measured for 7 days. The heartbeat rate of stimulated zebrafish was recorded 2 days postfertilization (dpf) [25,26].

2.10.3. Cell Death, Intracellular Lipid Peroxidation, and ROS Analysis

Cell death was evaluated using acridine orange staining (7 μg/mL) and intracellular ROS levels were measured using DCFH2-DA (20 μg/mL) at 3 dpf. The lipid peroxidation levels of PM-stimulated zebrafish were measured by diphenyl-1-pyrenylphosphine (DPPP) staining (25 μg/mL) [25].

2.11. *Statistical Analyses*

Quantifiable data are expressed as the means± standard deviation (SD), based on at least three independent evaluations (n = 3). One-way analysis of variance was used to compare mean values. *p*-values < 0.05 ($p < 0.05$), * and 0.01 ($p < 0.01$), ** were considered statistically significant.

3. Results

3.1. *Characterization of CRHF2 Fraction of C. racemosa and PM*

Based on the bioassay-guided evaluations of the five hexane fractions, CRHF2 was selected for further studies because of its superior activity. GC/MS analysis of CRHF2 identified two sterol compounds, clionasterol and phytol. Between the two, clionasterol in CRHF2 was identified as a key compound (Figure 1b). The chemical structure of clionasterol is shown in Figure 1c.

3.2. *Protective Effect of Five Hexane Fractions of C. racemosa against PM-Induced ROS Generation*

The cell viability and intracellular ROS in PM-induced cells were evaluated to investigate the antioxidant potential of hexane fractions isolated from the seaweed. As shown in Figure 2a, the cell viability of PM-stimulated HaCaT cells was drastically decreased compared to that in nontreated cells; however, treatment with the hexane fractions (25, 50, 100, and 200 μg/mL) increased cell viability in a dose-dependent manner. Similarly, PM significantly stimulated intracellular ROS generation in cells, whereas treatment with *C. racemosa* hexane fractions remarkably and dose-dependently downregulated ROS production and attenuated the effects of PM on HaCaT cells in treated cells (Figure 2b). However, among the five fractions, the CRHF2 fraction showed excellent ROS scavenging activity. Therefore, CRHF2 was used in further experiments.

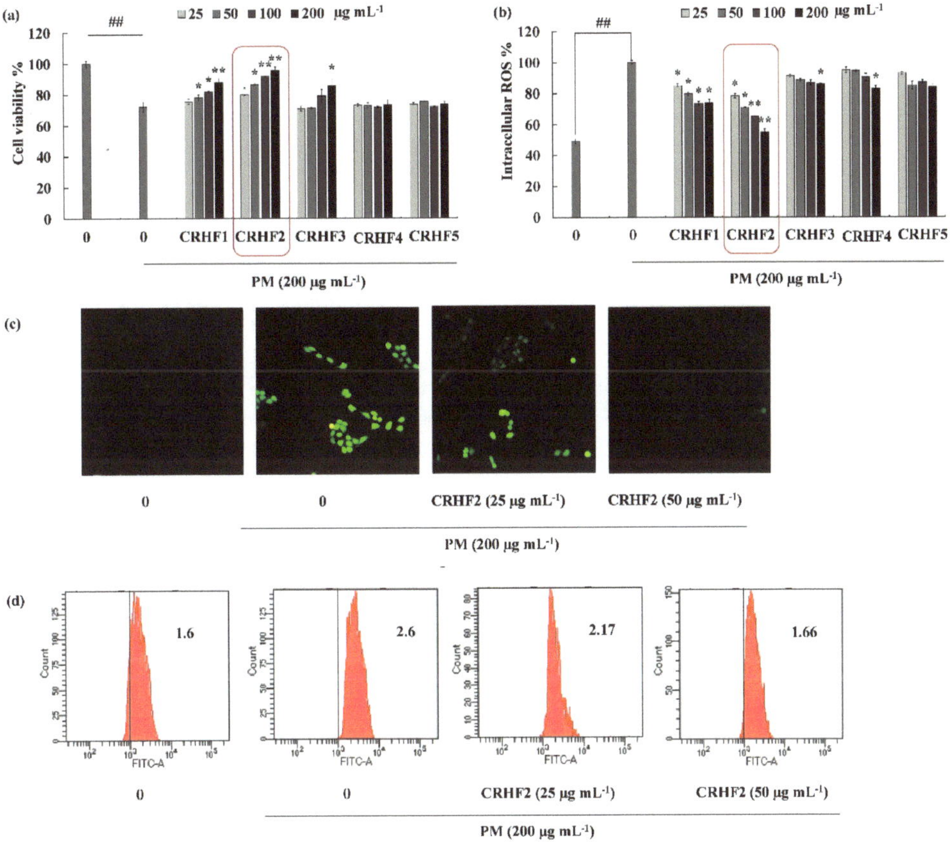

Figure 2. CRHF2 protects PM-induced oxidative stress in HaCaT keratinocytes. Effect of column fractions on (**a**) cell viability and (**b**) intracellular ROS levels, (**c**) fluorescence microscopy images, and (**d**) flow cytometry results of DCFH2-DA stained cells. The values in the histogram represent the mean intracellular ROS levels. Results are represented as mean ± SD of triplicate determinants (n = 3); * $p < 0.05$, ** $p < 0.01$. (# denotes significance compared to control while * represents significance compared to PM-treated group).

Images for DCFH2-DA staining of cells showed antioxidative activity of CRHF2 in PM-stimulated HaCaT cells. According to the fluorescence microscopy images, an amplified intensity of green fluorescence was observed in PM-stimulated cells compared to that in the untreated group (Figure 2c). Upon treatment with CRHF2, the fluorescence intensity dose-dependently decreased. The superior antioxidant activity of CRHF2 was confirmed using FACS and DCFH2-DA fluoroprobe. Flow cytometry analysis of DCFH2-DA-stained cells indicated a peak shift towards higher intensity (X axis-FITC-A) due to PM stimulation of cells compared to that in the nontreated group (Figure 2d). However, this effect was reversed with increasing concentrations of CRHF2.

3.3. CRHF2 Attenuated PM-Induced Apoptotic Body Formation

Hoechst 33342 staining and nuclear double staining with ethidium bromide and acridine orange were used in studying the nuclear morphology of the cells. Hoechst staining revealed an increase in chromatin condensation and DNA fragmentation in PM-stimulated cells compared to those in nontreated cells, representing a higher intensity in the nuclei region (Figure 3a). After dose-dependent treatment with CRHF2, the number of apoptotic bodies was significantly decreased, resulting in the decreased fluorescence intensity. This was proven with the percentage of apoptotic cell measurement with ImageJ software. As shown in Figure 3b, PM stimulation of HaCaT cells increased in the late apoptotic stages, as indicated with by orange-colored nuclei fragments upon double nuclear staining using ethidium bromide and acridine orange double staining (Figure 3b). Treatment with CRHF2 suppressed the appearance of apoptotic bodies in a dose-dependent manner, demonstrating its protective effect.

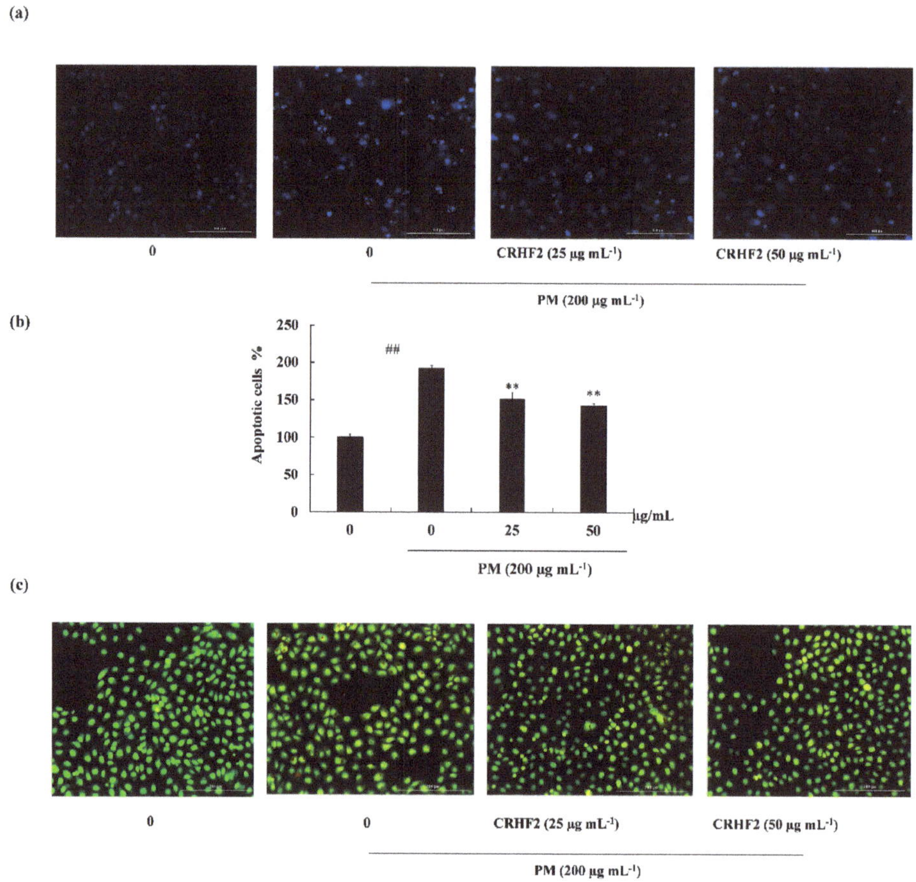

Figure 3. CRHF2 attenuates PM-stimulated apoptotic body formation and DNA damage. Evaluation of nuclear morphology via (**a**) Hoechst 33342 staining fluorescence images (**b**) apoptotic cell percentage, and (**c**) ethidium bromide and acridine orange-involved double staining. Experiments were triplicated to confirm their repeatability (n = 3), ** $p < 0.01$ (# denotes significance compared to control, while * denotes significance compared to PM-treated groups).

3.4. Effects of CRHF2 on Sub-G1 DNA Content and Late Apoptotic Event

Flow cytometric analysis was performed to quantify apoptosis. As shown in Figure 4a, the sub-G1 DNA content in the nontreated control and the PM-stimulated group were 2.55 ± 0.66% and 76.11 ± 5.21%, respectively, indicating an increase in the percentage of cells in the late apoptotic stage. However, this effect was improved by dose-dependent treatment with CRHF2, along with a substantial reduction in the proportion of sub-G1 phase cells.

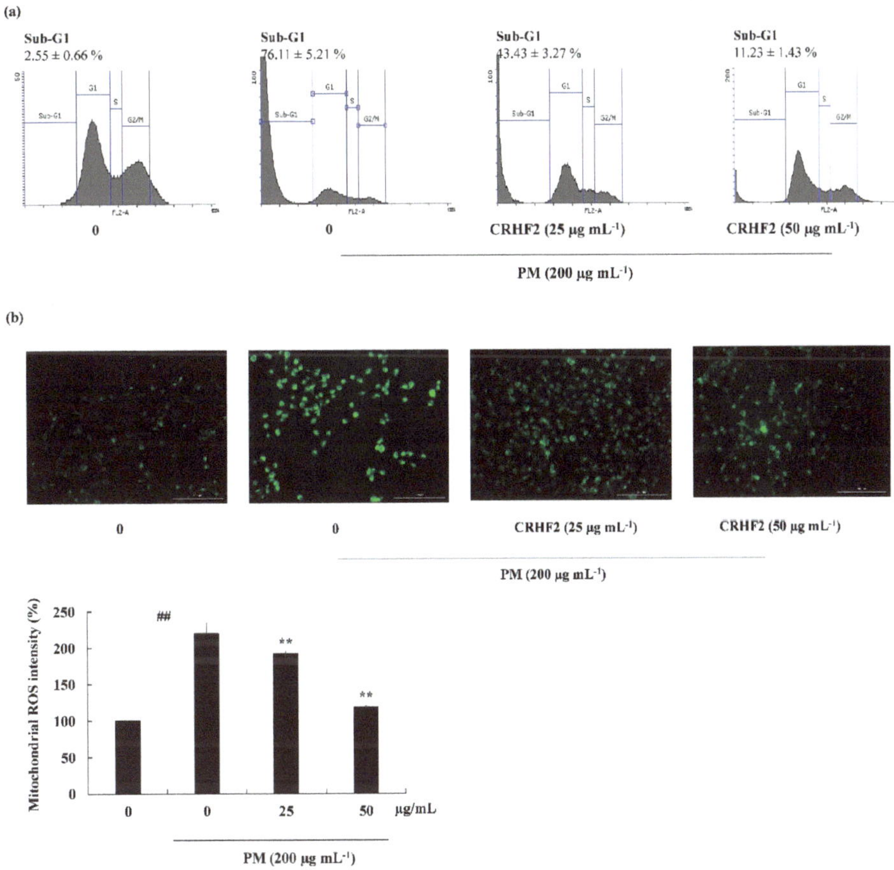

Figure 4. Cell-cycle analysis and mitochondrial ROS analysis in CRHF2-treated cells. (**a**) Cell-cycle analysis of sub-G1 cell population. PM affected cell-cycle progression and it was recovered with CHRF2 treatment. Cells were treated with CRHF2 and induced with PM. Harvested cells were evaluated via flow cytometer. (**b**) Mitochondrial ROS analysis by DHR123 staining. Triplicated independent experiments (n = 3) were involved to confirm the repeatability. Results are represented as mean ± SD; ** $p < 0.01$. (# denotes significance compared to control while * represents significance compared to PM-treated group).

3.5. Mitochondrial ROS Production was Ameliorated by the Treatment of PM-Induced Cells with CRHF2

DHR123 is a cell-permeable fluorogenic indicator of ROS levels, particularly in the mitochondria. The ROS generated in mitochondria oxidized DHR123 into rhodamine 123 and emitted green-color fluorescence (Figure 4b). Mitochondrial ROS is a major determinant of cell apoptosis. Based on the results obtained, PM stimulation resulted in higher fluorescent intensity, indicating overexpression of mitochondrial ROS; however, dose-dependent treatment with CRHF2 decreased fluorescent intensities, thus inhibiting the ROS generation.

3.6. Effect of CRHF2 on Mitochondria-Mediated Apoptosis Signaling in HaCaT Cells

To investigate the mechanism by which CRHF2 inhibits the expression of mitochondria-mediated apoptosis-related proteins, Western blot analysis was conducted. The results revealed that PM stimulation caused an immediate increase in the expression levels of Bax, Caspase-3, p53, and Caspase-9, and cleaved P poly (ADP-ribose) polymerase (PARP)-like apoptotic proteins (Figure 5). Correspondingly, the levels of antiapoptotic protein Bcl-xL were downregulated. However, concentration-dependent treatment with CRHF2 downregulated apoptotic protein expressions and upregulated the antiapoptotic protein Bcl-xL, thus exhibiting its protective effects.

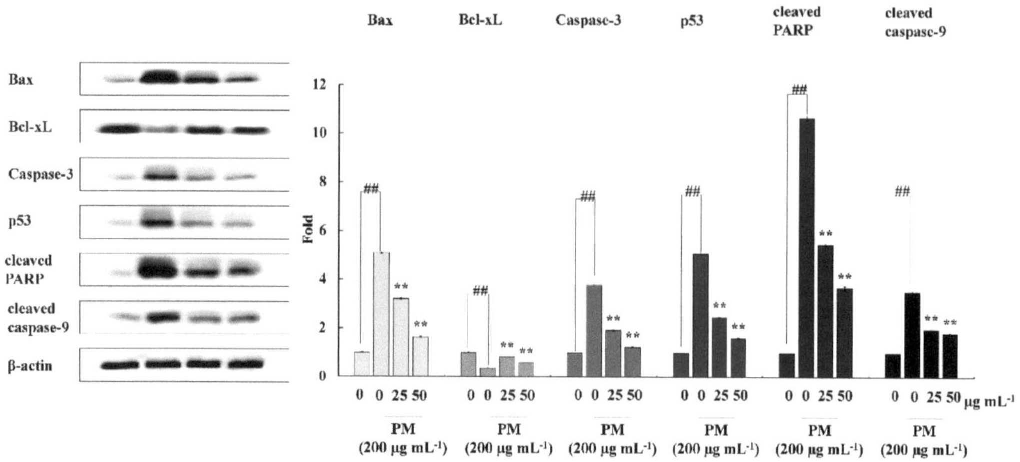

Figure 5. CRHF2 protects HaCaT keratinocytes from PM-stimulated apoptosis by inhibiting apoptotic proteins. Experiments were triplicated (n = 3). Results are represented as mean ± SD; ** $p < 0.01$. (# denotes significance compared to control while * represents significance compared to PM-treated group).

3.7. Protective Effect of CRHF2 on PM-Induced Zebrafish Embryo Survival and Heartbeat Rate

The protective effect of CRHF2 on PM-induced toxicity in zebrafish was evaluated with 25, 50, and 100 μg/mL concentration ranges of CRHF2 due to there being no toxic effect on the embryos. As shown in Figure 6a, the survival rate of embryos was decreased with the treatment of PM compared to the nontreated group. Preincubation with CRHF2 remarkably and dose-dependently increased the survival rate of embryos. Additionally, the heartbeat rate of zebrafish induced with PM was considerably increased compared to the nontreated group (Figure 6b). It was remarkably recovered to normal levels with the treatment of CRHF2.

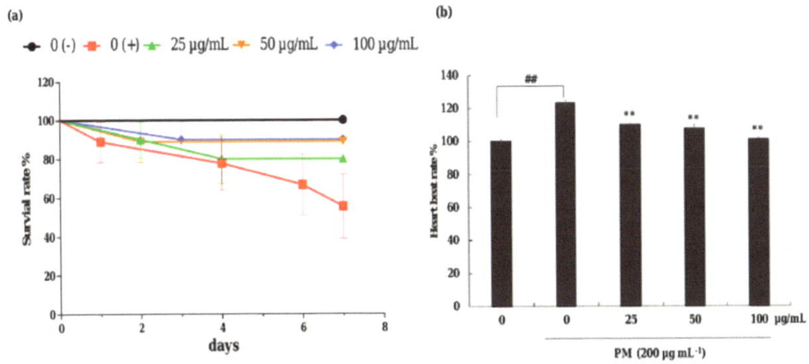

Figure 6. Effects of CRHF2 on PM-induced alterations in survival rate and heart beat rate of zebrafish. (**a**) survival rate, (**b**) heartbeat rate. Experiments were triplicated (n = 3). Results are represented as mean ± SD; ** $p < 0.01$. (# denotes significance compared to control while * represents significance compared to PM treated group 0 (+)).

3.8. Potential of CRHF2 on Inhibiting PM-Induced ROS Accumulation, Lipid Peroxidation and Cell Death in Zebrafish

As shown in Figure 7a, the ROS production in PM-induced zebrafish was increased compared to that in nontreated fish. However, CRHF2 treatment at 25, 50, and 100 µg/mL decreased the intracellular ROS levels. Additionally, PM stimulation resulted in significant cell death and lipid peroxidation in zebrafish (Figure 7b,c). Similarly, the results showed a decline in lipid peroxidation and cell death in zebrafish following dose-dependent CRHF2 treatment.

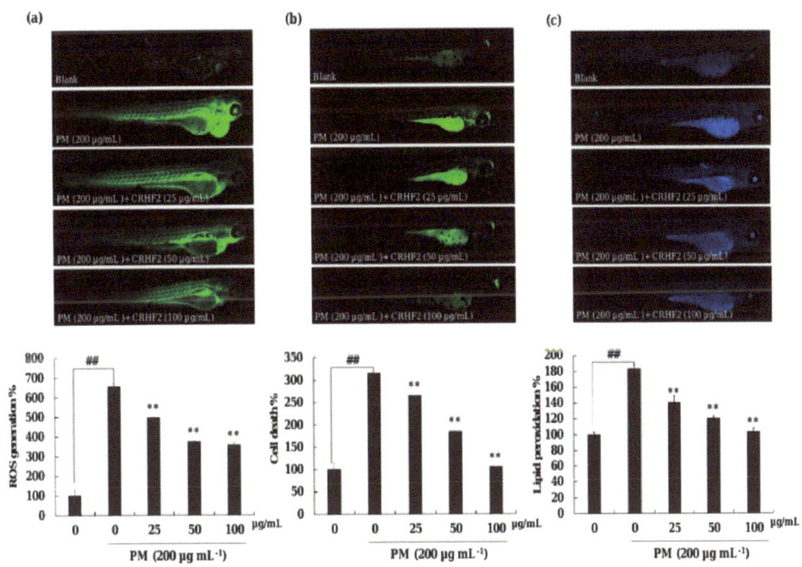

Figure 7. Protective effect of CRHF2 on PM-induced damage in zebrafish. (**a**) ROS production, (**b**) cell death, and (**c**) lipid peroxidation. Experiments were triplicated (n = 3). Results are represented as mean ± SD; ** $p < 0.01$. (# denotes significance compared to control while * represents significance compared to PM-treated group).

4. Discussion

With increasing environmental pollution, airborne PM has been identified as the main source of air pollution worldwide. Industrial expansion, coal-burning power plants, vehicle traffic, sandstorms, and other natural events contribute to high levels of PM in the environment [27]. Fine-dust PM has been identified as a major cause of skin diseases, including oxidative stress, inflammation, androgenic alopecia, extrinsic aging, and skin cancers [28]. PM damages the skin by penetrating keratinocytes and causing oxidative stress, which leads to cell apoptosis [28]. Therefore, it is necessary to identify potential treatment strategies for counteracting PM-induced diseases in humans. Recent trends in medicine and cosmeceuticals have identified natural compounds extracted from seaweeds, as safe treatment methods owing to their natural origin. The increasing requirement for potential new substances for the treatment of diseases has highlighted the applicability of these bioactive compounds in the pharmacology, medicine, and cosmeceuticals sectors [29]. Hence, this study was conducted to test the antioxidant and antiapoptotic activities of clionasterol-rich fraction of *C. racemosa* extract against PM-induced oxidative stress, thereby reducing skin damage. *C. racemosa* is a known source of several bioactive constituents such as squalene, having antioxidant and anti-inflammatory activities [30]. The importance of sesqiterpene caulerpenyne isolated from *C. racemosa* was reported earlier [31]. In the present study, fractionation of CRE using hexane resulted in five fractions. Among these, CRHF2 was proven to have superior activity. GC/MS analysis identified clionasterol as the key enriched compound in CRHF2. Several studies have reported the bioactivity of clionasterol [32,33]. In this study, the HaCaT cell culture system was used as an in vitro model and the zebrafish model was used as an in vivo model.

Reactive oxygen species (ROS) represent the major agents of oxidative stress which can be beneficial or deleterious to cells. ROS are small, short-lived, and highly reactive molecules that are important in the regulation of normal physiological functions at low doses. Excess ROS generation in living cells has detrimental effects on proteins, DNA, and membranes [34–36]. ROS can be involved in the initiation of oxidative processes as well as in the development of skin diseases. Even though harmful effects of ROS are attenuated by endogenous antioxidants, increased or prolonged presence of ROS interferes with the ROS defense mechanisms in the body. This leads to the activation of cellular responses, which results in the development of numerous skin disorders [33].

The present study findings showed that PM stimulation decreased the HaCaT cell viability, whereas CRHF2 reversed the effect of PM in limiting cell viability. Moreover, intracellular ROS levels increased with PM stimulation; however, they were considerably downregulated with dose-dependent CRHF2 treatment, proving excellent ROS scavenging activity. These findings are similar to those of several previous studies conducted to determine the effect of PM-induced oxidative stress, confirming that PM induces ROS generation in living cells, whereas it is reversed by the treatment with seaweed extracts [37,38].

Elevated ROS levels rework the cellular redox potential and trigger apoptosis [39]. In the present study, PM-induced apoptosis in HaCaT cells was detected using Hoechst 33342 staining, ethidium bromide and acridine double staining, and cell-cycle analysis using flow cytometry. In line with previous studies, Hoechst staining revealed that HaCaT cells exhibit a clear apoptotic morphology of fragmentation and condensation of the nucleus in PM-stimulated cells, which decreased with dose-dependent CRHF2 treatment. Apoptotic nuclei were identified based on the binding of Hoechst to DNA. Therefore, viable cells are indicated as homogeneously stained, round, intact nuclei, whereas apoptotic cells are indicated as fragmented and chromatin-condensed cells. In addition, nuclear double staining allows for the characterization of viable cells and cells undergoing early and late apoptosis. A homogenous nucleus is stained green, and a fragmented nucleus is stained green and orange [40]. In our study, double staining indicated the presence of necrotic cells in PM-stimulated HaCaT cells and indicated that a high number of cells had a clear apoptotic morphology with fragmented nuclei. Pretreatment with CRHF2 ameliorated PM-induced apoptosis by decreasing the number of apoptotic cells. This treatment also restored chromatin condensation and fragmentation in a dose-dependent manner. This phenomenon has also been demonstrated in an earlier study on antiapoptotic effect of Eckol isolated from *Ecklonia cava* against PM-induced cells [41]. To investigate the cell-cycle phase, flow cytometry was performed using propidium iodide staining. Similar to results obtained in an earlier study, prominent sub-G1 accumulation was observed in PM-stimulated HaCaT cells [42]. However, a reduction in sub-G1 phase cell population after CRHF2 treatment clarifies its antioxidant potential.

PM triggers apoptosis by the following two main pathways; the mitochondria-mediated intrinsic pathway and cell surface receptor-mediated extrinsic pathway. The mitochondria-mediated apoptosis pathway is a major signaling pathway resulting in apoptosis due to its sensitivity to a high number of death stimuli. Increased ROS generation in mitochondria results in mitochondrial damage as well as the activation of mitochondria-mediated apoptosis. Therefore, the effects of PM on mitochondrial ROS generation were also tested in this study, using the DHR123 staining method. Excessive mitochondrial ROS results in the activation of mitochondrial-mediated apoptosis pathway [43]. Mitochondria-derived ROS targets nearby structures such as mitochondrial DNA (mtDNA), further increasing ROS generation, leading to loss of mitochondrial membrane potential. This results in apoptosis through the mitochondrial pathway. In our study, compared to nontreated cells, PM stimulation increased ROS generation in mitochondria; however, CRHF2 regulated the mitochondria in a stable state.

The mitochondria-mediated apoptosis pathway is controlled by a complex network of signaling cascades. Therefore, Western blot analysis was carried out to assess the effect of CRHF2 on mitochondria-mediated apoptosis pathway protein expression. PM stimulation initiates this pathway by the permeabilization of the outer mitochondrial membrane. This is controlled by Bcl-2 family proteins, which include antiapoptotic proteins (Bcl-2 and Bcl-xL) and proapoptotic proteins such as Bax [44]. The activation of Bax and inhibition of Bcl-xL promote the opening of mitochondrial permeability transition holes, thus causing the release of cytochrome C into the cytoplasm. This activates caspase-9 and caspase-3 protein expression, resulting in apoptosis [45]. The permeabilization of the mitochondrial membrane by PM also results in the activation of p53, leading to apoptosis [46]. PARP inhibits its catalytic activity, which ultimately provokes cell apoptosis. In our study, PM treatment resulted in the alteration of Bax and Bcl-xL protein levels in cells, and it ultimately increased the expression levels of caspase-3, caspase-9, p53, and cleaved PARP. Similar results were observed in a study on the effect of PM on endothelial cells [41]. Our study findings suggest that PM initiates mitochondria-mediated apoptosis protein expression, leading to cell damage. Results showed that dose-dependent treatment of CRHF2 inhibits the expression of proapoptotic proteins, attenuating the negative effects of PM in skin keratinocytes. Similar results were obtained in a study on the antiapoptotic activity of dioxinodehydrockol, a phlorotannin isolated from *E. cava* [47].

The zebrafish model is becoming popular as an animal model for testing human disease conditions. Therefore, in this study, it was used as an in vivo model to investigate the antioxidant activity of CRHF2. ROS production levels measured in zebrafish indicated that CRHF2 treatment resulted in declining intensities, revealing its protective effect against PM-induced ROS generation. Moreover, DPPP staining indicated the intensity of lipid peroxidation due to PM stimulation in zebrafish. ROS stimulated the lipid peroxidation, resulting in the disruption of membrane lipid bilayer and membrane-bound receptor activities. This leads to an increase in tissue permeability [24]. Lipid peroxidation results in protein inactivation by increasing unsaturated aldehydes through crosslinking [9]. Moreover, results indicated that cell death in zebrafish was caused by the accumulation of important macromolecules, such as lipid peroxidases, due to ROS generation. However, CRHF2 protected zebrafish against PM-induced oxidative stress through ROS scavenging. The results obtained in this study are well consistent with those of previous studies on polysaccharide extracts from *Hizikia fusiforme* and *Padina boryana* [25,48]. Our results indicate that *C. racemosa* extract possesses strong antioxidant activity, as demonstrated by the decline in cell death, ROS production, and lipid peroxidation in the zebrafish model.

5. Conclusions

In conclusion, the present study demonstrated that PM induced oxidative stress-mediated cell apoptosis, and the hexane fraction isolated from *C. racemosa*, CRHF2, attenuated PM-induced skin damage by inhibiting ROS generation and activating the mitochondrial-mediated apoptosis pathway in HaCaT cells. The apoptosis was inhibited by regulating apoptosis pathway proteins and DNA damage in PM-induced cells. Furthermore, CRHF2 suppressed ROS production, cell death, and lipid peroxidation in PM-treated zebrafish, suggesting its potential use in attenuating oxidative stress in vivo. These results suggest that *C. racemosa* possesses strong antioxidant and antiapoptotic activities and could be a promising ingredient in the pharmaceutical and cosmeceutical industries.

Author Contributions: Conceptualization, N.M.L. and T.U.J.; methodology, N.M.L. and D.P.N.; software, N.M.L., D.P.N., T.U.J., H.H.A.C.K.J., H.-G.L. and Y.-S.K.; validation, N.M.L., D.P.N. and T.U.J.; formal analysis, N.M.L.; investigation, N.M.L. and D.P.N.; resources, T.U.J., D.P.N., H.H.A.C.K.J., N.M.L., H.-G.L. and Y.-S.K.; data curation, D.P.N. writing—original draft preparation, N.M.L.; writing—review and editing, Y.-J.J.; visualization, N.M.L.; supervision, Y.-J.J.; project administration, Y.-J.J.; funding acquisition, Y.-J.J. All authors have read and agreed to the published version of the manuscript.

Funding: This research was supported by the 2022 scientific promotion program funded by Jeju National University.

Institutional Review Board Statement: Animal experiments were conducted in accordance with the experimental animal guidelines provided by Jeju National University Animal center and were authorized by the Animal Care Use Committee (IACUC) of Jeju National University (protocol 2020-0049).

Informed Consent Statement: Not applicable.

Data Availability Statement: Data are contained within the article.

Conflicts of Interest: The authors declare no conflict of interest.

References

1. Magnani, N.D.; Muresan, X.M.; Belmonte, G.; Cervellati, F.; Sticozzi, C.; Pecorelli, A.; Miracco, C.; Marchini, T.; Evelson, P.; Valacchi, G. Skin damage mechanisms related to airborne particulate matter exposure. *Toxicol. Sci.* **2016**, *149*, 227–236. [CrossRef] [PubMed]
2. Piao, M.J.; Ahn, M.J.; Kang, K.A.; Ryu, Y.S.; Hyun, Y.J.; Shilnikova, K.; Zhen, A.X.; Jeong, J.W.; Choi, Y.H.; Kang, H.K.; et al. Particulate matter 2.5 damages skin cells by inducing oxidative stress, subcellular organelle dysfunction, and apoptosis. *Arch. Toxicol.* **2018**, *92*, 2077–2091. [CrossRef] [PubMed]
3. Han, X.; Liang, W.; Zhang, Y.; Sun, L.; Liang, W. Effect of atmospheric fine particles on epidermal growth factor receptor mRNA expression in mouse skin tissue. *Genet. Mol. Res.* **2016**, *15*, 1–5. [CrossRef] [PubMed]
4. Li, Q.; Kang, Z.; Jiang, S.; Zhao, J.; Yan, S.; Xu, F.; Xu, J. Effects of ambient fine particles PM2. 5 on human HaCaT cells. *Int. J. Environ. Res. Public Health* **2017**, *14*, 72. [CrossRef]
5. Liu, J.; Zhang, J.; Ren, L.; Wei, J.; Zhu, Y.; Duan, J.; Jing, L.; Sun, Z.; Zhou, X. Fine particulate matters induce apoptosis via the ATM/P53/CDK2 and mitochondria apoptosis pathway triggered by oxidative stress in rat and GC-2spd cell. *Ecotoxicol. Environ. Saf.* **2019**, *180*, 280–287. [CrossRef]
6. Gao, Z.-X.; Song, X.-L.; Li, S.-S.; Lai, X.-R.; Yang, Y.-L.; Yang, G.; Li, Z.-J.; Cui, Y.-H.; Pan, H. Assessment of DNA damage and cell senescence in corneal epithelial cells exposed to airborne particulate matter (PM2. 5) collected in Guangzhou, China. *Investig. Ophthalmol. Vis. Sci.* **2016**, *57*, 3093–3102. [CrossRef]
7. Zhang, J.; Liu, J.; Ren, L.; Wei, J.; Duan, J.; Zhang, L.; Zhou, X.; Sun, Z. PM2. 5 induces male reproductive toxicity via mitochondrial dysfunction, DNA damage and RIPK1 mediated apoptotic signaling pathway. *Sci. Total Environ.* **2018**, *634*, 1435–1444. [CrossRef]
8. Birben, E.; Sahiner, U.M.; Sackesen, C.; Erzurum, S.; Kalayci, O. Oxidative Stress and Antioxidant Defense. *World Allergy Organ. J.* **2012**, *5*, 9–19. [CrossRef]
9. Kang, M.-C.; Kim, S.Y.; Kim, Y.T.; Kim, E.-A.; Lee, S.-H.; Ko, S.-C.; Wijesinghe, W.; Samarakoon, K.W.; Kim, Y.-S.; Cho, J.H.; et al. In vitro and in vivo antioxidant activities of polysaccharide purified from aloe vera (*Aloe barbadensis*) gel. *Carbohydr. Polym.* **2014**, *99*, 365–371. [CrossRef]

10. Sundarraj, S.; Thangam, R.; Sreevani, V.; Kaveri, K.; Gunasekaran, P.; Achiraman, S.; Kannan, S. γ-Sitosterol from *Acacia nilotica* L. induces G2/M cell cycle arrest and apoptosis through c-Myc suppression in MCF-7 and A549 cells. *J. Ethnopharmacol.* **2012**, *141*, 803–809. [CrossRef]
11. Cerqueira, F.; Watanadilok, R.; Sonchaeng, P.; Kijjoa, A.; Pinto, M.; van Ufford, H.Q.; Kroes, B.; Beukelman, C.; Nascimento, M.S.J. Clionasterol: A potent inhibitor of complement component C1. *Planta Med.* **2003**, *69*, 174–176. [CrossRef]
12. Dzeha, T.; Jaspars, M.; Tabudravu, J. Clionasterol, a triterpenoid from the Kenyan marine green macroalga Halimeda macroloba. *West. Indian Ocean J. Mar. Sci.* **2003**, *2*, 157–161. [CrossRef]
13. Chaturvedi, A.K.; Patel, M.K.; Mishra, A.; Tiwari, V.; Jha, B. The SbMT-2 gene from a halophyte confers abiotic stress tolerance and modulates ROS scavenging in transgenic tobacco. *PLoS ONE* **2014**, *9*, e111379. [CrossRef]
14. Matanjun, P.; Mohamed, S.; Mustapha, N.M.; Muhammad, K.; Ming, C.H. Antioxidant activities and phenolics content of eight species of seaweeds from north Borneo. *J. Appl. Phycol.* **2008**, *20*, 367. [CrossRef]
15. Nagahawatta, D.P.; Sanjeewa, K.K.A.; Jayawardena, T.U.; Kim, H.-S.; Yang, H.-W.; Jiang, Y.; Je, J.-G.; Lee, T.-K.; Jeon, Y.-J. Drying seaweeds using hybrid hot water Goodle dryer (HHGD): Comparison with freeze-dryer in chemical composition and antioxidant activity. *Fish. Aquat. Sci.* **2021**, *24*, 19–31. [CrossRef]
16. Kim, H.-H.; Kim, H.-S.; Ko, J.-Y.; Kim, C.-Y.; Lee, J.-H.; Jeon, Y.-J.; Sciences, A. A single-step isolation of useful antioxidant compounds from *Ishige okamurae* by using centrifugal partition chromatography. *Fish. Aquat. Sci.* **2016**, *19*, 22. [CrossRef]
17. Mori, I.; Sun, Z.; Ukachi, M.; Nagano, K.; McLeod, C.W.; Cox, A.G.; Nishikawa, M. Development and certification of the new NIES CRM 28: Urban aerosols for the determination of multielements. *Anal. Bioanal. Chem.* **2008**, *391*, 1997–2003. [CrossRef]
18. Fernando, I.P.S.; Jayawardena, T.U.; Sanjeewa, K.K.A.; Wang, L.; Jeon, Y.-J.; Lee, W.W. Anti-inflammatory potential of alginic acid from *Sargassum horneri* against urban aerosol-induced inflammatory responses in keratinocytes and macrophages. *Ecotoxicol. Environ. Saf.* **2018**, *160*, 24–31. [CrossRef]
19. Wang, L.; Cui, Y.R.; Yang, H.-W.; Lee, H.G.; Ko, J.-Y.; Jeon, Y.-J.; Sciences, A. A mixture of seaweed extracts and glycosaminoglycans from sea squirts inhibits α-MSH-induced melanogenesis in B16F10 melanoma cells. *Fish. Aquat. Sci.* **2019**, *22*, 11. [CrossRef]
20. Fernando, I.P.S.; Jayawardena, T.U.; Kim, H.-S.; Vaas, A.P.J.P.; De Silva, H.I.C.; Nanayakkara, C.M.; Abeytunga, D.T.U.; Lee, W.; Ahn, G.; Lee, D.-S.; et al. A keratinocyte and integrated fibroblast culture model for studying particulate matter-induced skin lesions and therapeutic intervention of fucosterol. *Life Sci.* **2019**, *233*, 116714. [CrossRef]
21. Hao, W.; Zhang, X.; Zhao, W.; Chen, X. Psoralidin induces autophagy through ROS generation which inhibits the proliferation of human lung cancer A549 cells. *PeerJ* **2014**, *2*, e555. [CrossRef]
22. Fernando, I.S.; Sanjeewa, K.A.; Ann, Y.-S.; Ko, C.-i.; Lee, S.-H.; Lee, W.W.; Jeon, Y.-J. Apoptotic and antiproliferative effects of Stigmast-5-en-3-ol from Dendronephthya gigantea on human leukemia HL-60 and human breast cancer MCF-7 cells. *Toxicol. Vitr.* **2018**, *52*, 297–305. [CrossRef]
23. Wijesinghe, W.; Jeon, Y.J.; Ramasamy, P.; Wahid, M.E.A.; Vairappan, C.S. Anticancer activity and mediation of apoptosis in human HL-60 leukaemia cells by edible sea cucumber (*Holothuria edulis*) extract. *Food Chem.* **2013**, *139*, 326–331. [CrossRef]
24. Chaudhary, A.K.; Yadav, N.; Bhat, T.A.; O'Malley, J.; Kumar, S.; Chandra, D. A potential role of X-linked inhibitor of apoptosis protein in mitochondrial membrane permeabilization and its implication in cancer therapy. *Drug Discov. Today* **2016**, *21*, 38–47. [CrossRef]
25. Jayawardena, T.U.; Wang, L.; Sanjeewa, K.K.A.; Kang, S.I.; Lee, J.-S.; Jeon, Y.-J. Antioxidant Potential of Sulfated Polysaccharides from *Padina boryana*; Protective Effect against Oxidative Stress in In Vitro and In Vivo Zebrafish Model. *Mar. Drugs* **2020**, *18*, 212. [CrossRef]
26. Wang, L.; Lee, W.; Jayawardena, T.U.; Cha, S.-H.; Jeon, Y.-J. Dieckol, an algae-derived phenolic compound, suppresses airborne particulate matter-induced skin aging by inhibiting the expressions of pro-inflammatory cytokines and matrix metalloproteinases through regulating NF-κB, AP-1, and MAPKs signaling pathways. *Food Chem. Toxicol.* **2020**, *146*, 111823. [CrossRef]
27. Lee, J.-H.; Ko, J.-Y.; Oh, J.-Y.; Kim, E.-A.; Kim, C.-Y.; Jeon, Y.-J. Evaluation of phlorofucofuroeckol-A isolated from *Ecklonia cava* (Phaeophyta) on anti-lipid peroxidation in vitro and in vivo. *ALGAE* **2015**, *30*, 313–323. [CrossRef]
28. Soeur, J.; Belaïdi, J.-P.; Chollet, C., Denat, L.; Dimitrov, A.; Jones, C.; Perez, P.; Zanini, M.; Zobiri, O.; Mezzache, S.; et al. Photo-pollution stress in skin: Traces of pollutants (PAH and particulate matter) impair redox homeostasis in keratinocytes exposed to UVA1. *J. Dermatol. Sci.* **2017**, *86*, 162–169. [CrossRef] [PubMed]
29. Takamatsu, S.; Hodges, T.W.; Rajbhandari, I.; Gerwick, W.H.; Hamann, M.T.; Nagle, D.G. Marine natural products as novel antioxidant prototypes. *J. Nat. Prod.* **2003**, *66*, 605–608. [CrossRef] [PubMed]
30. Fernando, I.P.S.; Sanjeewa, K.K.A.; Samarakoon, K.W.; Lee, W.W.; Kim, H.S.; Jeon, Y.J. Squalene isolated from marine macroalgae *Caulerpa racemosa* and its potent antioxidant and anti-inflammatory activities. *J. Food Biochem.* **2018**, *42*, e12628. [CrossRef]
31. Dumay, O.; Pergent, G.; Pergent-Martini, C.; Amade, P. Variations in Caulerpenyne Contents in *Caulerpa taxifolia* and *Caulerpa racemosa*. *J. Chem. Ecol.* **2002**, *28*, 343–352. [CrossRef]
32. Marsan, M.-P.; Muller, I.; Milon, A. Ability of clionasterol and poriferasterol (24-epimers of sitosterol and stigmasterol) to regulate membrane lipid dynamics. *Chem. Phys. Lipids* **1996**, *84*, 117–121. [CrossRef]
33. Valentine, F.R.; Bergmann, W. Contributions to the study of marine products. VIII. the sterol of sponges: Clionasterol and poriferasterol. *J. Org. Chem.* **1941**, *6*, 452–461. [CrossRef]
34. Krutmann, J.; Liu, W.; Li, L.; Pan, X.; Crawford, M.; Sore, G.; Seite, S. Pollution and skin: From epidemiological and mechanistic studies to clinical implications. *J. Dermatol. Sci.* **2014**, *76*, 163–168. [CrossRef]

35. Halliwell, B. Free radicals and antioxidants—quo vadis? *Trends Pharmacol. Sci.* **2011**, *32*, 125–130. [CrossRef]
36. Fernando, I.P.S.; Jayawardena, T.U.; Kim, H.-S.; Lee, W.W.; Vaas, A.P.J.P.; De Silva, H.I.C.; Abayaweera, G.S.; Nanayakkara, C.M.; Abeytunga, D.T.U.; Lee, D.-S.; et al. Beijing urban particulate matter-induced injury and inflammation in human lung epithelial cells and the protective effects of fucosterol from *Sargassum binderi* (Sonder ex J. Agardh). *Environ. Res.* **2019**, *172*, 150–158. [CrossRef]
37. Diao, P.; He, H.; Tang, J.; Xiong, L.; Li, L. Natural compounds protect the skin from airborne particulate matter by attenuating oxidative stress. *Biomed. Pharmacother.* **2021**, *138*, 111534. [CrossRef]
38. Fernando, I.P.S.; Sanjeewa, K.K.A.; Lee, H.G.; Kim, H.-S.; Vaas, A.P.J.P.; De Silva, H.I.C.; Nanayakkara, C.M.; Abeytunga, D.T.U.; Lee, W.W.; Lee, D.-S.; et al. Characterization and cytoprotective properties of *Sargassum natans* fucoidan against urban aerosol-induced keratinocyte damage. *Int. J. Biol. Macromol.* **2020**, *159*, 773–781. [CrossRef]
39. Dagher, Z.; Garçon, G.; Billet, S.; Gosset, P.; Ledoux, F.; Courcot, D.; Aboukais, A.; Shirali, P. Activation of different pathways of apoptosis by air pollution particulate matter (PM2.5) in human epithelial lung cells (L132) in culture. *Toxicology* **2006**, *225*, 12–24. [CrossRef]
40. Wang, L.; Jayawardena, T.U.; Yang, H.-W.; Lee, H.G.; Kang, M.-C.; Sanjeewa, K.K.A.; Oh, J.Y.; Jeon, Y.-J. Isolation, Characterization, and Antioxidant Activity Evaluation of a Fucoidan from an Enzymatic Digest of the Edible Seaweed, *Hizikia fusiforme*. *Antioxidants* **2020**, *9*, 363. [CrossRef]
41. Zhen, A.X.; Hyun, Y.J.; Piao, M.J.; Fernando, P.D.S.M.; Kang, K.A.; Ahn, M.J.; Yi, J.M.; Kang, H.K.; Koh, Y.S.; Lee, N.H.; et al. Eckol Inhibits Particulate Matter 2.5-Induced Skin Keratinocyte Damage via MAPK Signaling Pathway. *Mar. Drugs* **2019**, *17*, 444. [CrossRef] [PubMed]
42. Riccardi, C.; Nicoletti, I. Analysis of apoptosis by propidium iodide staining and flow cytometry. *Nat. Protoc.* **2006**, *1*, 1458. [CrossRef] [PubMed]
43. Orrenius, S.; Gogvadze, V.; Zhivotovsky, B. Calcium and mitochondria in the regulation of cell death. *Biochem. Biophys. Res. Commun.* **2015**, *460*, 72–81. [CrossRef] [PubMed]
44. Zheng, J.H.; Viacava Follis, A.; Kriwacki, R.W.; Moldoveanu, T. Discoveries and controversies in BCL-2 protein-mediated apoptosis. *FEBS J.* **2016**, *283*, 2690–2700. [CrossRef]
45. Li, Z.; Jo, J.; Jia, J.-M.; Lo, S.-C.; Whitcomb, D.J.; Jiao, S.; Cho, K.; Sheng, M. Caspase-3 Activation via Mitochondria Is Required for Long-Term Depression and AMPA Receptor Internalization. *Cell* **2010**, *141*, 859–871. [CrossRef]
46. Wang, W.; Deng, Z.; Feng, Y.; Liao, F.; Zhou, F.; Feng, S.; Wang, X. PM2.5 induced apoptosis in endothelial cell through the activation of the p53-bax-caspase pathway. *Chemosphere* **2017**, *177*, 135–143. [CrossRef]
47. Ryu, B.; Ahn, B.-N.; Kang, K.-H.; Kim, Y.-S.; Li, Y.-X.; Kong, C.-S.; Kim, S.-K.; Kim, D.G. Dioxinodehydroeckol protects human keratinocyte cells from UVB-induced apoptosis modulated by related genes Bax/Bcl-2 and caspase pathway. *J. Photochem. Photobiol. B Biol.* **2015**, *153*, 352–357. [CrossRef]
48. Wang, L.; Oh, J.Y.; Kim, H.S.; Lee, W.; Cui, Y.; Lee, H.G.; Kim, Y.-T.; Ko, J.Y.; Jeon, Y.-J. Protective effect of polysaccharides from Celluclast-assisted extract of *Hizikia fusiforme* against hydrogen peroxide-induced oxidative stress in vitro in Vero cells and in vivo in zebrafish. *Int. J. Biol. Macromol.* **2018**, *112*, 483–489. [CrossRef]

antioxidants

MDPI

Article

3-Bromo-4,5-dihydroxybenzaldehyde Protects Keratinocytes from Particulate Matter 2.5-Induced Damages

Ao-Xuan Zhen [1], Mei-Jing Piao [2], Kyoung-Ah Kang [2], Pincha-Devage-Sameera-Madushan Fernando [1], Herath-Mudiyanselage-Udari-Lakmini Herath [1], Suk-Ju Cho [3,*] and Jin-Won Hyun [1,2,*]

1 Department of Biochemistry, College of Medicine, Jeju National University, Jeju 63243, Republic of Korea; zhenaoxuan705@stu.jejunu.ac.kr (A.-X.Z.); sameera@stu.jejunu.ac.kr (P.-D.-S.-M.F.); lakmini@stu.jejunu.ac.kr (H.-M.-U.-L.H.)

2 Jeju Research Center for Natural Medicine, Jeju National University, Jeju 63243, Republic of Korea; mjpiao@jejunu.ac.kr (M.-J.P.); legna07@jejunu.ac.kr (K.-A.K.)

3 Department of Anesthesiology, Jeju National University Hospital, College of Medicine, Jeju National University, Jeju 63241, Republic of Korea

* Correspondence: sukjucho@jejunu.ac.kr (S.-J.C.); jinwonh@jejunu.ac.kr (J.-W.H.); Tel.: +82-64-717-2062 (S.-J.C.); +82-64-754-3838 (J.-W.H.)

Abstract: Cellular senescence can be activated by several stimuli, including ultraviolet radiation and air pollutants. This study aimed to evaluate the protective effect of marine algae compound 3-bromo-4,5-dihydroxybenzaldehyde (3-BDB) on particulate matter 2.5 ($PM_{2.5}$)-induced skin cell damage in vitro and in vivo. The human HaCaT keratinocyte was pre-treated with 3-BDB and then with $PM_{2.5}$. $PM_{2.5}$-induced reactive oxygen species (ROS) generation, lipid peroxidation, mitochondrial dysfunction, DNA damage, cell cycle arrest, apoptotic protein expression, and cellular senescence were measured using confocal microscopy, flow cytometry, and Western blot. The present study exhibited $PM_{2.5}$-generated ROS, DNA damage, inflammation, and senescence. However, 3-BDB ameliorated $PM_{2.5}$-induced ROS generation, mitochondria dysfunction, and DNA damage. Furthermore, 3-BDB reversed the $PM_{2.5}$-induced cell cycle arrest and apoptosis, reduced cellular inflammation, and mitigated cellular senescence in vitro and in vivo. Moreover, the mitogen-activated protein kinase signaling pathway and activator protein 1 activated by $PM_{2.5}$ were inhibited by 3-BDB. Thus, 3-BDB suppressed skin damage induced by $PM_{2.5}$.

Keywords: particulate matter 2.5; 3-bromo-4,5-dihydroxybenzaldehyde; reactive oxygen species; skin damage

Citation: Zhen, A.-X.; Piao, M.-J.; Kang, K.-A.; Fernando, P.-D.-S.-M.; Herath, H.-M.-U.-L.; Cho, S.-J.; Hyun, J.-W. 3-Bromo-4,5-dihydroxybenzaldehyde Protects Keratinocytes from Particulate Matter 2.5-Induced Damages. *Antioxidants* **2023**, *12*, 1307. https://doi.org/10.3390/antiox12061307

Academic Editor: Yasuhiro Yoshida

Received: 16 May 2023
Revised: 13 June 2023
Accepted: 15 June 2023
Published: 20 June 2023

1. Introduction

Fine particulate matter 2.5 ($PM_{2.5}$) causes air pollution from various sources, such as coal burning, transport, and anthropogenic emissions [1]. Approximately 90% of human beings face health risks from pollution, which violates the WHO Air Quality Guidelines [2]. $PM_{2.5}$ induces damage in vitro and in vivo to the bronchial epithelium, human endothelial cells, and macrophage-like cells [3–9]. The effects of air pollutants on the human skin have become a global concern recently [10]. Moreover, skin directly exposed to $PM_{2.5}$ can result in acute and chronic reactions. Recently, many studies, including ours, have outlined the potential mechanism by which $PM_{2.5}$ triggers excessive formation of reactive oxidative species (ROS), leading to skin inflammation and senescence [11–15]. $PM_{2.5}$ induces ROS generation, inflammatory cytokines, and apoptosis, and it promotes skin aging by interacting with p53, nuclear factor kappa B (NF-κB), interleukin-1 beta (IL-1β), IL-6, and caspase-3 [14,16,17].

3-Bromo-4,5-dihydroxybenzaldehyde (3-BDB), a natural marine compound from red algae (*Rhodomela confervoides*, *Polysiphonia morrowii*, and *Polysiphonia urceolata*), is known to have free radical scavenging, anticancer, and antibacterial properties [18–20]. We previously demonstrated that 3-BDB exerted antioxidant effects in keratinocytes by regulating nuclear factor and erythroid 2-like 2 (Nrf2) pathways. It also protects skin cells from ultraviolet B by inhibiting the generation of ROS [21–23]. Moreover, it inhibits macrophage infiltration, thereby improving cardiac function, preventing myocardial ischemia, and suppressing allergic inflammation [24–26]. However, little is known about the effects of 3-BDB on skin damage (senescence and apoptosis) caused by $PM_{2.5}$.

Therefore, we aimed to elucidate the effect of 3-BDB on $PM_{2.5}$-induced ROS generation, macromolecular damage, apoptosis, and senescence of skin cells in vitro and in vivo.

2. Materials and Methods

2.1. Sample Preparation

3-Bromo-4,5-dihydroxybenzaldehyde (3-BDB) was obtained from Matrix Scientific (Columbia, SC, USA). $PM_{2.5}$ (NIST particulate matter SRM 1650b) was purchased from Sigma-Aldrich Co., Ltd. (St. Louis, MO, USA). 3-BDB and $PM_{2.5}$ were dissolved in dimethyl sulfoxide (DMSO), and the DMSO concentration in the cell medium during treatment was maintained at <0.1%.

2.2. Cell Culture

The human HaCaT keratinocyte cell line was provided by Cell Lines Service (Heidelberg, Germany). They were cultured in Dulbecco's modified Eagle's medium (Life Technologies Co., Ltd., Grand Island, NY, USA), containing 10% heat-inactivated fetal calf serum (Life Technologies Co., Ltd.), and 1% antibiotic-antimycotic (Life Technologies Co., Ltd.) in a 37 °C incubator with a humidified atmosphere containing 5% CO_2.

2.3. Animal Experiment

We used HR-1 hairless male mice (OrientBio, Seongnam, Republic of Korea) for in vivo experiments following guidelines of the Jeju National University (Jeju, Republic of Korea) (permit number: 2017-0026). Moreover, mice were divided into four groups (n = 4 per group): phosphate buffered saline, $PM_{2.5}$ (100 µg/mL), 3-BDB (0.3 mM) + $PM_{2.5}$, and 3-BDB (3 mM) + $PM_{2.5}$. The dorsal portion of the skin of the mice was exposed to 3-BDB (0.3 mM or 3 mM) for 30 min before exposing them to $PM_{2.5}$. Then, they were covered with the nonwoven polyethylene pad (over a 1 cm^2 area), which dispersed $PM_{2.5}$ daily for 7 consecutive days. Finally, on day 7, the skin tissues were dissected for Western blot analysis [12].

2.4. ROS Scavenging Ability

We used 2′,7′-dichlorodihydrofluorescein diacetate (H_2DCFDA; Molecular Probes, Eugene, OR, USA) to measure the inhibition of $PM_{2.5}$-induced ROS by 3-BDB. Cells (1.0×10^5 cells/mL) were seeded into a 6-well plate. Cells were added to 10, 20, and 30 µM of 3-BDB or 1 mM of N-acetyl cysteine (NAC) for 1 h and then exposed to 50 µg/mL of $PM_{2.5}$ for 30 min. Cells were stained with H_2DCFDA (25 µM), and stained cells were detected using a FACSCalibur flow cytometer (Becton Dickinson, Mountain View, CA, USA). Similarly, cells were seeded into the chamber slides, and 30 µM of 3-BDB were treated for 1 h and then treated with $PM_{2.5}$ (50 µg/mL) for 30 min. Cells stained with H_2DCFDA were observed using an FV1200 laser scanning confocal microscope (Olympus, Tokyo, Japan).

2.5. Lipid Peroxidation Assay

We detected the suppression of $PM_{2.5}$-induced oxidation of lipids by 3-BDB using a diphenyl-1-pyrenylphosphine probe (DPPP, 2 µM; Molecular Probes). Cells were seeded into chamber slides, treated with 30 µM of 3-BDB for 1 h, and exposed to 50 µg/mL of $PM_{2.5}$ for another 24 h. Lipid peroxidation fluorescence was detected using a confocal microscope after DPPP staining.

2.6. Analysis of Mitochondria Function

We explored the mitochondrial calcium level and cell potential to access the inhibitory effect of 3-BDB on $PM_{2.5}$-induced mitochondrial dysfunction. For mitochondrial calcium detection, cells were treated with 30 μM of 3-BDB for 1 h and exposed to 50 μg/mL of $PM_{2.5}$ for another 24 h. The harvested cells were stained with Rhod-2 acetoxymethyl ester (Rhod-2 AM, 5 μM; Molecular Probes) and subjected to flow cytometry. We harvested cells stained with 5,5′,6,6′-tetrachloro-1,1′,3,3′-tetraethylbenzimidazolylcarbocyanine iodide (JC-1, 2 μM; Invitrogen, Carlsbad, CA, USA) to detect the mitochondrial membrane potential, and we captured the fluorescence using a flow cytometer or confocal microscope.

2.7. Detection of 8-Oxoguanine (8-OxoG)

8-OxoG is the most significant biomarker for oxidative DNA damage [27]. To detect 8-oxoG levels, we used avidin-tetramethylrhodamine isothiocyanate (TRITC) conjugate fluorescent dye (Sigma-Aldrich Co., Ltd.), which has an affinity to 8-oxoG [28]. Harvest cells in the chamber slide were treated with 30 μM of 3-BDB for 1 h and 50 μg/mL of $PM_{2.5}$ for another 24 h. Then, cells were stained with avidin-TRITC conjugate, and their fluorescence intensity was estimated using a 1.8.0 software program of image J under the confocal microscope [12].

2.8. Comet Assay

We performed a comet assay to assess the effect of 3-BDB on $PM_{2.5}$-induced DNA strand breaks. Cells (0.8×10^5 cells/mL) were seeded into the microtubes and treated with 3-BDB and/or $PM_{2.5}$ for 30 min. Harvested cells were fixed on the slides with 0.7% of agarose gel, immersed in lysis buffer (2.5 M NaCl, 100 mM Na_2-EDTA, 10 mM Tris, and 1% N-lauroylsarcosinate, pH 10) for 1 h, electrophoresed for 20 min, and then dried. Images of total fluorescence and the change in DNA tail length were recorded using ethidium bromide (10 μg/mL) under a fluorescence microscope equipped with Komet 5.5 software program of image analysis (Kinetic Imaging, Liverpool, UK). Fifty cells were counted per slide.

2.9. Detection of IL-1β and IL-6

The IL-1β and IL-6 concentrations in the culture medium were measured using a human Quantikine ELISA kit (R&D Systems, Minneapolis, MN, USA). Cells were treated with 30 μM of 3-BDB for 1 h. Then, they were incubated for 24 h with 50 μg/mL of $PM_{2.5}$ and then centrifuged at 3000 rpm for 15 min in the culture media. Cell-free supernatant was added to a 96-well plate coated with the primary antibodies against IL-1β and IL-6. The HRP-conjugated detection antibodies (100 μL) were then added and incubated for 1 h at 37 °C. After washing three times, the substrates were incubated for another 20 min. Finally, the stop solution was added to each well, and the absorbance of concentrations was measured at 450 nm using a SpectraMax i3x microplate reader (Molecular Devices, San Jose, CA, USA), which was performed immediately.

2.10. Western Blot

Cells were treated with 30 μM of 3-BDB for 1 h and then with 50 μg/mL $PM_{2.5}$ for 24 h, and mice skin tissues were treated with 3-BDB and $PM_{2.5}$ according to the above animal experiment method. Protein lysis buffers from the cells and mouse skin were loaded into a separating gel containing SDS-PAGE electrophoresis buffer. The target proteins were transferred onto membranes and shaken with primary and secondary antibodies sequentially. Finally, protein bands were obtained using the Amersham enhanced chemiluminescence, plus a Western blotting detection system (GE Healthcare, Buckinghamshire, UK). The primary antibodies used were as follows: actin (Sigma-Aldrich Co., Ltd.), c-Jun N-terminal kinase (JNK), p38 (Genetex Inc., Irvine, CA, USA), phospho-H2A.X, phospho-p53, caspase-9, caspase-3, mitogen-activated protein kinase kinases (MEK)1/2, phospho-MEK, phospho-extracellular regulated kinase (ERK), stress-activated ERK kinase

(SEK)1, phospho-SEK, phospho-JNK, phospho-p38, c-Fos, c-Jun, phospho-c-Jun (Cell Signaling Technology, Danvers, MA, USA), B-cell lymphoma protein (Bcl)-2, Bcl-2 associated X (Bax), ERK2 (Santa Cruz Biotechnology, Dallas, TX, USA), IL-1β, matrix metalloproteinase (MMP)-2, MMP-9 (Abcam, Cambridge, MA, USA), p53, IL-6 (Invitrogen), MMP-1 (Cusabio, Houston, TX, USA).

2.11. *Cell Cycle Analysis*

We performed a cell cycle analysis to evaluate the effect of 3-BDB on $PM_{2.5}$-induced cell cycle arrest. Cells were seeded into a 6-well plate, treated with 30 μM of 3-BDB for 1 h, and then with 50 μg/mL $PM_{2.5}$ for 24 h. Propidium iodide and RNase A (1:1000) were used to bind to cellular DNA. This analysis was performed using a flow cytometer.

2.12. *Hoechst 33342 Staining*

We utilized Hoechst 33342 (BIOMOL GmbH, Hamburg, Germany) to visualize the protection of $PM_{2.5}$-induced nuclei degradation by 3-BDB. Cells were seeded into a 60 mm culture dish and treated with 30 μM of 3-BDB for 1 h, followed by 50 μg/mL $PM_{2.5}$ for 24 h. Then, the cells were immersed in a medium with Hoechst 33342, a DNA-specific fluorescent dye (10 μM) for 15 min. Stained cells were visualized under a fluorescence microscope (Olympus, Tokyo, Japan).

2.13. *β-Galactosidase Staining Activity*

We used a cellular senescence detection kit (SPiDER-β-Gal, Dojindo Laboratories, Kumamoto, Japan) to detect the expression of the senescence-associated enzyme, β-galactosidase (SA-β-gal) [29]. Cells were seeded into chamber slides and treated with 30 μM of 3-BDB for 1 h, followed by $PM_{2.5}$ for another 24 h. After washing the chamber slides, the cells were stained with SPiDER-β-Gal solution and viewed under a confocal microscope.

2.14. *Statistical Analysis*

We performed statistical analyses among multiple groups by analyzing variance and Tukey's tests using Systat 3.5 software (Systat Software Inc., San Jose, CA, USA). All data are displayed as mean ± standard error. The *p*-values less than 0.05 were considered statistically significant.

3. Results

3.1. *Antioxidant Effect of 3-BDB against $PM_{2.5}$-Induced Intracellular ROS and Lipid Peroxidation*

We confirmed the ROS scavenging effect of 3-BDB at 10, 20, or 30 μM, and NAC (1 mM) was used as a positive control induced by $PM_{2.5}$ (Figure 1a). The results proved that 30 μM of 3-BDB, such as NAC, significantly prevented cells from $PM_{2.5}$-induced ROS. Next, confocal images confirmed that cells treated with 30 μM of 3-BDB contained lower ROS than the $PM_{2.5}$ group (Figure 1b). Thus, in the following trials, we used 30 μM as the optimal concentration of 3-BDB. Furthermore, to investigate ROS-induced lipid peroxidation, cells were subjected to 3-BDB and/or $PM_{2.5}$. Findings revealed that $PM_{2.5}$ caused lipid damage, whereas 3-BDB had an antagonistic effect (Figure 1c).

3.2. *Preventive Effect of 3-BDB against $PM_{2.5}$-Induced Mitochondrial Dysfunction*

As shown in Figure 2a, $PM_{2.5}$ increased mitochondrial calcium level, whereas the treatment with 3-BDB and $PM_{2.5}$ led to a decreased calcium level than $PM_{2.5}$. The mitochondrial membrane potential was analyzed to further assess mitochondrial dysfunction. The mitochondrial membrane potential depolarized by $PM_{2.5}$ was reversed after treatment with 3-BDB, as displayed by results from flow cytometry and confocal microscopic images (Figure 2b,c).

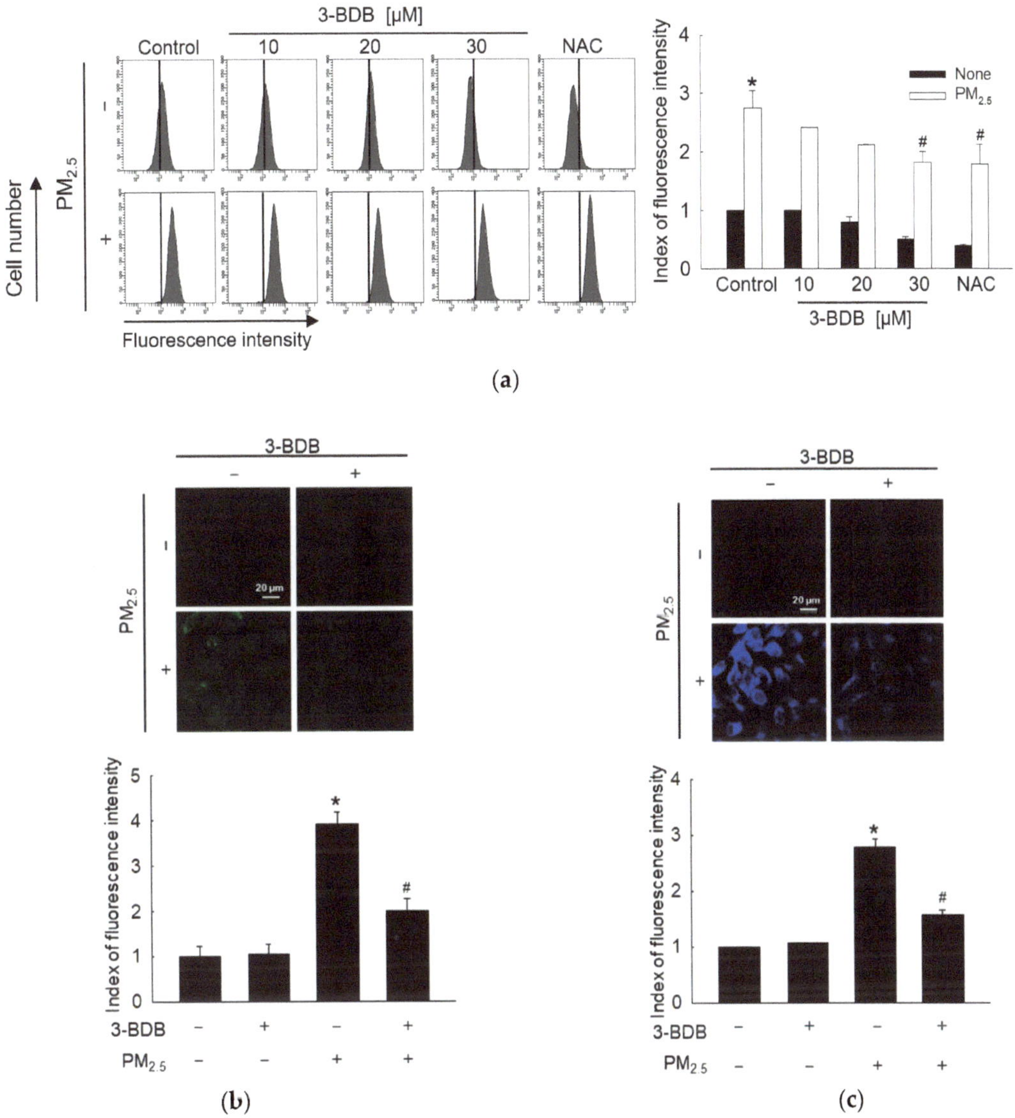

Figure 1. Inhibition of $PM_{2.5}$-induced ROS generation and lipid peroxidation were performed by 3-BDB in keratinocytes. (**a**) Cells were added to 10, 20, and 30 μM of 3-BDB or 1 mM of N-acetyl cysteine (NAC) for 1 h and then exposed to 50 μg/mL of $PM_{2.5}$ for 30 min. ROS were measured by a flow cytometer after H_2DCFDA staining. (**b**) Depletion of $PM_{2.5}$-induced ROS by 30 μM of 3-BDB was visualized by a confocal microscope after H_2DCFDA staining. (**c**) Prevention of $PM_{2.5}$-induced lipid peroxidation analysis by 3-BDB was performed using a confocal microscope after DPPP staining. (**a–c**) * $p < 0.05$ and # $p < 0.05$ compared to control cells and $PM_{2.5}$-exposed cells, respectively.

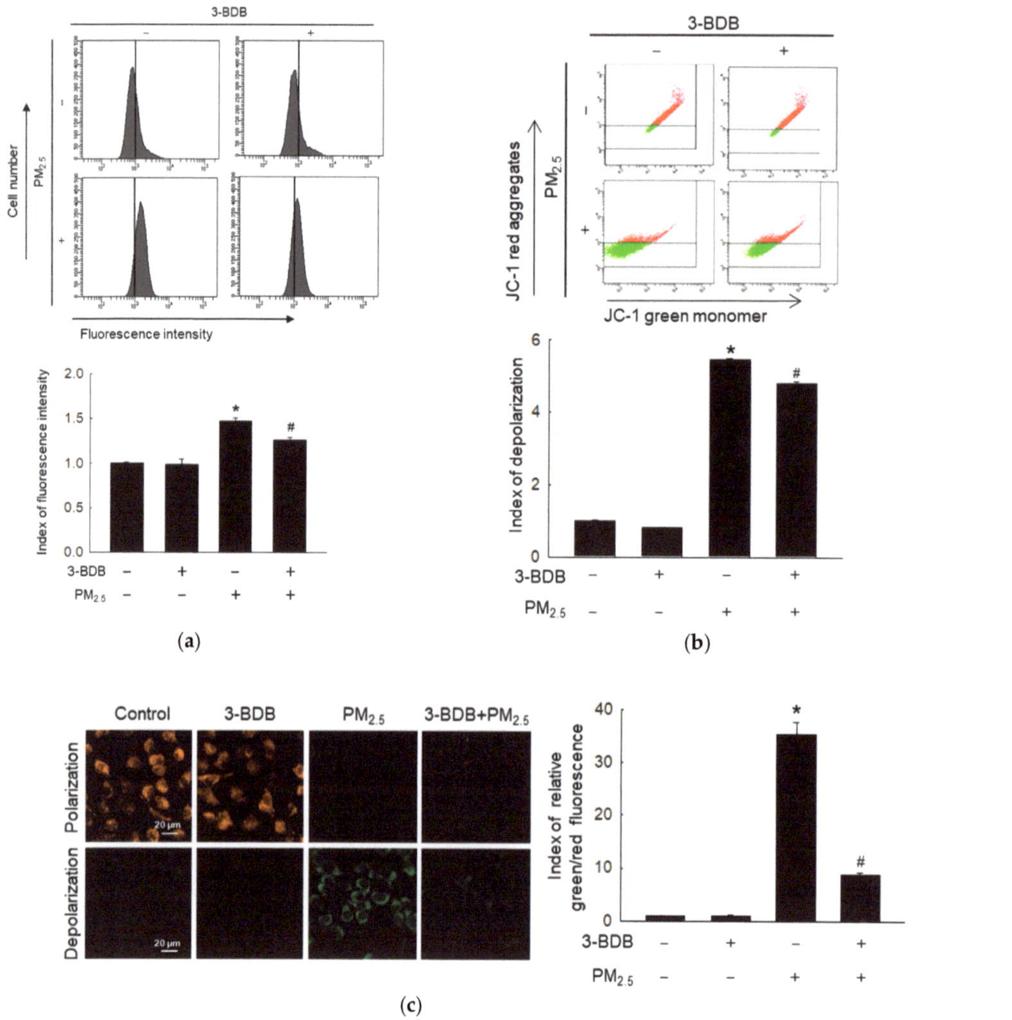

Figure 2. Prevention of $PM_{2.5}$-induced mitochondrial dysfunction was performed by 3-BDB in keratinocytes. Cells were treated with 30 μM of 3-BDB for 1 h and then exposed to 50 μg/mL of $PM_{2.5}$ for 24 h. (**a**) Rhod-2 AM was used to detect the mitochondrial calcium. (**b**,**c**) The mitochondrial membrane potential was obtained by (**b**) flow cytometry and (**c**) confocal microscopy by JC-1 staining. (**a**–**c**) * $p < 0.05$ and # $p < 0.05$ compared to control cells and $PM_{2.5}$-exposed cells, respectively.

3.3. Inhibitory Effect of 3-BDB against $PM_{2.5}$-Induced DNA Damage

Cells treated with 3-BDB and $PM_{2.5}$ possessed a lower level of 8-oxoG than those in the $PM_{2.5}$-treated group. This implies that 3-BDB inhibited $PM_{2.5}$-induced DNA oxidation (Figure 3a). Similar results confirmed that 3-BDB protected $PM_{2.5}$-induced DNA damage in the comet assay because 3-BDB reduced the DNA tail length induced by $PM_{2.5}$ in cells (Figure 3b). In addition, we detected phospho-H2A.X histone, a known indicator of DNA double-strand break [30], and a significant increase in phosphorylation of H2A.X in the $PM_{2.5}$ group in vitro and in vivo was observed; however, in the 3-BDB and $PM_{2.5}$ treatment group, a significant decrease in phospho H2A.X was observed (Figure 3c,d). ROS and ROS-induced DNA damage activate p53, a known tumor suppressor [31]. There was an increase in the phosphorylation of p53 in the $PM_{2.5}$ group; however, after treatment with 3-BDB, the level of activated p53 was lower in vitro and in vivo (Figure 3c,d). Furthermore, p53 controls the fate of cells when subjected to DNA damage, probably by arresting cell cycle progression [32,33]. Notably, the cell cycle analysis revealed that $PM_{2.5}$ arrested the cell cycle at the G_1 phase; however, 3-BDB attenuated it (Figure 3e).

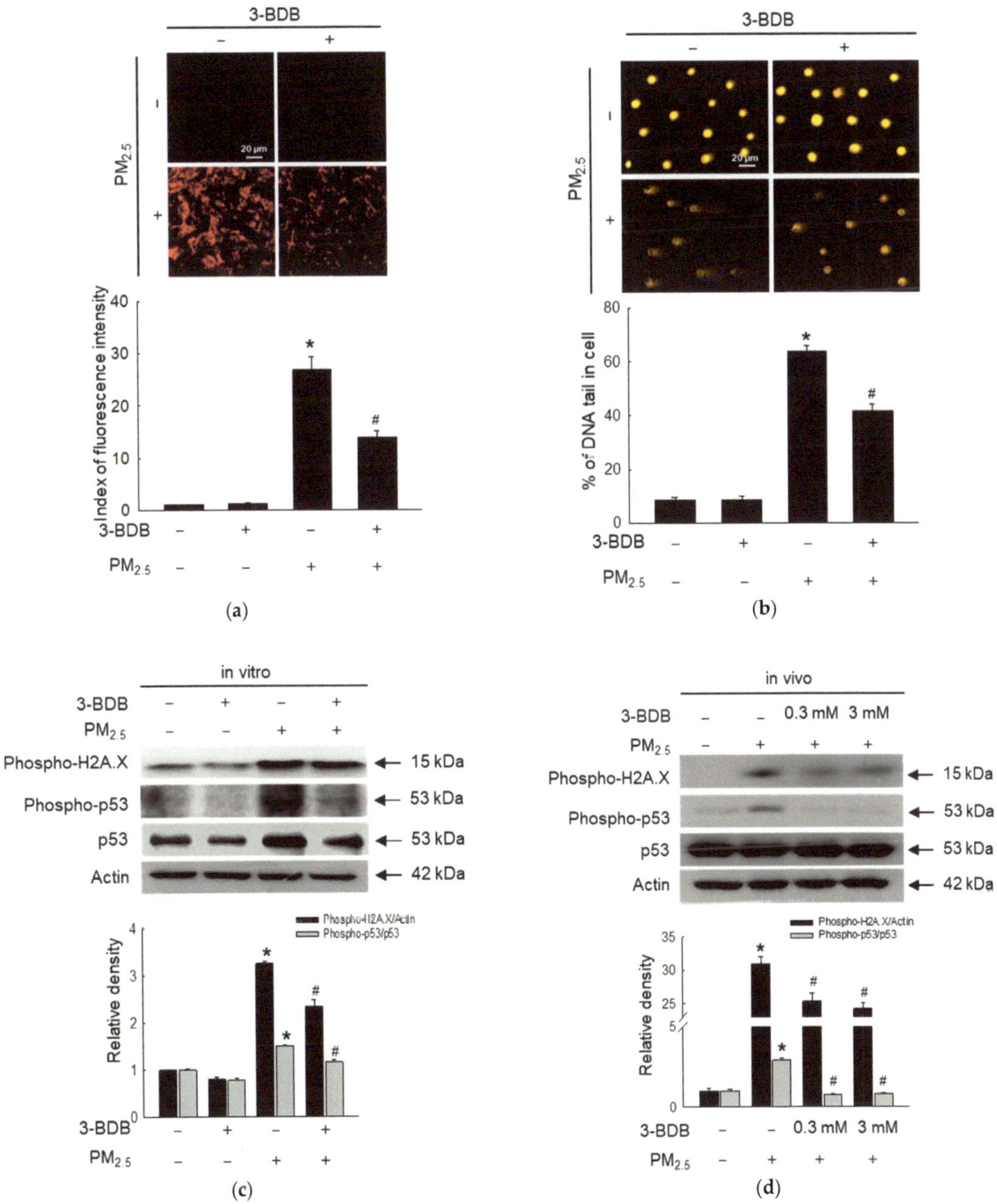

Figure 3. *Cont.*

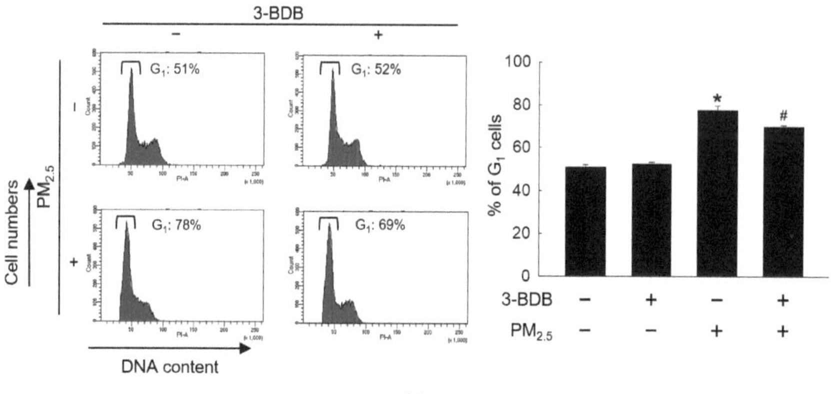

(e)

Figure 3. Reversibility of $PM_{2.5}$-induced DNA damage and cell cycle arrest was performed by 3-BDB. Cells were treated with 30 μM of 3-BDB for 1 h and then exposed to 50 μg/mL of $PM_{2.5}$ for 24 h. (**a**) Avidin-TRITC conjugate was used to detect the 8-oxoG. (**b**) A Comet assay was performed to analyze DNA damage. (**c**,**d**) The proteins were obtained from both (**c**) cells and (**d**) tissues, and phospho-H2A.X, phospho-p53, and p53 were examined by Western blot. (**e**) The checkpoint of the G_1 phase was measured by flow cytometry. (**a**–**e**) * $p < 0.05$ and # $p < 0.05$ compared to control groups and $PM_{2.5}$-exposed groups, respectively.

3.4. Anti-Apoptotic Effect of 3-BDB against $PM_{2.5}$-Induced Apoptosis

$PM_{2.5}$ decreased the anti-apoptotic protein, Bcl-2, and increased the pro-apoptotic protein, Bax; however, 3-BDB reversed these effects in vitro and in vivo (Figure 4a,b). 3-BDB also reversed $PM_{2.5}$-activated caspase-9 and caspase-3, the main markers of apoptosis-mediated cell death, in vitro and in vivo (Figure 4c,d). Nuclei integrity was visualized through Hoechst 33342 staining. Notably, $PM_{2.5}$ increased apoptotic bodies, but 3-BDB significantly decreased the number of apoptotic bodies (Figure 4e).

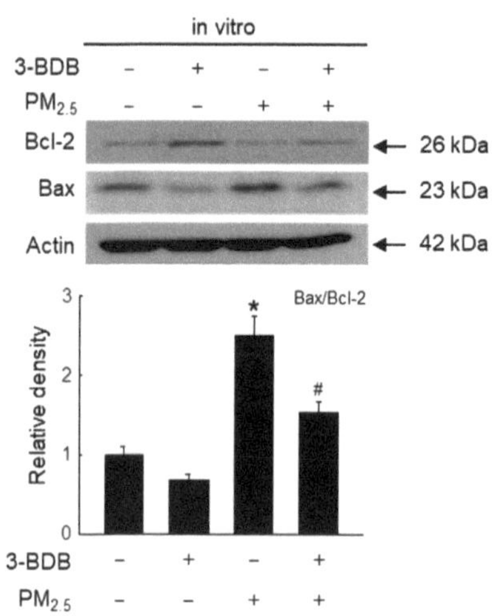

(a)

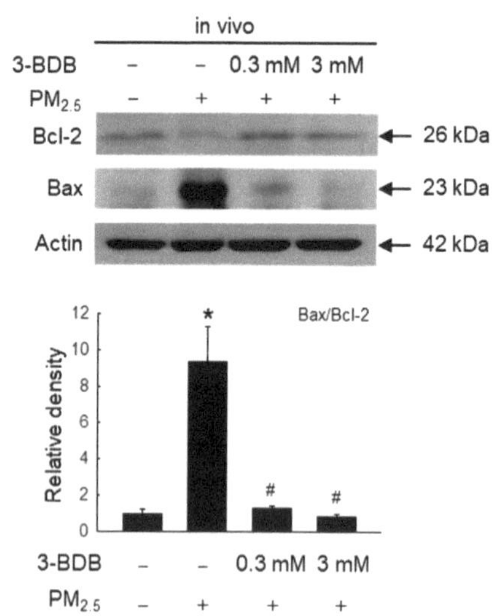

(b)

Figure 4. *Cont.*

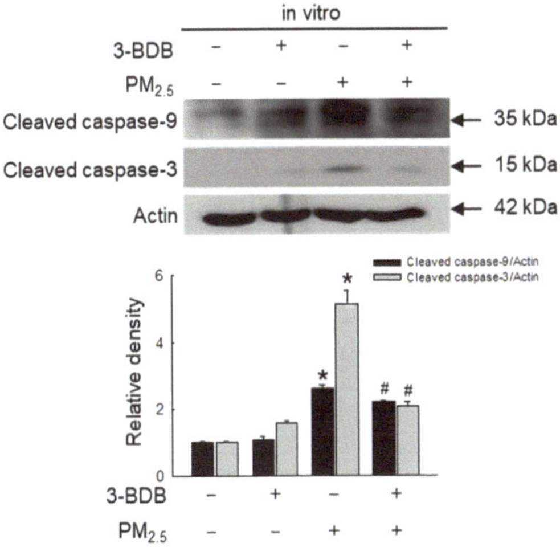

(c)

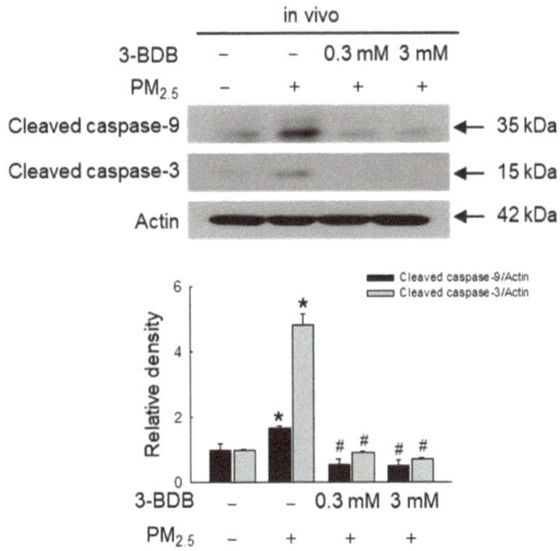

(d)

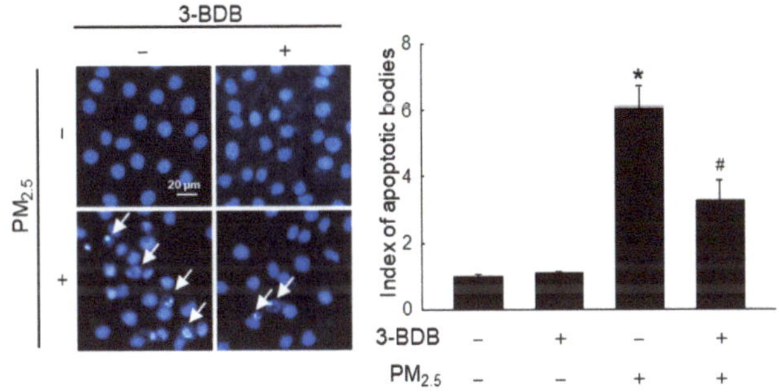

(e)

Figure 4. Reduction in $PM_{2.5}$-induced cell apoptosis was by 3-BDB in vitro and in vivo. Cells were treated with 30 μM of 3-BDB for 1 h and then exposed to 50 μg/mL of $PM_{2.5}$ for 24 h. Mice skin was treated with 3-BDB and $PM_{2.5}$ according to the animal experiment in Materials and Methods. (**a**–**d**) The proteins were isolated from (**a**,**c**) cells, (**b**,**d**) tissue, and Bcl-2, Bax, and cleaved caspase-9, and cleaved caspase-3 were detected by Western blot. (**e**) The apoptotic bodies were counted by using Hoechst 33342 staining. The arrows indicate the apoptotic bodies. (**a**–**e**) * $p < 0.05$ and # $p < 0.05$ compared to control groups and $PM_{2.5}$-exposed groups, respectively.

3.5. Inactivating Effect of 3-BDB against $PM_{2.5}$-Induced Activator Protein (AP)-1 via Mitogen-Activated Protein Kinase (MAPK) Signaling Pathway

AP-1 transcription factor is associated with MAPK-induced apoptosis [34]. Thus, we checked the expression levels of MAPK-related proteins, MEK, ERK, SEK, JNK, and p38. $PM_{2.5}$ induced a high level of activated MEK1/2, ERK1/2, SEK1, JNK, and p38, which were reversed by treatment with 3-BDB (Figure 5a). Besides, the transcription factor AP-1 (c-Jun and c-Fos) was also activated by $PM_{2.5}$, but it decreased in the 3-BDB- and $PM_{2.5}$-treated groups (Figure 5b).

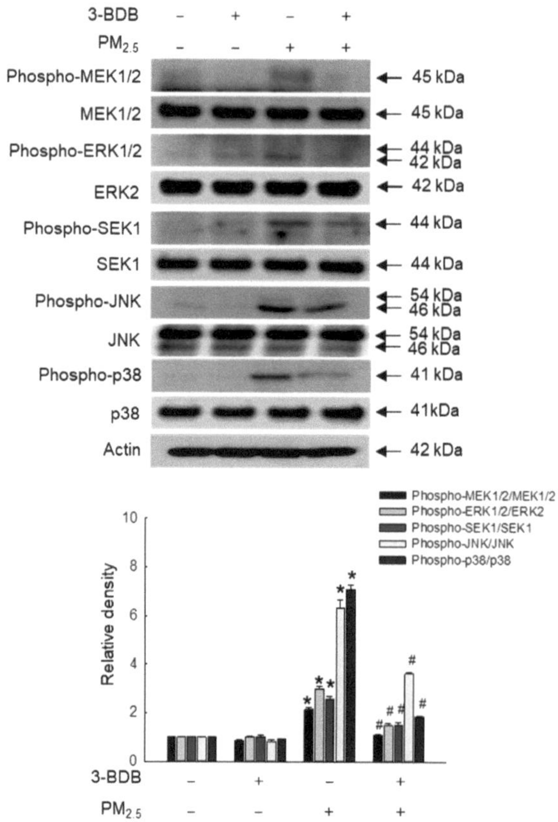

(**a**)

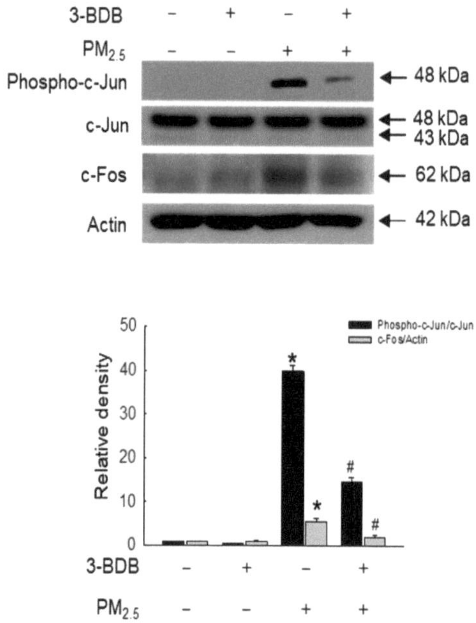

(**b**)

Figure 5. Inactivation of $PM_{2.5}$-induced MAPK signaling pathway, the transcription factor AP-1 was performed by 3-BDB. (**a**,**b**) Cells were treated with 30 µM of 3-BDB for 1 h, and then the cells were stimulated by $PM_{2.5}$ for 24 h, the proteins were separated from cells, and (**a**) phospho-MEK1/2, MEK1/2, phospho-ERK1/2, ERK2, phospho-SEK1, SEK1, phospho-JNK, JNK, phospho-p38, p38, as well as (**b**) c-Fos, phospho-c-Jun, and c-Jun expressions were detected by Western blot. (**a**,**b**) * $p < 0.05$ and # $p < 0.05$ compared to control cells and $PM_{2.5}$-exposed cells, respectively.

3.6. Antagonizing Effect of 3-BDB against $PM_{2.5}$-Induced Senescence

We mainly examined pro-inflammatory cytokines, senescence-related proteins, and markers. The levels of the pro-inflammatory cytokines IL-1β (Figure 6a) and IL-6 (Figure 6b) were induced in $PM_{2.5}$-exposed cells, while 3-BDB reduced levels of IL-1β and IL-6 (Figure 6a,b). Moreover, IL-1β and IL-6 protein levels were also higher in the $PM_{2.5}$ group than in the 3-BDB + $PM_{2.5}$-treated group in vitro and in vivo (Figure 6c,d). $PM_{2.5}$ also induced MMP-1, MMP-2, and MMP-9 expression; however, these effects were reversed by 3-BDB both in vitro and in vivo (Figure 6e,f). Finally, we examined senescent cells through staining with SA-β-gal [13]. The results showed that $PM_{2.5}$ generates higher fluorescence than the control group; however, 3-BDB notably inhibited the fluorescence intensity induced by $PM_{2.5}$ (Figure 6g).

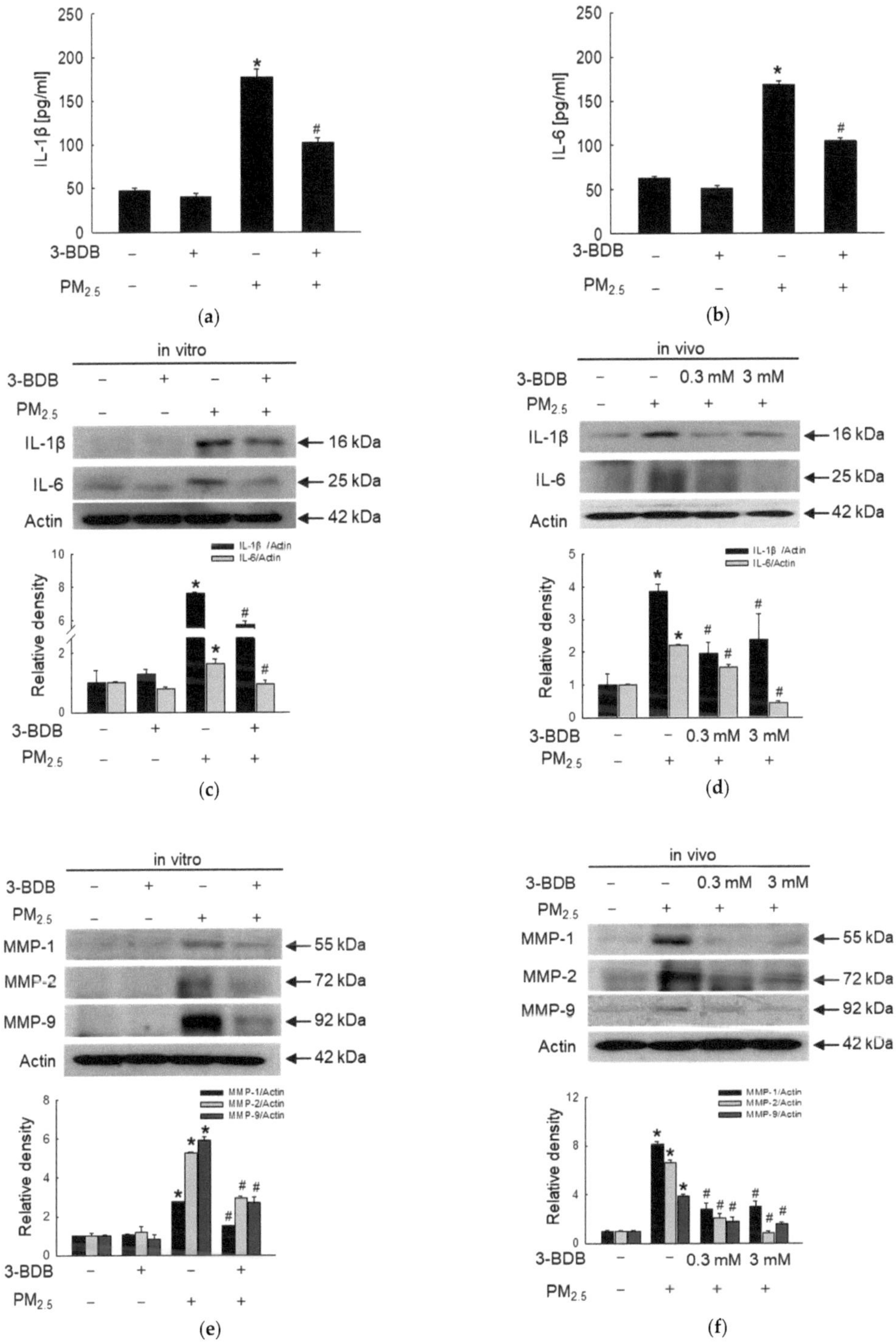

Figure 6. *Cont.*

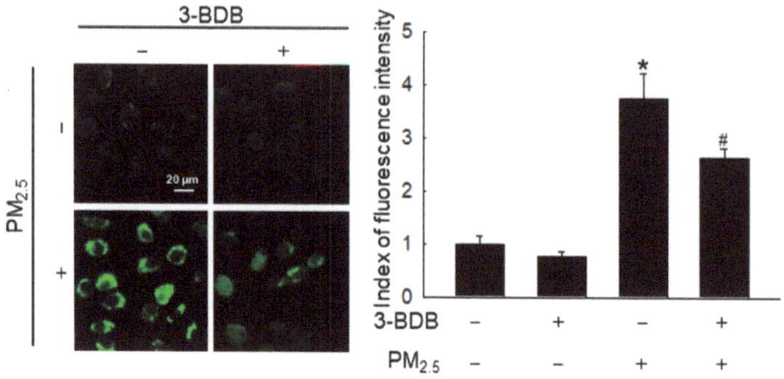

(g)

Figure 6. Inhibition of $PM_{2.5}$-induced pro-inflammatory cytokines and matrix metalloproteinases was performed by 3-BDB in vitro and in vivo. Cells were treated with 30 μM of 3-BDB for 1 h and then were exposed to 50 μg/mL of $PM_{2.5}$ for 24 h. Mice skin was treated with 3-BDB and $PM_{2.5}$ according to the animal experiment in Materials and Methods. (**a**,**b**) IL-1β and IL-6 concentrations in HaCaT cells were assessed using a human IL-1β and IL-6 ELISA kits, respectively. (**c**–**f**) From the proteins of (**c**,**e**) cells and (**d**,**f**) tissues, IL-1β, IL-6, MMP-1, MMP-2, and MMP-9 were examined by Western blot. (**g**) Senescence cells were available to visualize under a confocal microscope. (**a**–**g**) * $p < 0.05$ and # $p < 0.05$ compared to control groups and $PM_{2.5}$-exposed groups, respectively.

4. Discussion

Particulate matter ($PM_{2.5}$) is a strong matter of interest nowadays, as research is ongoing to determine its effect on human bodies. $PM_{2.5}$ possesses a different degree of toxicity, confirming that diesel engine combustion was more severe than biomass burning in the same mass [35]. One of the main organic components (polycyclic aromatic hydrocarbons, PAHs) from engine exhaust generates ROS, resulting in DNA damage [36]. Therefore, in this study, we used the recommended $PM_{2.5}$, mainly from diesel particulate matter, including PAHs and nitro-PAHs. Although, the skin can deal with different sources of ROS through a specific antioxidant mechanism; $PM_{2.5}$ overloaded it with high concentrations of ROS [37]. For ROS scavenging, we focused on the antioxidant compound 3-BDB, obtained from red algae, as it possesses a strong ability to protect against oxidative stress-related cell damage, including inflammation [21–24]. We previously showed that there was no cytotoxicity of 3-BDB at concentrations ranging from 10–30 μM in human HaCaT keratinocytes, and 3-BDB inhibited UVB-caused oxidative stress at 30 μM concentration [38]. In addition, $PM_{2.5}$ generated ROS [12–15]. Moreover, lipid peroxidation is vital for initiating the process of cell damage because lipids are prime targets of free radicals [39]. As shown in Figure 1, 3-BDB pretreatment alleviates both $PM_{2.5}$-induced cellular ROS generation and lipid peroxidation.

ROS are mainly generated from mitochondria, which are closely related to proton leaks [40]. However, oxidative stress via excessive ROS induces mitochondrial dysfunction associated with DNA damage via depolarization of mitochondrial membrane potential [41]. Mitochondrial calcium homeostasis is vital for proper mitochondrial function, but Ca^{2+} can also trigger the mitochondrial apoptosis pathway [42]. The increased ROS by $PM_{2.5}$ decreases mitochondrial action potential, causing apoptosis [12]. Furthermore, changes in the mitochondria are necessary for the senescence phenotype [43]. Thus, we examined mitochondrial calcium levels and membrane potential. Since 3-BDB inhibits ROS formation, it exerts protective effects on the mitochondria from calcium- and membrane-depolarization-induced dysfunction.

Oxidative stress leads to DNA oxidation and mutation, cancer, and senescence [29]. $PM_{2.5}$-induced oxidative stress causes DNA damage, which leads to cell cycle arrest in skin cells [44]. In this study, we noted less DNA damage in the 3-BDB and $PM_{2.5}$ groups than in the $PM_{2.5}$ group (Figure 3). One way to induce senescence is by activating the tumor suppressor, p53, which can be activated by oxidative stress or DNA damage [33]. p53 is a key factor for determining cell fate; under stress, it can maintain G_1 arrest to accelerate aging [31]. Our results were in agreement with the above explanation and demonstrated that $PM_{2.5}$ stimulated the activation of p53 and caused G_1 arrest; however, these effects were reversed by treatment with 3-BDB.

Our previous studies showed that $PM_{2.5}$ leads to cell apoptosis via ROS [12,14,37,45], and from the cell cycle analysis, $PM_{2.5}$ could induce aging and apoptosis through cell cycle arrest. Moreover, mitochondrial dysfunction is involved in apoptosis and aging [45,46]. We examined the effect of 3-BDB on $PM_{2.5}$-induced apoptosis. The results shown in Figure 4 show that 3-BDB reduces apoptotic bodies, as it inhibits ROS generation. Excessive ROS induced MMPs via the MAPK-transcription factor AP-1 signaling pathway [47,48]. Furthermore, ROS increases the secretion of pro-inflammatory cytokines, which are secreted at a high level in most senescent cells [49].

DNA damage has been regarded as an activator of the senescence-associated secretory phenotype (SASP), which is related to cell cycle arrest [50]. Two key SASP factors, IL-1β and IL-6, were detected at high levels in senescent cells [33]. Furthermore, we previously demonstrated that $PM_{2.5}$ increases levels of IL-1β and IL-6 [14]. MAPK induces the phosphorylation of NF-κB that promotes the secretion of pro-inflammatory cytokines and regulates AP-1 [51]. Collagen degradation is probably related to the formation of MMPs, especially in the epidermis and dermis [52]. Our previous study showed that $PM_{2.5}$ induces the production of MMPs (MMP-1, MMP-2, and MMP-9) and eventually induces cell senescence [13]. In the present study, we observed that $PM_{2.5}$ activated the MAPK signaling pathway and transcription factors (Figure 5), followed by the secretion of pro-inflammatory cytokines and MMPs; however, 3-BDB relieved cells from the stress condition induced by $PM_{2.5}$ (Figure 6). The senescence marker β-galactosidase is present in aged cells [29]. As shown in Figure 6e, β-galactosidase activity cells were stimulated by $PM_{2.5}$, but they were decreased via pretreatment with 3-BDB.

5. Conclusions

In summary, the inhibition of ROS generation by 3-BDB in human HaCaT keratinocytes and hairless mice reduces mitochondrial dysfunction and DNA damage response, which inhibits activation of the tumor suppressor p53 and cell cycle arrest. In addition, 3-BDB affects the inhibition of the MAPK signaling pathway and its regulated transcription factor, AP-1, reversing the formation of pro-inflammatory cytokines and MMPs, thereby inhibiting $PM_{2.5}$-induced senescence by 3-BDB (Figure 7). Notably, 3-BDB had a protective effect against $PM_{2.5}$-induced cellular damage and could be used as a preventive agent against air pollution-triggered skin damage.

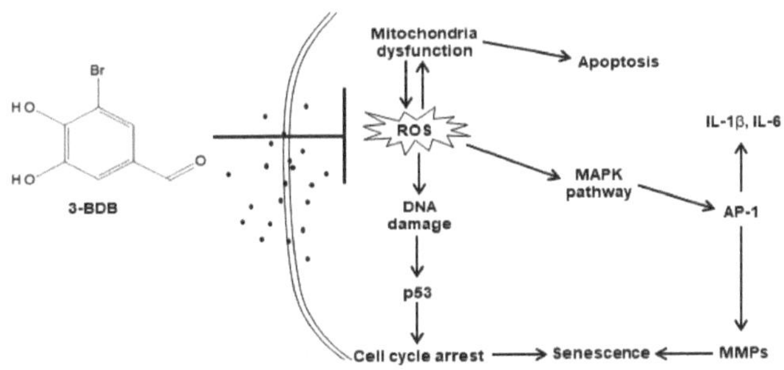

Figure 7. The schematic diagram for the protective effect of 3-BDB induced by $PM_{2.5}$ was exhibited. 3-BDB inhibited ROS generation induced by $PM_{2.5}$, which caused macromolecular damage, cell cycle arrest, and apoptosis. In addition, 3-BDB inhibited inflammatory cytokines release and MMPs expression through the MAPK signaling pathway, thus alleviating cell senescence through a complex intracellular mechanism.

Author Contributions: Experimental design, A.-X.Z., S.-J.C. and J.-W.H.; Data collection, A.-X.Z. and P.-D.-S.-M.F.; Formal analysis, A.-X.Z. and H.-M.-U.-L.H.; Writing and supervision, A.-X.Z., M.-J.P., K.-A.K., S.-J.C. and J.-W.H. All authors have read and agreed to the published version of the manuscript.

Funding: This work was funded by a research grant from Jeju National University Hospital in 2022.

Institutional Review Board Statement: Not applicable.

Informed Consent Statement: Not applicable.

Data Availability Statement: The data presented in this study are available upon request from the corresponding author.

Conflicts of Interest: The authors declare no conflict of interest.

References

1. Yin, S.; Wang, X.; Zhang, X.; Guo, M.; Miura, M.; Xiao, Y. Influence of biomass burning on local air pollution in mainland Southeast Asia from 2001 to 2016. *Environ. Pollut.* **2019**, *254*, 112949. [CrossRef]
2. Rembiesa, J.; Ruzgas, T.; Engblom, J.; Holefors, A. The impact of pollutants on skin and proper efficacy testing for anti-pollution claims. *Cosmetics* **2018**, *5*, 4. [CrossRef]
3. Han, X.; Zhuang, Y. PM2.5 induces autophagy-mediated cell apoptosis via PI3K/AKT/mTOR signaling pathway in mice bronchial epithelium cells. *Exp. Ther. Med.* **2021**, *21*, 1. [CrossRef] [PubMed]
4. Hu, T.; Zhu, P.; Liu, Y.; Zhu, H.; Geng, J.; Wang, B.; Yuan, G.; Peng, Y.; Xu, B. PM2.5 induces endothelial dysfunction via activating NLRP3 inflammasome. *Environ. Toxicol.* **2021**, *36*, 1886–1893. [CrossRef]
5. Zhou, J.; Zou, H.; Liu, Y.; Chen, Y.; Du, Y.; Liu, J.; Huang, Z.; Liang, L.; Xie, R.; Yang, Q. Acute cytotoxicity test of PM2.5, NNK and BPDE in human normal bronchial epithelial cells: A comparison of a co-culture model containing macrophages and a mono-culture model. *Toxicol. In Vitro* **2022**, *85*, 105480. [CrossRef]
6. Lin, C.I.; Tsai, C.H.; Sun, Y.L.; Hsieh, W.Y.; Lin, Y.C.; Chen, C.Y.; Lin, C.S. Instillation of particulate matter 2.5 induced acute lung injury and attenuated the injury recovery in ACE2 knockout mice. *Int. J. Biol. Sci.* **2018**, *14*, 253–265. [CrossRef] [PubMed]
7. Tang, W.; Du, L.; Sun, W.; Yu, Z.; He, F.; Chen, J.; Li, X.; Yu, L.; Chen, D. Maternal exposure to fine particulate air pollution induces epithelial-to-mesenchymal transition resulting in postnatal pulmonary dysfunction mediated by transforming growth factor-β/Smad3 signaling. *Toxicol. Lett.* **2017**, *267*, 11–20. [CrossRef] [PubMed]
8. Duan, S.; Wang, N.; Huang, L.; Zhao, Y.; Shao, H.; Jin, Y.; Zhang, R.; Li, C.; Wu, W.; Wang, J.; et al. NLRP3 inflammasome activation is associated with PM2.5 -induced cardiac functional and pathological injury in mice. *Environ. Toxicol.* **2019**, *34*, 1246–1254. [CrossRef]
9. Zhang, J.; Fulgar, C.C.; Mar, T.; Young, D.E.; Zhang, Q.; Bein, K.J.; Cui, L.; Castañeda, A.; Vogel, C.F.A.; Sun, X.; et al. TH17-induced neutrophils enhance the pulmonary allergic response following BALB/c exposure to house dust mite allergen and fine particulate matter from California and China. *Toxicol. Sci.* **2018**, *164*, 627–643. [CrossRef]
10. Ngoc, L.T.N.; Park, D.; Lee, Y.; Lee, Y.C. Systematic review and meta-analysis of human skin diseases due to particulate matter. *Int. J. Environ. Res. Public Health* **2017**, *14*, 1458. [CrossRef]
11. Kim, K.E.; Cho, D.; Park, H.J. Air pollution and skin diseases: Adverse effects of airborne particulate matter on various skin dis eases. *Life Sci.* **2016**, *152*, 126–134. [CrossRef] [PubMed]
12. Piao, M.J.; Ahn, M.J.; Kang, K.A.; Ryu, Y.S.; Hyun, Y.; Shilnikova, K.; Zhen, A.X.; Jeong, J.W.; Choi, Y.H.; Kang, H.K.; et al. Partic ulate matter 2.5 damages skin cells by inducing oxidative stress, subcellular organelle dysfunction, and apoptosis. *Arch. Toxicol.* **2018**, *92*, 2077–2091. [CrossRef] [PubMed]
13. Hyun, Y.J.; Piao, M.J.; Kang, K.A.; Zhen, A.X.; Madushan Fernando, P.D.S.; Kang, H.K.; Ahn, Y.S.; Hyun, J.W. Effect of fer-mented fish oil on fine particulate matter-induced skin aging. *Mar. Drugs* **2019**, *17*, 61. [CrossRef]
14. Ryu, Y.S.; Kang, K.A.; Piao, M.J.; Ahn, M.J.; Yi, J.M.; Hyun, Y.M.; Kim, S.H.; Ko, M.K.; Park, C.O.; Hyun, J.W. Particulate matter induces inflammatory cytokine production via activation of NFκB by TLR5-NOX4-ROS signaling in human skin keratinocyte and mouse skin. *Redox Biol.* **2019**, *21*, 101080. [CrossRef]
15. Ryu, Y.S.; Kang, K.A.; Piao, M.J.; Ahn, M.J.; Yi, J.M.; Bossis, G.; Hyun, Y.M.; Park, C.O.; Hyun, J.W. Particulate matter-induced senescence of skin keratinocytes involves oxidative stress-dependent epigenetic modifications. *Exp. Mol. Med.* **2019**, *51*, 1–14. [CrossRef]
16. Zhang, Y.; Zheng, L.; Tuo, J.; Liu, Q.; Zhang, X.; Xu, Z.; Liu, S.; Sui, G. Analysis of $PM_{2.5}$-induced cytotoxicity in human HaCaT cells based on a microfluidic system. *Toxicol. In Vitro* **2017**, *43*, 1–8. [CrossRef]
17. Nguyen, L.T.H.; Nguyen, U.T.; Kimb, Y.H.; Shin, H.M.; Yang, I.J. Astragali radix and its compound formononetin ameliorate diesel particulate matter-induced skin barrier disruption by regulation of keratinocyte proliferation and apoptosis. *J. Ethnopharmacol.* **2019**, *228*, 132–141. [CrossRef]

18. Cho, S.H.; Heo, S.J.; Yang, H.W.; Ko, E.Y.; Jung, M.S.; Cha, S.H.; Ahn, G.; Jeon, Y.J.; Kim, K.N. Protective effect of 3-Bromo-4,5-dihydroxybenzaldehyde from Polysiphonia morrowii harvey against hydrogen peroxide-induced oxidative stress in vitro and in vivo. *J. Microbiol. Biotechnol.* **2019**, *29*, 1193–1203. [CrossRef] [PubMed]
19. Kim, S.Y.; Kim, S.R.; Oh, M.J.; Jung, S.J.; Kang, S.Y. In vitro antiviral activity of red alga, Polysiphonia morrowii extract and its bromophenols against fish pathogenic infectious hematopoietic necrosis virus and infectious pancreatic necrosis virus. *J. Microbiol.* **2011**, *49*, 102–106. [CrossRef]
20. Jayasinghe, A.M.K.; Han, E.J.; Kirindage, K.G.I.S.; Fernando, I.P.S.; Kim, E.A.; Kim, J.; Jung, K.; Kim, K.N.; Heo, S.J.; Ahn, G. 3-bromo-4,5-dihydroxybenzaldehyde isolated from Polysiphonia morrowii suppresses TNF-α/IFN-γ-stimulated inflammation and deterioration of skin barrier in HaCaT keratinocytes. *Mar. Drugs* **2022**, *20*, 563. [CrossRef]
21. Ryu, Y.S.; Fernando, P.D.S.M.; Kang, K.A.; Piao, M.J.; Zhen, A.X.; Kang, H.K.; Koh, Y.S.; Hyun, J.W. Marine compound 3-bromo-4,5-dihydroxybenzaldehyde protects skin cells against oxidative damage via the Nrf2/HO-1 pathway. *Mar. Drugs* **2019**, *17*, 234. [CrossRef] [PubMed]
22. Kim, K.C.; Hyun, Y.J.; Hewage, S.R.K.M.; Piao, M.J.; Kang, K.A.; Kang, H.K.; Koh, Y.S.; Ahn, M.J.; Hyun, J.W. 3-Bromo-4,5-dihy droxybenzaldehyde enhances the level of reduced glutathione via the Nrf2-mediated pathway in human keratinocytes. *Mar. Drugs* **2017**, *15*, 291. [CrossRef]
23. Piao, M.J.; Kang, K.A.; Ryu, Y.S.; Shilnikova, K.; Park, J.E.; Hyun, Y.J.; Zhen, A.X.; Kang, H.K.; Koh, Y.S.; Ahn, M.J.; et al. The red algae compound 3-bromo-4,5-dihydroxybenzaldehyde protects human keratinocytes on oxidative stress-related molecules and pathways activated by UVB irradiation. *Mar. Drugs* **2017**, *15*, 268. [CrossRef]
24. Ji, N.; Lou, H.; Gong, X.; Fu, T.; Ni, S. Treatment with 3-bromo-4,5-dihydroxybenzaldehyde improves cardiac function by inhib iting macrophage infiltration in mice. *Korean Circ. J.* **2018**, *48*, 933–943. [CrossRef] [PubMed]
25. Qin, S.G.; Tian, H.Y.; Wei, J.; Han, Z.H.; Zhang, M.J.; Hao, G.H.; Liu, X.; Pan, L.F. 3-Bromo-4,5-dihydroxybenzaldehyde protects against myocardial ischemia and reperfusion injury through the Akt-PGC1α-Sirt3 pathway. *Front. Pharmacol.* **2018**, *9*, 722. [CrossRef]
26. Kang, N.J.; Han, S.C.; Kang, H.J.; Ko, G.; Yoon, W.J.; Kang, H.K.; Yoo, E.S. Anti-inflammatory effect of 3-bromo-4,5-dihydroxyben zaldehyde, a component of Polysiphonia morrowii, in vivo and in vitro. *Toxicol. Res.* **2017**, *33*, 325–332. [CrossRef] [PubMed]
27. Chiorcea-Paquim, A.M. 8-oxoguanine and 8-oxodeoxyguanosine biomarkers of oxidative DNA damage: A review on HPLC-ECD determination. *Molecules* **2022**, *27*, 1620. [CrossRef]
28. Conners, R.; Hooley, E.; Clarke, A.R.; Thomas, S.; Brady, R.L. Recognition of oxidatively modified bases within the biotin-binding site of avidin. *J. Mol. Biol.* **2006**, *357*, 263–274. [CrossRef]
29. Rinnerthaler, M.; Bischof, J.; Streubel, M.K.; Trost, A.; Richter, K. Oxidative stress in aging human skin. *Biomolecules* **2015**, *5*, 545–589. [CrossRef]
30. Drigeard Desgarnier, M.C.; Rochette, P.J. Enhancement of UVB-induced DNA damage repair after a chronic low-dose UVB pre-stimulation. *DNA Repair* **2018**, *63*, 56–62. [CrossRef]
31. Reed, S.M.; Quelle, D.E. p53 acetylation: Regulation and consequences. *Cancers* **2014**, *7*, 30–69. [CrossRef] [PubMed]
32. Keyvanloo Shahrestanaki, M.; Bagheri, M.; Ghanadian, M.; Aghaei, M.; Jafari, S.M. *Centaurea cyanus* extracted 13-O-acetylsolsti tialin A decrease Bax/Bcl-2 ratio and expression of cyclin D1/Cdk-4 to induce apoptosis and cell cycle arrest in MCF-7 and MDA-MB-231 breast cancer cell lines. *J. Cell. Biochem.* **2019**, *120*, 18309–18319. [CrossRef]
33. Kumari, R.; Jat, P. Mechanisms of cellular senescence: Cell cycle arrest and senescence associated secretory phenotype. *Front Cell Dev. Biol.* **2021**, *9*, 645593. [CrossRef] [PubMed]
34. Yue, J.; López, J.M. Understanding MAPK signaling pathways in apoptosis. *Int. J. Mol. Sci.* **2020**, *21*, 2346. [CrossRef] [PubMed]
35. Park, M.; Joo, H.S.; Lee, K.; Jang, M.; Kim, S.D.; Kim, I.; Borlaza, L.J.S.; Lim, H.; Shin, H.; Chung, K.H.; et al. Differential toxicities of fine particulate matters from various sources. *Sci. Rep.* **2018**, *8*, 17007. [CrossRef]
36. Quezada-Maldonado, E.M.; Sánchez-Pérez, Y.; Chirino, Y.I.; García-Cuellar, C.M. Airborne particulate matter induces oxidative damage, DNA adduct formation and alterations in DNA repair pathways. *Environ. Pollut.* **2021**, *287*, 117313. [CrossRef]
37. Zhen, A.X.; Piao, M.J.; Hyun, Y.J.; Kang, K.A.; Ryu, Y.S.; Cho, S.J.; Kang, H.K.; Koh, Y.S.; Ahn, M.J.; Kim, T.H.; et al. Purpurogallin protects keratinocytes from damage and apoptosis induced by Ultraviolet B radiation and particulate matter 2.5. *Biomol. Ther.* **2019**, *27*, 395–403. [CrossRef]
38. Hyun, Y.J.; Piao, M.J.; Zhang, R.; Choi, Y.H.; Chae, S.; Hyun, J.W. Photo-protection by 3-bromo-4, 5-dihydroxybenzaldehyde against ultraviolet B-induced oxidative stress in human keratinocytes. *Ecotoxicol. Environ. Saf.* **2012**, *83*, 71–78. [CrossRef]
39. Stockwell, B.R. Ferroptosis turns 10: Emerging mechanisms, physiological functions, and therapeutic applications. *Cell* **2022**, *185*, 2401–2421. [CrossRef]
40. Cadenas, S. Mitochondrial uncoupling, ROS generation and cardioprotection. *Biochim. Biophys. Acta Bioenerg.* **2018**, *1859*, 940–950. [CrossRef]
41. Park, C.; Cha, H.J.; Hong, S.H.; Kim, G.Y.; Kim, S.; Kim, H.S.; Kim, B.W.; Jeon, Y.J.; Choi, Y.H. Protective effect of phloroglucinol on oxidative stress-induced DNA damage and apoptosis through activation of the Nrf2/HO-1 signaling pathway in HaCaT human keratinocytes. *Mar. Drugs* **2019**, *17*, 225. [CrossRef] [PubMed]
42. Sterea, A.M.; El Hiani, Y. The role of mitochondrial calcium signaling in the pathophysiology of cancer cells. *Adv. Exp. Med. Biol.* **2020**, *1131*, 747–770. [PubMed]

43. Correia-Melo, C.; Marques, F.D.; Anderson, R.; Hewitt, G.; Hewitt, R.; Cole, J.; Carroll, B.M.; Miwa, S.; Birch, J.; Merz, A.; et al. Mitochondria are required for pro-ageing features of the senescent phenotype. *EMBO J.* **2016**, *35*, 724–742. [CrossRef] [PubMed]
44. Herath, H.M.U.L.; Piao, M.J.; Kang, K.A.; Zhen, A.X.; Fernando, P.D.S.M.; Kang, H.K.; Yi, J.M.; Hyun, J.W. Hesperidin exhibits protective effects against PM2. 5-mediated mitochondrial damage, cell cycle arrest, and cellular senescence in human HaCaT keratinocytes. *Molecules* **2022**, *27*, 4800. [CrossRef] [PubMed]
45. Zhen, A.X.; Piao, M.J.; Hyun, Y.J.; Kang, K.A.; Madushan Fernando, P.D.S.; Cho, S.J.; Ahn, M.J.; Hyun, J.W. Diphlorethohydroxycarmalol attenuates fine particulate matter-induced subcellular skin dysfunction. *Mar. Drugs* **2019**, *17*, 95. [CrossRef]
46. Chistiakov, D.A.; Sobenin, I.A.; Revin, V.V.; Orekhov, A.N.; Bobryshev, Y.V. Mitochondrial aging and age-related dysfunction of mitochondria. *Biomed. Res. Int.* **2014**, *2014*, 238463. [CrossRef]
47. Kwon, K.R.; Alam, M.B.; Park, J.H.; Kim, T.H.; Lee, S.H. Attenuation of UVB-induced photo-aging by polyphenolic-rich spathol obus suberectus stem extract via modulation of MAPK/AP-1/MMPs signaling in human keratinocytes. *Nutrients* **2019**, *11*, 1341. [CrossRef]
48. Lee, Y.H.; Seo, E.K.; Lee, S.T. Skullcapflavone II inhibits degradation of type I collagen by suppressing MMP-1 transcription in human skin fibroblasts. *Int. J. Mol. Sci.* **2019**, *20*, 2734. [CrossRef]
49. Zhou, Q.; Wang, W.; Wu, J.; Qiu, S.; Yuan, S.; Fu, P.L.; Qian, Q.R.; Xu, Y.Z. Ubiquitin-specific protease 3 attenuates interleukin-1β-mediated chondrocyte senescence by deacetylating forkhead box O-3 via sirtuin-3. *Bioengineered* **2022**, *13*, 2017–2027. [CrossRef]
50. Hernandez-Segura, A.; Nehme, J.; Demaria, M. Hallmarks of cellular senescence. *Trends Cell Biol.* **2018**, *28*, 436–453. [CrossRef]
51. Kim, H.; Woo, S.M.; Choi, W.R.; Kim, H.; Yi, C.; Kim, K.; Cheng, J.; Yang, S.H.; Suh, J. Scopoletin downregulates MMP-1 expression in human fibroblasts via inhibition of p38 phosphorylation. *Int. J. Mol. Med.* **2018**, *42*, 2285–2293. [CrossRef] [PubMed]
52. Park, K.H.; Kim, J.; Jung, S.; Sung, K.H.; Son, Y.K.; Bae, J.M.; Kim, B.H. Alleviation of ultraviolet B-induced photoaging by 7-MEGATM 500 in hairless mouse skin. *Toxicol. Res.* **2019**, *35*, 353–359. [CrossRef] [PubMed]

antioxidants

MDPI

Article

Inflammatory Cytokines and Chemokines Are Synergistically Induced in a ROS-Dependent Manner by a Co-Culture of Corneal Epithelial Cells and Neutrophil-like Cells in the Presence of Particulate Matter

Zirui Zeng [1], Yasuhiro Yoshida [2,*], Duo Wang [3], Yuri Fujii [2], Mengyue Shen [2,4], Tatsuya Mimura [5] and Yoshiya Tanaka [1,*]

1 The First Department of Internal Medicine, School of Medicine, University of Occupational and Environmental Health, Japan, 1-1 Iseigaoka, Yahatanishi-ku, Kitakyushu 807-8555, Japan

2 Department of Immunology and Parasitology, School of Medicine, University of Occupational and Environmental Health, Japan, 1-1 Iseigaoka, Yahatanishi-ku, Kitakyushu 807-8555, Japan; z201078@info.uoeh-u.ac.jp (Y.F.)

3 Department of Radiobiology and Hygiene Management, Institute of Industrial Ecological Sciences, University of Occupational and Environmental Health, Japan, 1-1 Iseigaoka, Yahatanishi-ku, Kitakyushu 807-8555, Japan

4 Department of Medical Teaching, West China Center of Medical Sciences of Sichuan University, Chengdu 610041, China

5 Department of Ophthalmology, Teikyo University School of Medicine, Tokyo 173-0003, Japan

* Correspondence: freude@med.uoeh-u.ac.jp (Y.Y.); tanaka@med.uoeh-u.ac.jp (Y.T.); Tel.: +81-936917431 (Y.Y.)

Abstract: Ocular exposure to particulate matter (PM) causes local inflammation; however, the influence of neutrophils on PM-induced ocular inflammation is still not fully understood. In this study, we constructed a system to investigate the role of PM in ocular inflammation using a co-culture of human corneal epithelial cells (HCE-T) and differentiation-induced neutrophils (dHL-60). To investigate whether HCE-T directly endocytosed PM, we performed a holographic analysis, which showed the endocytosis of PM in HCE-T. The cytokines and chemokines produced by HCE-T were measured using an ELISA. HCE-T treated with PM produced IL-6 and IL-8, which were inhibited by N-Acetyl-L-cysteine (NAC), suggesting the involvement of ROS. Their co-culture with dHL-60 enhanced their production of IL-6, IL-8, and MCP-1. This suggests an inflammatory loop involving intraocular corneal epithelial cells and neutrophils. These cytokines and chemokines are mainly regulated by NF-κB. Therefore, this co-culture system was examined in the presence of an IKK inhibitor known to downregulate NF-κB activity. The IKK inhibitor dramatically suppressed the production of these factors in co-culture supernatants. The results suggest that the inflammatory loop observed in the co-culture is mediated through ROS and the transcription factor NF-κB. Thus, the co-culture system is considered a valuable tool for analyzing complex inflammations.

Keywords: corneal epithelial cells; neutrophils; particulate matter; inflammation; reactive oxygen

Citation: Zeng, Z.; Yoshida, Y.; Wang, D.; Fujii, Y.; Shen, M.; Mimura, T.; Tanaka, Y. Inflammatory Cytokines and Chemokines Are Synergistically Induced in a ROS-Dependent Manner by a Co-Culture of Corneal Epithelial Cells and Neutrophil-like Cells in the Presence of Particulate Matter. *Antioxidants* **2024**, *13*, 467. https://doi.org/10.3390/antiox13040467

Academic Editor: Alessandra Napolitano

Received: 29 February 2024
Revised: 31 March 2024
Accepted: 12 April 2024
Published: 16 April 2024

1. Introduction

The cornea is a transparent tissue located at the front of the eyeball; however, since there are no blood vessels in the cornea, there are no hematopoietic immune cells within the corneal tissue unless an injury occurs. Therefore, it is clinically important to understand how the avascular cornea responds to inflammations resulting from injury and infection. When the cornea receives an infection or physical injury, a wound-healing response immediately begins, and inflammatory cells such as neutrophils infiltrate the cornea [1]. While these inflammatory cells play an important role in eliminating microbes [2] and restoring corneal transparency, their persistent presence within the corneal stroma ultimately leads to corneal opacity. During corneal inflammation, inflammatory cytokines are released from cells [3,4].

Neutrophils produce IL-6 and TNF-α after the endocytosis of PM [5]. Neutrophils play an important role in the immune system; however, their short lifespan makes primary cultured neutrophils unsuitable for use in repeated in vitro studies. Treatment with dimethyl sulfoxide (DMSO) or all-trans retinoic acid (ATRA) induces their differentiation into neutrophil-like cells [6,7]. In a previous study, we successfully established a cell line that mimics pro-inflammatory neutrophils using the human cell line HL-60 [8]. We observed that these established cells produce inflammatory cytokines and chemokines such as IL-8 and MCP-1 when they endocytose particles such as PM2.5. In addition to PM2.5, it has also been reported that a similar phenomenon occurs with the endocytosis of Staphylococcus aureus-derived particles (bio-particles) [9]. Although these particles can cause ocular damage, there are no clear studies on the subsequent responses of ocular epithelial cells and neutrophils, and whether these further exacerbate inflammation.

Previous research has demonstrated that the damage to human corneal epithelial cells induced by PM2.5 follows a pattern dependent on both time and dosage [10] and induces significant ROS elevation [11] and corneal toxicity by triggering pyroptosis in human corneal epithelial cells [12]. PM2.5 can induce DNA damage and cell senescence in corneal epithelial cells [13], but there are no in-depth studies on the relationship between PM-induced ocular inflammation and neutrophils.

The transcription factor Nuclear factor-κB (NF-κB) is pivotal in regulating immune responses. The classical form of NF-κB within cells is a heterodimer composed of a p50 subunit and a p65 subunit. It undergoes rapid activation by IKK [14] in response to diverse stimuli, including pathogens, stress signals, and pro-inflammatory cytokines [15,16]. NF-κB translocates to the nucleus of the cell, where it stimulates the transcription of specific inflammatory factors, including IL-6, IL-8, and MCP-1 [17].

To ethically reduce the usage of animals in experiments, there is a growing emphasis on embracing the 3 Rs, the principles of Refinement, Reduction, and Replacement [18,19]. An in vitro experimental system employing corneal epithelial cell lines has emerged as a valuable tool for understanding ocular events [20]. It is important to recognize that although these models are valuable tools, they are currently unable to completely replace animal experiments. Since we have previously reported the establishment of inflammatory-induced neutrophils, we wondered whether it would be possible to construct a new inflammatory system in the eye using differentiating neutrophils and a corneal epithelial cell line.

In this study, we constructed a system to investigate ocular inflammation and the role of PM using a co-culture of human corneal epithelial cells (HCE-T) and differentiation-induced neutrophils (dHL-60).

2. Materials and Methods

2.1. Materials and Cells

Fluorescent particulate matter (PM, sicastar®-redF, particles of different sizes (0.1, 0.3, and 1 μm)) was purchased from Corefront (Waltham, MA, USA). Biological particulate matter was obtained from pHrodo™ Red. Staphylococcus aureus Bioparticles™ Conjugate (BioPM), for phagocytosis (A10010), was purchased from Invitrogen (Carlsbad, CA, USA). The IKK Inhibitor VII (401486) was purchased from Calbiochem (Darmstadt, Germany), and N-Acetyl-L-cysteine (A8199) was purchased from SIGMA-ALDRICH (St. Louis, MI, USA). These inhibitors were dissolved in dimethyl sulfoxide (DMSO) and diluted in PBS before treatment. HCE-T and HL-60 cells were purchased from RIKEN BioResource Center (Tsukuba, Japan).

2.2. Cell Cultures, Cells' Differentiation into Neutrophils, and Flow Cytometry

HCE-T was incubated in Roswell Park Memorial Institute (RPMI) 1640 medium (Nissui, Tokyo, Japan) including 10% fetal bovine serum (Scrana Europe GmbH, Brandenburg, Germany), L-glutamine (2 mM, Wako Pure Chemical Industries, Tokyo, Japan), and a penicillin–streptomycin solution (Gibco, New York, NY, USA) and maintained in a humidified incubator with 5% CO_2 at 37 °C. HCE-T (3 × 10^4 cells/well/200 μL, in a 48-well plate in Figure 2 and 3 × 10^4 cells/well/500 μL in a 24-well plate for the co-culture and ROS assay system) were spread. HL-60 cells were differentiated into neutrophils as previously described [8]. Briefly, HL-60 cells were plated in a 6-well plate at a density of 5 × 10^5 cells/well/2 mL with 1.3% (*v*/*v*) dimethyl sulfoxide (DMSO, Nacalai Tesque, Kyoto, Japan) or 1 μM all-trans retinoic acid (ATRA, R2625, Sigma-Aldrich, St. Louis, MO, USA). The cell differentiation conditions were renewed after 3 days of their 6-day differentiation period. To identify whether HL-60 differentiated into neutrophil-like cells (CD11b-positive), surface markers were analyzed by flow cytometry. The cells were stained with APC-conjugated anti-CD11b (20-0112-U100, Cytek, Fremont, CA, USA) and PerCP Cy5.5-conjugated anti-human CD14 (325622, Biolegend, San Diego, CA, USA) antibodies and incubated for 30 min at 4 °C. The cells were washed and analyzed using a CytoFLEX flow cytometer (Beckman Coulter, Brea, CA, USA). Representative data are shown in Supplementary Figure S1.

2.3. Holotomography

HCE-T were treated with PM (1 μm) for 24 h in a specific dish (central glass-bottom TomoDish, Tomocube, 901002-02, Sinseong-ro, Yuseong-gu, Daejeon, 34109, Republic of Korea). Three-dimensional holotomographic images, based on the refractive indexes of the materials, were analyzed following the manufacturer's instructions using a 3D optical diffraction tomography (3D-ODT) device. Digital staining, based on RI and 3D images, was reconstructed.

2.4. Enzyme-Linked Immunosorbent Assay (ELISA)

Enzyme-linked immunosorbent assay kits for human IL-1β, IL-6, IL-8, MCP-1, and TNF-α were purchased from BioLegend (San Diego, CA, USA). Cells were treated with PM (20 μg/mL) or LPS (1 μg/mL) for either 24 or 48 h, with or without inhibitors for 24 h. After treatment, culture supernatants were collected and analyzed for their cytokine levels using a previously described method [21].

2.5. Co-Culture HCE-T with HL-60 Using a Transwell

After HCE-T (3 × 10^4) were adhered to a 24-well plate, HCE-T were co-cultured with DMSO or ATRA differentiated-HL-60 (DHL-60, AHL-60, respectively), or original HL-60 (3 × 10^4 or 10 × 10^4) in the presence of PBS, LPS (1 μg/mL), or BioPM (20 μg/mL) for 24 h. The cell culture supernatant was used for the ELISA.

To separate HCE-T and HL-60 cells, a Transwell chamber (Transwell porous cell culture insert, 6.4 mm membrane diameter, 3 μm pore size, polyethylene terephthalate membrane, Corning Caster Corp [Corning, NY, USA]) was used. HCE-T (3 × 10^4) were seeded into a 24-well plate. The medium was then replaced, with/without DHL-60, and treated with PBS or BioPM while within the Transwell chamber. After 24 h of incubation, supernatants from both the 24-well plate and the Transwell chamber were collected. The combined supernatants from both compartments were used for the ELISA analysis. Figures 3 and 6 use the same experimental setup, with the only difference being the presence of the Transwell chamber.

2.6. Measurement of ROS Production

HCE-T (3 × 10^4) were treated with PBS or NAC (10 mM). After 24 h, the cells were stained with Oxivision Green and observed via a fluorescence microscope. The area of the green fluorescent region was calculated and normalized by the number of cells confirmed in the bright field.

DCFH was used for the measurement of ROS production following the manufacturer's protocol. HCE-T (3 × 10^4) were stained with photo-oxidation-resistant DCFH-DA working solution and incubated for 30 min, and fluorescence signals were measured using a plate reader.

2.7. Western Blotting

Cells were lysed using RIPA buffer to obtain whole-cell extracts [22]. Equal amounts of protein (10 μg) were separated by electrophoresis. The density of each band was measured using ImageJ software (bundled with 64-bit Java 1.8.0_112; National Institute of Health, Bethesda, MD, USA). The expression levels of target proteins were standardized against β-actin.

2.8. Reporter Assay

HCE-T (3×10^4) were transfected using Lipofectamine 2000 Reagent (Invitrogen, Carlsbad, CA, USA) at a ratio of 3 μL of reagent per microgram of DNA, following previously established protocols [15]. In brief, the NF-κB luciferase reporter construct was transfected into cells cultured in 24-well plates at 90% confluency. Later, 24 h post-transfection, the cells were treated with either LPS or an IKK inhibitor for a duration of 6 h. Following stimulation, cell lysates were prepared using Passive Lysis Buffer (Promega, Madison, WI, USA), and their luciferase activity was quantified using a luminometer (Fluoroskan™ FLMicroplate Fluorometer and Luminometer, Thermo Scientific, Waltham, MA, USA).

2.9. Ultraviolet (UV) Exposure

HCE-T (3×10^4) were seeded into 24-well plate and incubated for 24 h. The medium RPMI was replaced with PBS. The cells were then exposed to UV-A for 0.5 h (2.6 J/s/m^2). Then, the PBS was replaced with RPMI medium. After 24 h, culture supernatants were harvested and their IL-6 production was measured using an ELISA kit (Biolegend).

2.10. Statistics

The results are presented as mean ± standard deviation. Each column represents the mean level for its respective group. Statistical analyses were conducted using Fisher's least significant difference (LSD) test following a one-way analysis of variance. Statistical significance was defined as $p < 0.05$. A t-test was performed, as seen in Figures 4B and 5B.

3. Results

3.1. HCE-T's Ability to Take Up PM Is Analyzed Using Holotomography Technology

It has been reported that HCE-T can take up particles [23]. To investigate precisely whether HCE-T can take up particles in our system, HCE-T were treated with particles, and then a tomography analysis was performed. After the photo was taken, it was reconstructed so that the particles were represented in yellow and the nucleus represented in blue, showing that HCE-T could endocytose particles and exist around the nucleus (Figure 1). We have also included a 3D movie (Supplementary Video S1) in the Supplementary Materials.

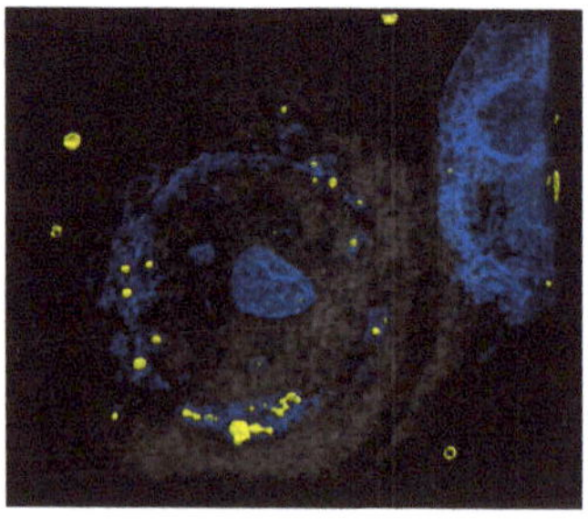

Figure 1. HCE-T takes up PM. At 3 h after HCE-T's treatment with PM (1 μm), a 3D optical diffraction tomography (3D-ODT) device was employed to capture 3D images of the PM endocytosed by HCE-T. Yellow: 1 μm PM, Blue: nucleus.

3.2. HCE-T Produces IL-6 and IL-8 Due to PM Treatment

Generally, cells produce cytokines during the endocytosis of PM, triggering or enhancing an inflammatory response. We have previously reported that neutrophils have a preferable particle size for endocytosis [5]. Next, H-CET were treated with several particles and the cytokine levels in the supernatant were measured. As shown in Figure 2, IL-6 was induced by several PMs (0.1, 0.3, and BioPM) at 24 h, but not IL-8. However, IL-8 was detected in BioPM-treated cells at 48 h. In contrast, the IL-1β, TNF-α, and MCP-1 levels were below the detection level. These results suggest that HCE-T endocytose PM and produce inflammatory factors.

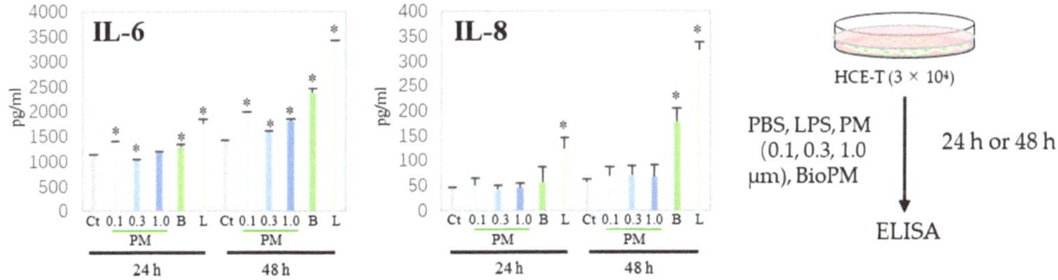

Figure 2. HCE-T produce IL-6 and IL-8 upon PM treatment. HCE-T (3 × 10^4) were stimulated with PBS; LPS (1 µg/mL); 0.1, 0.3, and 1.0 µm beads (20 µg/mL); and BioPM (20 µg/mL) for 24 h and 48 h. IL-6 and IL-8 production was measured using ELISA. Representative analyses from 3 independent experiments are shown. Ct, control; B, bioparticulate matter (BioPM); L, LPS. * $p < 0.05$ vs. each time point Ct.

3.3. Co-Culture with Differentiated HL-60 Cells Enhances the IL-6, IL-8, and MCP-1 Production in HCE-T

When inflammation arises, neutrophils are recruited to the site of inflammation. In our previous study, we demonstrated that HL-60 cells can be differentiated into neutrophil-like cells by DMSO and ATRA. DHL-60, AHL-60, undifferentiated HL-60, and HCE-T were co-cultured to investigate their effects on recruited neutrophils. IL-6 and IL-8 production was further increased compared to that of mono-cultured cells (Figure 3). No production of TNF-α, IL-1β, or IL-17A was observed even after co-culturing. On the other hand, an enhanced production of MCP-1 was observed in the co-cultures. These results imply that neutrophils can amplify the inflammatory response in the cornea. In the following experiments, DMSO-differentiated HL-60 is used for the co-culture system because we reported that DMSO-induced neutrophils more closely mimic the properties and data of murine proinflammatory neutrophils than ATRA-induced ones [8], and their response to BioPM is higher than that of ATRA-differentiated HL-60.

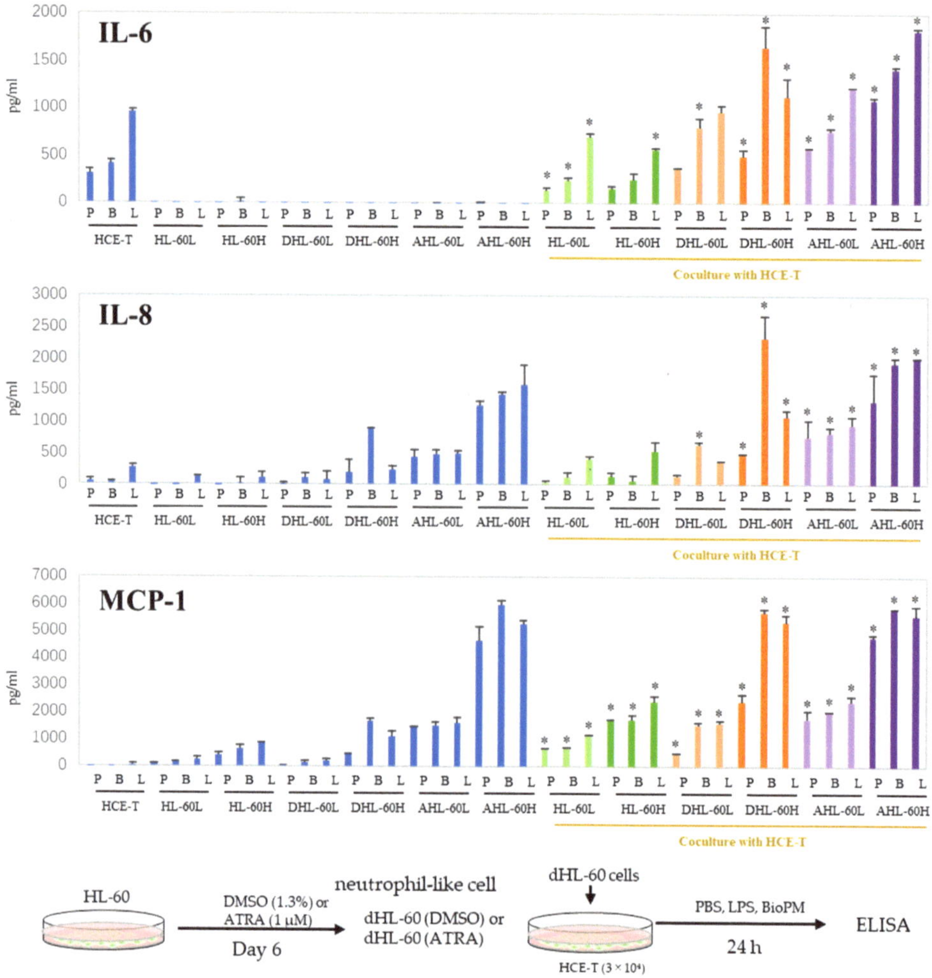

Figure 3. Co-culture with HL-60 enhances IL-6, IL-8, and MCP-1 production in HCE-T. HL-60 was differentiated into neutrophils by DMSO (1.3%) and ATRA (1 μM) over 6 days. HCE-T (3×10^4) were co-cultured with DMSO- or ATRA-differentiated and original HL-60 (3×10^4 or 10×10^4) in the presence of PBS (P), LPS (L, 1 μg/mL), or BioPM (B, 20 μg/mL) for 24 h. Cytokine and chemokine levels were analyzed using ELISA. Representative analyses from 3 independent experiments are shown. DHL-60: DMSO-differentiated HL-60, AHL-60: ATRA-differentiated HL-60. (HL-60L: 3×10^4, HL-60H: 10×10^4). Blue column: single culture, green column: co-culture with undifferentiated HL-60, orange column: co-culture with DMSO-differentiated HL-60, purple column: co-culture with ATRA-differentiated HL-60. * $p < 0.05$ vs. each comparable HCE-T treatment.

3.4. *ROS Are Involved in the Production of Cytokines and Chemokines in Co-Culture Systems*

NAC is widely used as a scavenger of ROS production. To examine whether ROS are associated with the production of enhanced soluble factors, NAC was applied to our co-culture system. Surprisingly, HCE-T continuously produce ROS (Figure 4A, green fluoresce cells), which are dramatically suppressed by NAC (Figure 4A,B). As expected, there was a significant decrease in the production of IL-6, IL-8, and MCP-1 in the presence of NAC (Figure 4C). These results indicate that ROS are involved in the production of inflammatory cytokines and chemokines, regardless of whether they are produced by a co-culture or not. This suggests a connection between the inflammation observed in our system and ROS.

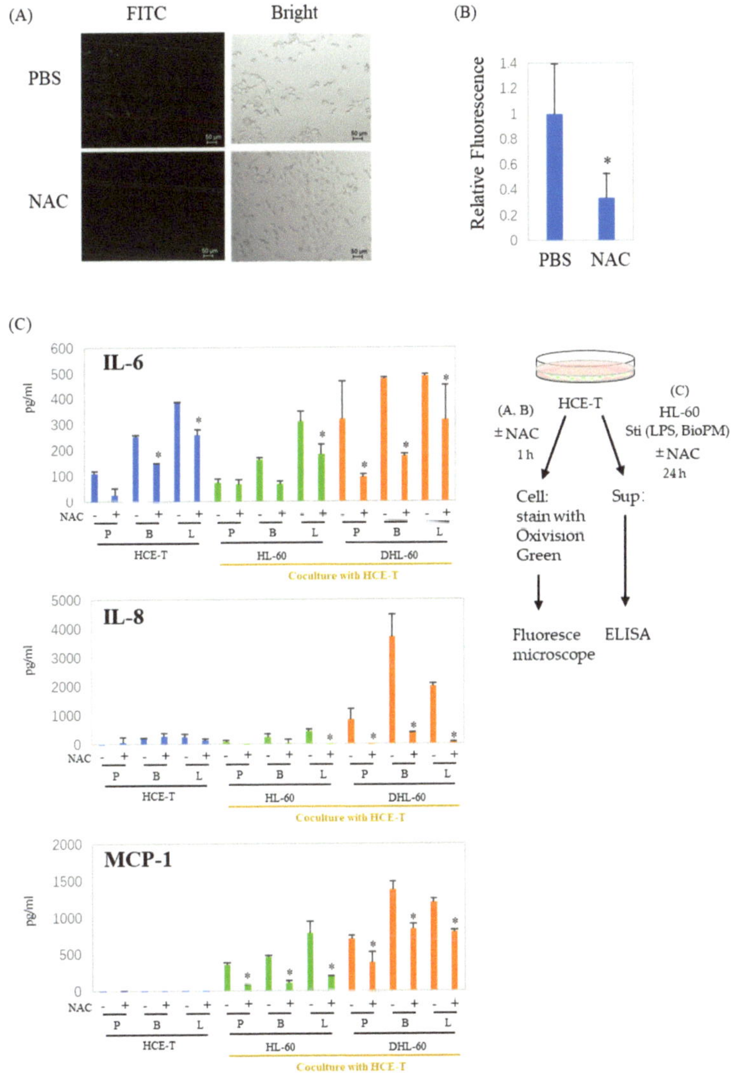

Figure 4. ROS are involved in the production of cytokines and chemokines in co-culture systems. (**A**,**B**) HCE-T (1×10^4) were treated with PBS or NAC (10 mM) for 1 h, and cells were stained with Oxivision Green and analyzed by fluoresce microscope. A representative photo is shown in (**A**). (**B**) The area of the green fluorescent region was calculated and normalized by the number of cells confirmed in the bright field. The graph shows the average ± SD of the four regions. The PBS group is represented as 1. * $p < 0.05$, vs. PBS. (**C**) HCE-T (3×10^4) were co-cultured with DMSO-differentiated and original HL-60 cells (10×10^4) in the presence of PBS (P), LPS (L, 1 μg/mL), or BioPM (B, 20 μg/mL), with or without NAC (10 mM). IL-6, IL-8, and MCP-1 levels were analyzed by ELISA. Representative analyses from 2 independent experiments are shown. DHL-60: DMSO-differentiated HL-60 cells. * $p < 0.05$ vs. the without-NAC sample of each group.

3.5. IL-6, IL-8, and MCP-1 Production Are Inhibited by an IKK Inhibitor after HCE-T Are Co-Cultured with Neutrophil-like Cells

NF-κB is a pivotal regulator of inflammatory cytokine production. Our previous investigations have demonstrated that the inhibition of IKK suppresses endocytosis [9]. To investigate the role of NF-κB in inflammatory cytokine production, an IKK inhibitor was applied to the co-culture system. As shown in Figure 5A, the IKK inhibitor inhibits p65 phosphorylation. In addition, transcription activity is also inhibited by the IKK inhibitor (Figure 5B). ELISA data show a dramatic reduction in IL-6 production in HCE-T at both 24 h (Figure 5D, blue column) and 48 h (Supplementary Figure S2A). This was also demonstrated by the fact that other NF-κB inhibitors similarly inhibited stimulus-induced IL-6 production (Supplementary Figure S2B). Furthermore, the ROS production of DHL-60 was also attenuated by the IKK inhibitor (Figure 5C). In addition, the enhanced production of inflammatory cytokines and chemokines in the co-culture system, in the presence or absence of BioPM, was also suppressed by the IKK inhibitor. (Figure 5D, green and orange column). To investigate the cells contributing to enhanced cytokine production in the co-cultures, DHL-60 were pretreated with an IKK inhibitor and co-cultured. When DHL-60 were pretreated with an IKK inhibitor, their IL-6, IL-8, and MCP-1 production in response to BioPM were dramatically reduced in co-culture (Figure 5E, black column). Additionally, when HCE-T were pre-treated solely with an IKK inhibitor before co-culturing, there was a notable decrease in their production of IL-8 and MCP-1, even with the subsequent absence of the inhibitor (Figure 5E).

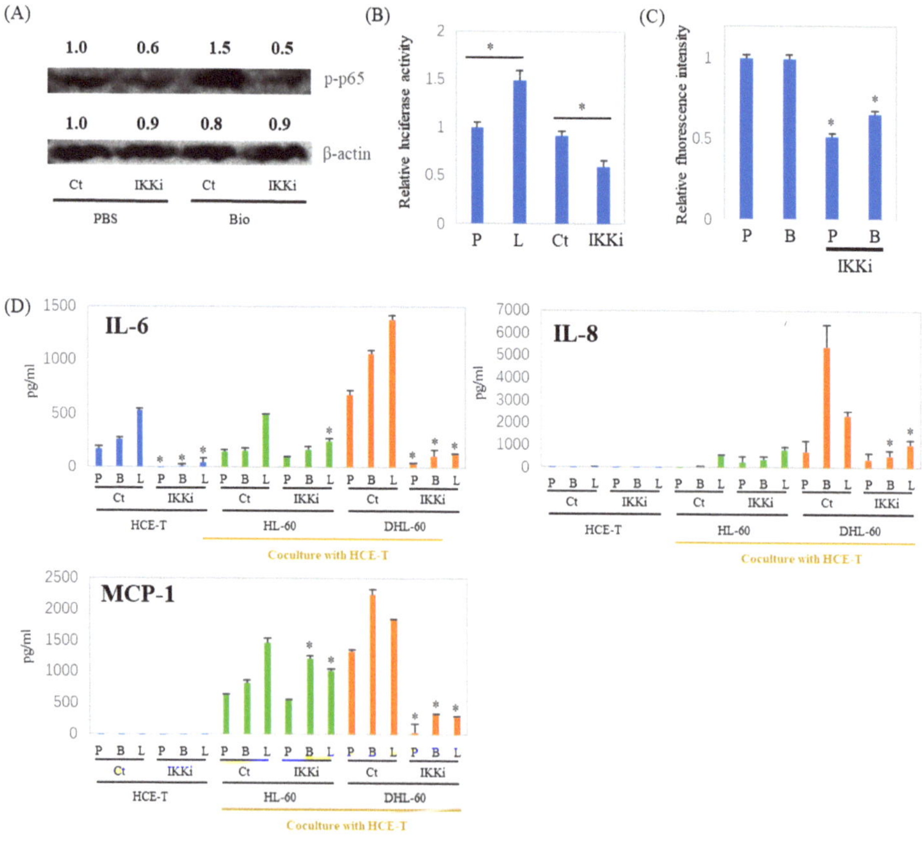

Figure 5. *Cont.*

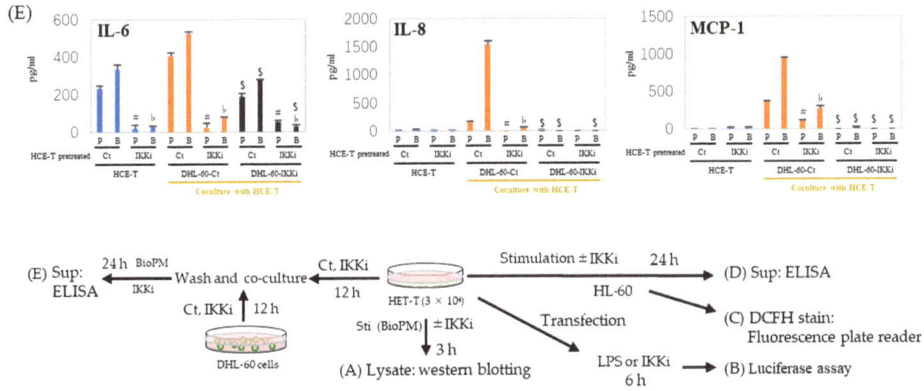

Figure 5. IL-6, IL-8, and MCP-1 production in a co-culture with neutrophil-like cells is inhibited by an IKK inhibitor. (**A**) HCE-T (3×10^4) were treated with BioPM, in the presence or absence of IKK Inhibitor (IKKi, 10 μM), for 3 h. Cell lysate was used for Westeren blotting. Phosphorylated p65 at Serine 536 (p-p65) was detected by immunoblotting, with beta-actin serving as the loading control. The relative density of the bands is indicated by the numbers above each band. The densities from the control band (PBS treatment) were utilized as the reference (1.0) to determine the relative band intensities. (**B**) HCE-T (3×10^4) were transfected using a luciferase reporter construct. Cells were treated with LPS or IKK inhibitor 24 h after transfection for 6 h and cell lysates were used for luciferase assay. (**C**,**D**) HCE-T (3×10^4) were co-cultured with DMSO-differentiated or original HL-60 (10×10^4) in the presence of PBS (P), BioPM (B, 20 μg/mL), or LPS (L, 1 μg/mL) with or without IKK Inhibitor (IKKi, 10 μM) for 24 h, DHL-60 were harvested and ROS production was assessed by DCFH stain (**C**), the supernatants were used for ELISA (**D**). (**E**) HCE-T (3×10^4) were pre-treated with the IKK inhibitor (IKKi, 10 μM) for 12 h and co-cultured with DHL-60 (Ct) or IKKi pre-treated DHL-60 (10×10^4) for 24 h. Cytokine levels were analyzed by ELISA. DHL-Ct: DHL-60 pre-treatment with control for 12 h before co-culture, DHL-IKKi: DHL-60 pre-treatment with IKKi for 12 h before co-culture. Ct, control (0.1% DMSO), DHL-60H: DMSO-differentiated HL-60. Representative analyses from 2 independent experiments are shown. * $p < 0.05$ vs. Ct of each group. # $p < 0.05$, pretreated HCE-T Ct vs. IKKi, PBS stimulation. ♭ $p < 0.05$, pretreated HCE-T Ct vs. IKKi, BioPM stimulation. $ $p < 0.05$, DHL-60 pretreated Ct vs. IKKi.

3.6. Cell–Cell Interactions Affect Cytokines and Chemokines' Production

In the co-culture of HCE-T with HL-60, there are cell–cell interactions. We used a transwell to separate HCE-T and HL-60 to demonstrate whether this cell–cell interaction affects our co-culture system. There is reduction after the transwell is used, as shown in the results of IL-6 and MCP-1 production (Figure 6A). This decrease in IL-6 production was more effectively suppressed when DHL-60 was pretreated with an IKK inhibitor (Figure 6B). IL-8 and MCP-1 production were completely inhibited by pre-treatment with IKKi of DHL-60 (Supplementary Figure S3). Therefore, cell–cell interactions can be important in the production of cytokines and chemokines in co-culture systems.

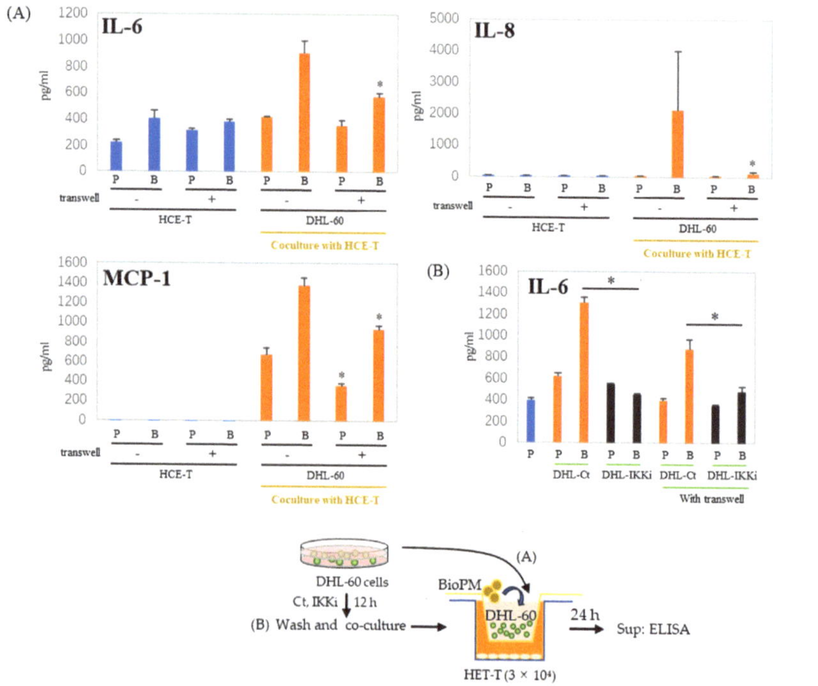

Figure 6. Production of cytokines and chemokines was decreased by the insertion of a transwell. (**A**) HCE-T (3×10^4) were co-cultured with DHL-60 (10×10^4) in the presence of PBS (P) or BioPM (B, 20 μg/mL), with or without a transwell. (**B**) HCE-T (3×10^4) were co-cultured with pre-treated DHL-60, in the presence or absence of a transwell. * $p < 0.05$ vs. Bio-treated Ct. Cytokine and chemokine levels were analyzed by ELISA. Representative analyses from 2 independent experiments are shown. DHL-60: DMSO-differentiated HL-60. DHL-Ct: DHL-60 pre-treatment with control for 12 h before co-culture, DHL-IKKi: DHL-60 pre-treatment with IKKi for 12 h before co-culture. Ct, control (0.1% DMSO). * $p < 0.05$, vs. without a transwell.

4. Discussion

In this study, we proposed a cell line system that mimics the situation in which corneal epithelial cells are damaged by particulate matter and immune cells are recruited. We demonstrated that when both cells endocytose particles, they produce cytokines and chemokines that indicate inflammation. Furthermore, we showed that recruited neutrophils significantly amplified the inflammatory response through NF-κB and ROS. Figure 7 shows these interactions as a simple illustration.

The air pollution caused by PM2.5 is a serious concern in numerous Asian countries [24]. PM2.5, a critical element of atmospheric pollution, can attack a variety of different organs in the body [25]. In particular, eye epithelial cells that are subject to direct airborne contact with pollutants may become more susceptible to inflammation [26,27]. As we previously showed that PM induces inflammation [28], PM was also thought to be involved in inducing inflammation in the cornea. Nagai et al. showed that HCE-T can uptake nanoparticle beads, suggesting that PM2.5 can be phagocytized into HCE-T and damage cells [29]. Holotomography technology provides label-free 4D quantitative imaging solutions for imaging and cell analyses. Importantly, it allows for subcellular organelles to be observed in live cells without fixation, transfection, or antibody staining [30]. In this study, we demonstrated that HCE-T can take up 1 μm diameter PE-conjugated beads without staining them, using holotomography technology. To our knowledge, this is the first direct evidence that HCE-T take up PM without any treatment being used for the analysis.

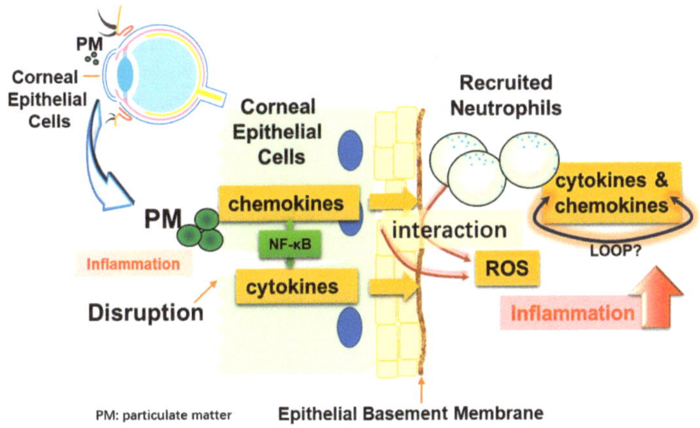

Figure 7. Diagrammatic representation. Corneal epithelial cells may be damaged by physical stimulation. These cells can directly take up particles (PM), resulting in inflammation. At this time, blood vessels are regenerated and neutrophils, which are immune cells, are recruited. Inflammation is exacerbated by the interaction of the recruited neutrophils with particulate matter, and cell–cell interactions occur between corneal epithelial cells and neutrophils.

In this study, we present a system that mimics the situation in which immune-competent cells flow into the cornea after an injury occurs in the cornea, which lacks a vascular system. We showed that corneal epithelial cells alone induce inflammation in response to PM, but when neutrophils influx or exist there, inflammation is synergistically exacerbated. At this time, considering that the effect was limited only by adding cells that had not yet differentiated into neutrophils, it is expected that the inflammatory cells that flowed in made a large contribution to this inflammatory response. In particular, the dramatic change in chemokine production (IL-8) in the co-culture system suggests an exacerbated inflammatory loop caused by cells that subsequently migrate further.

Consistent with previous reports [31], we demonstrated that ROS are involved in cytokine and chemokine production in this co-culture system. PM 2.5 can induce DNA damage in corneal epithelial cells, probably by promoting ROS formation [13]. IL-6, ROS, NF-κB, and UV are closely related, and UV irradiation can easily occur in the eye. It has been reported that UV-A exposure induces mitochondrial damage, ROS production, and NF-κB activation in HCE-T, and decreases the cell barrier function [32,33]. In addition, Benko et al. demonstrated that UV exposure increases IL-6 production [34]. We also briefly examined the cellular response to UV exposure. Our results suggest that UV exposure increases basal-level IL-6 production from HCE-T (Supplementary Figure S4). Since IL-6 is a senescence biomarker, this suggests that UV induces cell damage or senescence. However, there are still no detailed studies using our system regarding the mechanisms by which HCE-T undergo cellular senescence after UV exposure and the inflammation that is exacerbated by injury. In the future, we hope to use our system to obtain information about these relationships.

Following endocytosis, neutrophils promptly generate inflammatory cytokines [5,35]. The NF-κB signaling pathway primarily influences the production of IL-6 and IL-8 by enhancing the transcriptional activity of these genes [17,36]. In our previous paper, IKK inhibitor VII, which inhibits NF-κB signaling, completely abrogated the production of these cytokines in these cells before their stimulation with PM [5,9]. Therefore, our observations regarding the effects of IKK inhibitors on the co-culture system in this study suggest that IKK inhibitors are probably partially affecting differentiated HL-60. Furthermore, since the phosphorylation of the p65 of NF-κB is inhibited, it is expected that HCE-T are also influenced by IKK inhibitors and affect cytokine production. In other words, the phenomenon of enhanced inflammation observed in this co-culture is thought to be due to the effect of the IKK inhibitor on both cells.

Here, we demonstrated that cell–cell contact was partially responsible for the phenomena observed in this co-culture system. Furthermore, regarding the production of IL-6, we may be able to make some guesses about the producing cells. As shown in Figure 3, DHL-60 does not produce much IL-6, which is consistent with Klein M.B. et al.'s comment [37]. Therefore, the enhanced IL-6 production observed in our co-culture is likely derived from HCE-T. In Figure 6, the presence of a transwell suppresses this enhancement. Interestingly, when HCE-T were pre-treated solely with an IKK inhibitor before co-culture, there was a notable decrease in the production of IL-8 and MCP-1. This indicates that HCE-T also play a role in their augmented production during co-culture, although HL-60 cells seem to be the primary source of these cytokines. This suggests that cell–cell contact influences HCE-T-derived IL-6 production. Furthermore, the antibody neutralization of IL-6 impacts the production of IL-8 and MCP-1 in this co-culture, indicating that an inflammatory loop is involved in this system (Supplementary Figure S5). Even in the AHL-60 co-culture system, IL-6 and IL-8 production tends to be attenuated in transwells.

Although a disruption of immune cells (neutrophils) was not evident under the microscope, it is thought that Neutrophil Extracellular Traps (NETs) [38,39], one of the characteristics of neutrophils, are formed within this system. NETs have been identified as triggers for a self-limited inflammatory reaction [40]. At this time, endogenous contents may be released, which may secondarily affect epithelial cells. In fact, our paper also demonstrated that differentiated neutrophils exerted a NET-like phenomenon [8]. Furthermore, the observed transwell-inhibited experimental data did not display a complete elimination of the effect of the co-culture. In fact, when the co-culture was observed using a fluorescence microscope, the presence of the transwell was seen to reduce cell–cell interactions (Supplementary Figure S6). In the co-culture system, it is expected that neutrophils will phagocytose particles and release soluble factors immediately. These influences may still persist. Additionally, while the pore size of the transwell used in this study was 3 microns, when considering the motility of neutrophils, it may be more effective to use a transwell with a smaller pore size.

Ocular epithelial cells serve as the primary physical barrier against foreign substances. This barrier is susceptible to weakening or breakdown under various influences. Factors such as dry eye and cellular aging contribute to this vulnerability. The system demonstrated in this study is deemed optimal for replicating these phenomena.

5. Conclusions

In this study, we investigated how PM causes ocular inflammation and how pro-inflammatory neutrophils are involved in this process. We were able to demonstrate that HCE-T can directly endocytose PM without using antibodies. However, cytokine production, which is a sign of inflammation, was dramatically enhanced by HCE-T's co-culture with neutrophils, and cell-to-cell contact was important for this inflammatory loop due to the suppressive effect of the transwell on cytokine production. It was also suggested that these inflammatory responses are mediated through ROS and NF-κB signaling pathways. These findings highlight the importance of studying PM-induced ocular inflammation in co-culture systems to understand complex inflammatory mechanisms. The in vitro system using corneal epithelial cells and neutrophils that we have demonstrated here will be of great value in predicting the ocular inflammation caused by various substances in the future.

Supplementary Materials: The following supporting information can be downloaded at https://www.mdpi.com/article/10.3390/antiox13040467/s1, Video S1: HCE-T take up PM. Figure S1: HL-60 differentiated into neutrophil-like cells. Figure S2: NF-κB inhibitors inhibit cytokine production. Figure S3: IL-8 and MCP-1 production were inhibited in pre-treated DHL-60 by IKKi. Figure S4: UV-A exposure has effects on the production of IL-6. Figure S5: Cytokines' production was inhibited by anti-IL-6 antibody in co-culture. Figure S6: Photo from HCE-T co-culture with DHL-60.

Author Contributions: Conceptualization, Z.Z. and Y.Y.; methodology, Z.Z. and Y.F.; validation, Y.T.; formal analysis, Z.Z.; investigation, Z.Z. and M.S.; data curation, D.W. and Y.T.; writing—original draft preparation, Z.Z.; writing—review and editing, Y.Y.; and funding acquisition, T.M., D.W. and Y.Y. All authors have read and agreed to the published version of the manuscript.

Funding: This research was funded by a Grant-in-Aid for Scientific Research (B) [grant number 20H04347 to T. Mimura], a Grant-in-Aid for Challenging Research (Exploratory) [grant number 20K21738 to Y. Yoshida], a Grant-in-Aid for Scientific Research (C) [grant number 22K09060 to D. Wang], a UOEH Grant-in-Aid for Priority Research in the field of Occupational Medicine [grant number 2022-8 to D. Wang], and Grants-in-Aid of The Internatinoal Research Fund for Subsidy of Kyushu University School of Medicine Alumni, Japan [to D. Wang and Y. Yoshida].

Institutional Review Board Statement: Not applicable.

Data Availability Statement: The data presented in this study are available upon request from the corresponding author.

Acknowledgments: We appreciate the help of all laboratory members in the management of our laboratory and teaching technologies.

Conflicts of Interest: The authors declare no conflicts of interest.

References

1. Rajaiya, J.; Zhou, X.; Barequet, I.; Gilmore, M.S.; Chodosh, J. Novel model of innate immunity in corneal infection. *In Vitro Cell. Dev. Biol. Anim.* **2015**, *51*, 827–834. [CrossRef]
2. Daheshia, M.; Kanangat, S.; Rouse, B.T. Production of key molecules by ocular neutrophils early after herpetic infection of the cornea. *Exp. Eye Res.* **1998**, *67*, 619–624. [CrossRef] [PubMed]
3. Hazlett, L.D. Corneal response to Pseudomonas aeruginosa infection. *Prog. Retin. Eye Res.* **2004**, *23*, 1–30. [CrossRef]
4. Chintakuntlawar, A.V.; Astley, R.; Chodosh, J. Adenovirus type 37 keratitis in the C57BL/6J mouse. *Investig. Ophthalmol. Vis. Sci.* **2007**, *48*, 781–788. [CrossRef] [PubMed]
5. Miyake, T.; Wang, D.; Matsuoka, H.; Morita, K.; Yasuda, H.; Yatera, K.; Kanazawa, T.; Yoshida, Y. Endocytosis of particulate matter induces cytokine production by neutrophil via Toll-like receptor 4. *Int. Immunopharmacol.* **2018**, *57*, 190–199. [CrossRef] [PubMed]
6. Pelletier, M.; Savoie, A.; Girard, D. Activation of human neutrophils by the air pollutant sodium sulfite (Na(2)SO(3)): Comparison with immature promyelocytic HL-60 and DMSO-differentiated HL-60 cells reveals that Na(2)SO(3) is a neutrophil but not a HL-60 cell agonist. *Clin. Immunol.* **2000**, *96*, 131–139. [CrossRef] [PubMed]
7. Verdon, R.; Gillies, S.L.; Brown, D.M.; Henry, T.; Tran, L.; Tyler, C.R.; Rossi, A.G.; Stone, V.; Johnston, H.J. Neutrophil activation by nanomaterials in vitro: Comparing strengths and limitations of primary human cells with those of an immortalized (HL-60) cell line. *Nanotoxicology* **2021**, *15*, 1–20. [CrossRef]
8. Wang, D.; Sennari, Y.; Shen, M.; Morita, K.; Kanazawa, T.; Yoshida, Y. ERK is involved in the differentiation and function of dimethyl sulfoxide-induced HL-60 neutrophil-like cells, which mimic inflammatory neutrophils. *Int. Immunopharmacol.* **2020**, *84*, 106510. [CrossRef]
9. Wang, D.; Zeng, Z.; Shen, M.; Okazaki, R.; Miyata, H.; Yonezawa, T.; Yoshida, Y. ATP Consumption Is Coupled with Endocytosis in Exudated Neutrophils. *Int. J. Mol. Sci.* **2023**, *24*, 9039. [CrossRef]
10. Fu, Q.; Lyu, D.; Zhang, L.; Qin, Z.; Tang, Q.; Yin, H.; Lou, X.; Chen, Z.; Yao, K. Airborne particulate matter (PM2.5) triggers autophagy in human corneal epithelial cell line. *Environ. Pollut.* **2017**, *227*, 314–322. [CrossRef]
11. Lee, M.; Lee, S.Y.; Bae, Y.S. Emerging roles of neutrophils in immune homeostasis. *BMB Rep.* **2022**, *55*, 473–480. [CrossRef] [PubMed]
12. Niu, L.; Li, L.; Xing, C.; Luo, B.; Hu, C.; Song, M.; Niu, J.; Ruan, Y.; Sun, X.; Lei, Y. Airborne particulate matter (PM(2.5)) triggers cornea inflammation and pyroptosis via NLRP3 activation. *Ecotoxicol. Environ. Saf.* **2021**, *207*, 111306. [CrossRef] [PubMed]
13. Gao, Z.X.; Song, X.L.; Li, S.S.; Lai, X.R.; Yang, Y.L.; Yang, G.; Li, Z.J.; Cui, Y.H.; Pan, H.W. Assessment of DNA Damage and Cell Senescence in Corneal Epithelial Cells Exposed to Airborne Particulate Matter (PM2.5) Collected in Guangzhou, China. *Investig. Ophthalmol. Vis. Sci.* **2016**, *57*, 3093–3102. [CrossRef] [PubMed]
14. Guha, M.; Mackman, N. LPS induction of gene expression in human monocytes. *Cell Signal.* **2001**, *13*, 85–94. [CrossRef] [PubMed]

15. Yoshida, Y.; Kumar, A.; Koyama, Y.; Peng, H.; Arman, A.; Boch, J.A.; Auron, P.E. Interleukin 1 activates STAT3/nuclear factor-kappaB cross-talk via a unique TRAF6- and p65-dependent mechanism. *J. Biol. Chem.* **2004**, *279*, 1768–1776. [CrossRef] [PubMed]
16. Li, Q.; Verma, I.M. NF-kappaB regulation in the immune system. *Nat. Rev. Immunol.* **2002**, *2*, 725–734. [CrossRef]
17. Bao, Y.; Cao, X. The immune potential and immunopathology of cytokine-producing B cell subsets: A comprehensive review. *J. Autoimmun.* **2014**, *55*, 10–23. [CrossRef] [PubMed]
18. Bert, B.; Dorendahl, A.; Leich, N.; Vietze, J.; Steinfath, M.; Chmielewska, J.; Hensel, A.; Grune, B.; Schonfelder, G. Rethinking 3R strategies: Digging deeper into AnimalTestInfo promotes transparency in in vivo biomedical research. *PLoS Biol.* **2017**, *15*, e2003217. [CrossRef]
19. Schechtman, L.M. Implementation of the 3Rs (refinement, reduction, and replacement): Validation and regulatory acceptance considerations for alternative toxicological test methods. *ILAR J.* **2002**, *43* (Suppl. S1), S85–S94. [CrossRef]
20. Abdalkader, R.K.; Fujita, T. Corneal epithelium models for safety assessment in drug development: Present and future directions. *Exp. Eye Res.* **2023**, *237*, 109697. [CrossRef]
21. Song, Y.; Okazaki, R.; Yoshida, Y. Senescence-associated secretory phenotype and activation of NF-kappaB in splenocytes of old mice exposed to irradiation at a young age. *Dev. Comp. Immunol.* **2021**, *122*, 104124. [CrossRef]
22. Yoshida, Y.; Liu, J.; Sugiura, T.; Ishidao, T.; Ueno, S.; Yanagita, H.; Fueta, Y.; Kunugita, N.; Hori, H.; Yamashita, U. The indoor air pollutant 2-ethyl-hexanol activates CD4 cells. *Chem. Biol. Interact.* **2009**, *177*, 137–141. [CrossRef] [PubMed]
23. Tan, G.; Li, J.; Yang, Q.; Wu, A.; Qu, D.Y.; Wang, Y.; Ye, L.; Bao, J.; Shao, Y. Air pollutant particulate matter 2.5 induces dry eye syndrome in mice. *Sci. Rep.* **2018**, *8*, 17828. [CrossRef] [PubMed]
24. Fang, D.; Wang, Q.; Li, H.; Yu, Y.; Lu, Y.; Qian, X. Mortality effects assessment of ambient PM2.5 pollution in the 74 leading cities of China. *Sci. Total Environ.* **2016**, *569*, 1545–1552. [CrossRef]
25. Karottki, D.G.; Spilak, M.; Frederiksen, M.; Gunnarsen, L.; Brauner, E.V.; Kolarik, B.; Andersen, Z.J.; Sigsgaard, T.; Barregard, L.; Strandberg, B.; et al. An indoor air filtration study in homes of elderly: Cardiovascular and respiratory effects of exposure to particulate matter. *Environ. Health* **2013**, *12*, 116. [CrossRef] [PubMed]
26. Mimura, T.; Ichinose, T.; Inoue, K.; Yoshida, Y.; Fujishima, H. Airborne Suspended Particulate Matter and the Prevalence of Allergic Conjunctivitis in Japan. *Cureus* **2024**, *16*, e53292. [CrossRef] [PubMed]
27. Stepp, M.A.; Menko, A.S. Immune responses to injury and their links to eye disease. *Transl. Res.* **2021**, *236*, 52–71. [CrossRef]
28. He, C.; Song, Y.; Ichinose, T.; He, M.; Morita, K.; Wang, D.; Kanazawa, T.; Yoshida, Y. Lipopolysaccharide levels adherent to PM2.5 play an important role in particulate matter induced-immunosuppressive effects in mouse splenocytes. *J. Appl. Toxicol.* **2018**, *38*, 471–479. [CrossRef]
29. Nagai, N.; Ogata, F.; Otake, H.; Nakazawa, Y.; Kawasaki, N. Energy-dependent endocytosis is responsible for drug transcorneal penetration following the instillation of ophthalmic formulations containing indomethacin nanoparticles. *Int. J. Nanomed.* **2019**, *14*, 1213–1227. [CrossRef]
30. Sung, M.; Kim, J.H.; Min, H.S.; Jang, S.; Hong, J.; Choi, B.K.; Shin, J.; Chung, K.S.; Park, Y.R. Three-dimensional label-free morphology of CD8 + T cells as a sepsis biomarker. *Light Sci. Appl.* **2023**, *12*, 265. [CrossRef]
31. Yu, D.; Cai, W.; Shen, T.; Wu, Y.; Ren, C.; Li, T.; Hu, C.; Zhu, M.; Yu, J. PM(2.5) exposure increases dry eye disease risks through corneal epithelial inflammation and mitochondrial dysfunctions. *Cell Biol. Toxicol.* **2023**, *39*, 2615–2630. [CrossRef]
32. Otsu, W.; Yako, T.; Sugisawa, E.; Nakamura, S.; Tsusaki, H.; Umigai, N.; Shimazawa, M.; Hara, H. Crocetin protects against mitochondrial damage induced by UV-A irradiation in corneal epithelial cell line HCE-T cells. *J. Pharmacol. Sci.* **2022**, *150*, 279–288. [CrossRef] [PubMed]
33. Ishida, K.; Yako, T.; Tanaka, M.; Otsu, W.; Nakamura, S.; Shimazawa, M.; Tsusaki, H.; Hara, H. Free-Radical Scavenger NSP-116 Protects the Corneal Epithelium against UV-A and Blue LED Light Exposure. *Biol. Pharm. Bull.* **2021**, *44*, 937–946. [CrossRef]
34. Benko, S.; Tozser, J.; Miklossy, G.; Varga, A.; Kadas, J.; Csutak, A.; Berta, A.; Rajnavolgyi, E. Constitutive and UV-B modulated transcription of Nod-like receptors and their functional partners in human corneal epithelial cells. *Mol. Vis.* **2008**, *14*, 1575–1583. [PubMed]
35. Cloutier, A.; Guindi, C.; Larivee, P.; Dubois, C.M.; Amrani, A.; McDonald, P.P. Inflammatory cytokine production by human neutrophils involves C/EBP transcription factors. *J. Immunol.* **2009**, *182*, 563–571. [CrossRef]
36. An, Z.; Li, J.; Yu, J.; Wang, X.; Gao, H.; Zhang, W.; Wei, Z.; Zhang, J.; Zhang, Y.; Zhao, J.; et al. Neutrophil extracellular traps induced by IL-8 aggravate atherosclerosis via activation NF-kappaB signaling in macrophages. *Cell Cycle* **2019**, *18*, 2928–2938. [CrossRef]
37. Klein, M.B.; Hu, S.; Chao, C.C.; Goodman, J.L. The agent of human granulocytic ehrlichiosis induces the production of myelosuppressing chemokines without induction of proinflammatory cytokines. *J. Infect. Dis.* **2000**, *182*, 200–205. [CrossRef] [PubMed]
38. Baz, A.A.; Hao, H.; Lan, S.; Li, Z.; Liu, S.; Chen, S.; Chu, Y. Neutrophil extracellular traps in bacterial infections and evasion strategies. *Front. Immunol.* **2024**, *15*, 1357967. [CrossRef] [PubMed]

39. Papayannopoulos, V. Neutrophil extracellular traps in immunity and disease. *Nat. Rev. Immunol.* **2018**, *18*, 134–147. [CrossRef]
40. Vaseruk, A.; Bila, G.; Bilyy, R. Nanoparticles for stimulation of neutrophil extracellular trap-mediated immunity. *Eur. J. Immunol.* **2024**, *54*, e2350582. [CrossRef]

antioxidants

Article

Oxidative Stress, Cytotoxic and Inflammatory Effects of Urban Ultrafine Road-Deposited Dust from the UK and Mexico in Human Epithelial Lung (Calu-3) Cells

Jessica Hammond [1,*], Barbara A. Maher [2,*], Tomasz Gonet [2,†], Francisco Bautista [3] and David Allsop [1,‡]

1 Division of Biomedical and Life Sciences, Faculty of Health and Medicine, Lancaster University, Lancaster LA1 4YQ, UK
2 Centre for Environmental Magnetism and Palaeomagnetism, Lancaster Environment Centre, Lancaster University, Lancaster LA1 4YQ, UK
3 Laboratorio Universitario de Geofísica Ambiental, Centro de Investigaciones en Geografía Ambiental, Universidad Nacional Autónoma de Mexico, Morelia 58190, Michoacan, Mexico
* Correspondence: jessica.hammond@lancaster.ac.uk (J.H.); b.maher@lancaster.ac.uk (B.A.M.)
† Present address: Jaguar Land Rover, Gaydon, Lighthorne Heath, Warwick CV35 0BJ, UK.
‡ Deceased author.

Abstract: Road-deposited dust (RD) is a pervasive form of particulate pollution identified (typically via epidemiological or mathematical modelling) as hazardous to human health. Finer RD particle sizes, the most abundant (by number, not mass), may pose greater risk as they can access all major organs. Here, the first in vitro exposure of human lung epithelial (Calu-3) cells to 0–300 μg/mL of the ultrafine (<220 nm) fraction of road dust (UF-RDPs) from three contrasting cities (Lancaster and Birmingham, UK, and Mexico City, Mexico) resulted in differential oxidative, cytotoxic, and inflammatory responses. Except for Cd, Na, and Pb, analysed metals were most abundant in Mexico City UF-RDPs, which were most cytotoxic. Birmingham UF-RDPs provoked greatest ROS release (only at 300 μg/mL) and greatest increase in pro-inflammatory cytokine release. Lancaster UF-RDPs increased cell viability. All three UF-RDP samples stimulated ROS production and pro-inflammatory cytokine release. Mass-based PM limits seem inappropriate given the location-specific PM compositions and health impacts evidenced here. A combination of new, biologically relevant metrics and localised regulations appears critical to mitigating the global pandemic of health impacts of particulate air pollution and road-deposited dust.

Keywords: air pollution; cytotoxicity; road-deposited dust; inflammation; ultrafine particles; reactive oxygen species; transition metals

check for updates

Citation: Hammond, J.; Maher, B.A.; Gonet, T.; Bautista, F.; Allsop, D. Oxidative Stress, Cytotoxic and Inflammatory Effects of Urban Ultrafine Road-Deposited Dust from the UK and Mexico in Human Epithelial Lung (Calu-3) Cells. *Antioxidants* **2022**, *11*, 1814. https://doi.org/10.3390/antiox11091814

Academic Editors: Yasuhiro Yoshida and Antonella Casola

Received: 24 July 2022
Accepted: 9 September 2022
Published: 14 September 2022

1. Introduction

Human exposure to outdoor, fine-grained airborne particulate matter ($PM_{2.5}$, with an aerodynamic diameter < 2.5 μm) was estimated to be responsible for an excess of 1.8 million deaths in urban areas in 2019 [1]; 99% of the world's population is exposed to high particulate pollution levels, i.e., above the World Health Organization annual mean limits of 15 $\mu g/m^3$ for PM_{10} (PM with an aerodynamic diameter <10 μm), and 5 $\mu g/m^3$ for $PM_{2.5}$ [2]. Epidemiological studies demonstrate significant associations between PM exposure and adverse health impacts, including pulmonary diseases [3], cardiovascular diseases (CVD) [4], brain tumours [5], and neurodegenerative diseases [6,7]. Road-deposited dust (RD) occurs when airborne PM, a mixture of organic and inorganic molecules from anthropogenic and natural sources, settles on/near road surfaces. RD can pose a substantial potential hazard to human health since it comprises an accumulating reservoir of deposited particulates, which can be re-suspended multiple times (e.g., through traffic-induced turbulence), providing multiple opportunities for inhalation/ingestion by all road-users and those living and/or working within close proximity to major roads. RD can further accumulate pollutants in situ, including carbonaceous compounds [8], heavy metals [9], and polyaromatic hydrocarbons (PAHs) [10].

The composition of PM, and thus of RD, is likely to vary significantly on local, national, and international scales. The cellular targets, toxic effects, and mechanisms of specific particle size fractions of PM and RD, and of PM and RD arising from different locations/sources, are currently imperfectly understood. Improved understanding of the specific, causal impacts of PM and RD, and of their differing components, would provide an evidenced rationale for legislative mitigation to reduce PM emissions; and may also be key in developing new therapeutic approaches to treat those already suffering adverse, PM- and RD-induced health outcomes.

Anthropogenic contributors to RD include not only diesel/petrol exhaust but also non-exhaust emissions (NEEs), such as brake, tyre, and road/asphalt wear, and industrial sources, e.g., combustion-derived emissions from factories, and space heating. Natural contributors include soil, endotoxins (bacteria), pollen, and aeolian dust [9,11–13]. As RD is often derived from diverse sources, its chemical and biological composition also varies, typically by location [13], but also with season and climate [14]. RD often contains a wide range of metal-bearing particulates, some attributable to specific sources; e.g., Ba is an additive in most brake pads [15], TiO_2 in road paint [16], and Pt and Pd are released from catalytic converters [17]. The presence in RD of redox-active transition metals and carcinogenic compounds (e.g., PAHs) is detrimental to human health [18]; exposure to PAHs in RD was associated with an incremental lifetime cancer risk (ILCR) of 9.9×10^{-4} in Taiwanese adults [10] (ICLR > 10^{-4} indicates high carcinogenic risk [19]). Several studies report a significant amount of strongly magnetic, iron-rich particles, such as magnetite (Fe_3O_4), in RD [11,20]. Magnetite nanoparticles (MNPs) are often associated with other metal elements such as Co, Cr, Cu, Mn, Ni, Pb, Zn [12], and PAHs [21], and magnetic methods are increasingly used for monitoring of airborne PM. MNPs are potential mediators of neurodegeneration; MNPs with a striking similarity to roadside MNPs have been found in human brain tissue [22], directly associated with key pathological markers of Alzheimer's disease (senile plaques and neurofibrillary tangles) [23,24], and may induce oxidative stress [25], leading eventually to cell death [26].

RD is estimated to comprise 25.7% of PM_{10} in Brazil [27], 55% of PM_{10} in India [28], and 24.6% of $PM_{2.5}$ in Lanzhou, China [29]. Conventionally, air quality is monitored by measuring PM mass concentrations (typically reported as the mass (μg) of PM_{10} and/or $PM_{2.5}$ per m^3 air). Such mass-based metrics are usually dominated by coarser PM. Conversely, in terms of particle number concentrations, ultrafine particles (UFPs, <1 μm) are both by far the dominant fraction, and currently unaccounted for when setting PM limits/guidelines [30]. UFPs frequently represent the majority of the particles to which humans are exposed [25]. Neither RD nor NEEs are restricted currently in terms of exposure limits or emission reductions, despite their abundance (NEEs reportedly form 60% of $PM_{2.5}$ by mass in the UK [31]), potential risk to human health, and contribution to PM when aerosolised.

Compared with larger PM size fractions, UFPs can disperse more widely in the environment [30] and their toxicity is reported to be greater [5,32], likely due to their high surface reactivity [33].

UFPs can penetrate further into the body; e.g., a multiple path particle dosimetry model suggests highest deposition of particles 10–100 nm in size in the alveolar region, regardless of their density. Alveolar deposition of larger particles (100 nm–1 μm) requires higher particle density (10 g/cm^3). There is also relatively high deposition of nanoparticles in the tracheobronchial tract, whilst large particles (<10 μm) deposit primarily in the extra-thoracic and tracheobronchial regions [34].

UFPs may enter the body via inhalation into the lungs [35] and olfactory nerve [36], ingestion [37], and/or dermal penetration [38]. If invading microbes, and foreign bodies such as UFPs, evade the thick protective mucus layer in the lungs, epithelial cells are the first cellular line of defence (Figure 1). The epithelial cells of the human respiratory system defend against incursion of inhaled particulates, primarily via physical barriers formed by cell adhesion proteins (e.g., E-cadherin) and tight junctions (e.g., occludin). Additional defence arises through the release of chemokines, cytokines, and growth factors, and production of reactive oxygen species (ROS) and nitrogen species and antimicrobial proteins [39]. Various antioxidants (e.g., glutathione) also have a protective effect, but decrease in abundance deeper into the respiratory tract (i.e., into the regions penetrated by UFPs) [40].

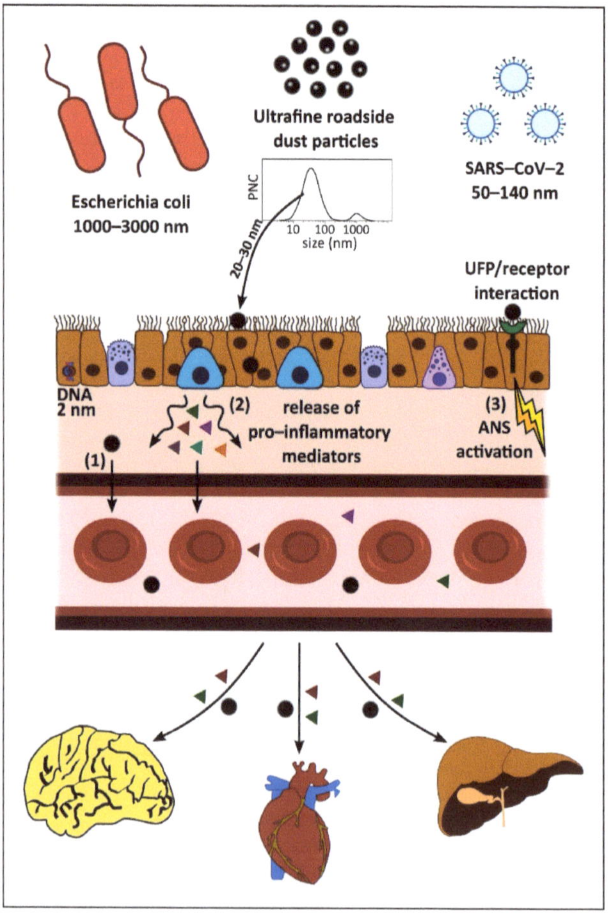

Figure 1. Fate of inhaled ultrafine road-deposited dust particles (UF-RDPs) in the human body.

Inhalation of UF-RDPs may occur when settled road-deposited dust (RD) particles become aerosolised due to wind conditions and/or traffic movement. RD contains a range of particle sizes as shown by the particle number concentration (PNC) graph (adapted from [12]). UFPs can become trapped in mucus or by cilia lining the airway epithelium but very small particles (~20–30 nm) may penetrate between and/or through cells. Such incursion is usually impeded by tight junctions between neighbouring cells, but these junctions can become damaged, and/or UFPs may increase paracellular permeability temporarily, allowing for transient opening of tight junctions and passage of particles [41]. UFPs may also pass transcellularly, travelling through epithelial cells lining the lungs, and can interact/damage cellular components during this passage [42]. The impact of UFP inhalation is not limited to the lungs but extends systemically via three possible mechanisms [43]: 1. incursion into the circulatory system, with subsequent travel to and deposition in extra-pulmonary organs, e.g., [36,44,45]; 2. stimulation of the release of pro-inflammatory mediators from the lungs, which enter the circulation and affect tissues beyond the lungs, e.g., [46]; and/or 3. interaction of pollutant UFPs with the nerves/receptors in the lung, which activates the autonomic nervous system (ANS) to affect a systemic change or response in the body [47]. Particles <200 nm can be transported to the brain directly [22,37,48] via the olfactory bulb [36], and may also be transported to the central nervous system (CNS) via other nerve pathways (trigeminal, vagus, neuroenteric).

UFP cytotoxicity is likely to arise via different pathways, including oxidative stress, damage to the cell membrane, altered gene expression, mitochondrial dysfunction, and/or DNA damage, including the inability to repair this damage [34]. Excess oxidative stress can cause a hierarchical oxidative response in cells: exposure to PM/RD generates the production of free radicals and/or ROS (e.g., via the Fenton reaction); altering the oxidant/antioxidant balance within the cell and stimulating expression of cytoprotective proteins. If the stress is prolonged and/or chronic, secretion of cytokines (e.g., IL-6, IL-8) is triggered, inducing an inflammatory response [49]. If the oxidative stress does not subside, the cell can undergo death via apoptosis or necrosis [50].

Since they are the first cells to encounter inhaled particulates, human lung epithelial cells have been used to model in vitro toxicity of air pollution, including RD particles, brake wear particles, tyre wear debris, and exhaust emissions, e.g., [14,51–55]. Cytotoxicity studies investigating RD using human cell lines are sparse [18,54–59] (Supplementary Table S1); with the majority of studies examining the effects of particles $\leq$2.5 µm. Exposing liver (HepG2) and skin (KERTr) cell lines to 66.7µg/100µL RD from Guangzhou, China decreased cell viability after 72 h; by 53.9% and 71.4%, respectively [54]. The decrease in viability was correlated to the sum of total metal (loids) present in the sample, with Zn, Mn, Cu, and Ni identified as major components [54]. The cytotoxicity of UF, traffic-related pollution particles has also been investigated in rat cell models [60,61]. Little is known about the biological effects in human cells of the UF fractions of RD, which are potentially hazardous owing to their small size, varied, often metal-rich, composition, and abundance in the environment.

To our knowledge, we report here the first investigations of the effects of the ultrafine fraction of RDPs in vitro. The aims of this study were to: (1) extract and characterise the ultrafine fraction (<220 nm) of road-deposited dust particles (RDPs) collected at heavily trafficked sites [62–64] in three contrasting cities; (2) examine the oxidative, cytotoxic, and pro-inflammatory responses of human bronchial epithelial (Calu-3) cells treated with UF-RDPs from these different locations; and (3) compare the cellular effects induced by UF-RDPs from these three cities; namely, a small UK city (Lancaster) and larger UK city (Birmingham), compared with the more highly polluted Mexico City.

2. Materials and Methods

2.1. Sampling Sites

The UK RD samples were collected within 0.5 m of heavily trafficked roads: the A6 at Cable Street (Lancaster, UK) [11] and A38, close to the Bristol Road Observation Site (Birmingham, UK). The Lancaster site is located near a taxi rank and opposite a bus station, where traffic queues are frequent. The Birmingham site is located close to two busy, traffic-lighted junctions, near the University of Birmingham entrance. Mexico City dust was collected from an area of 1 m^2 of the Constitución de la República Avenue. Sample site information is summarised in Table 1. The sampling sites were selected in order to compare sites with a range of $PM_{2.5}$ concentrations, within the same country (UK), and to compare to a densely populated city with greater air pollution levels (Mexico City, Table 1). We have previously reported the presence of UF magnetic nanoparticles (likely originating from PM, including RD) in human brains from Mexico City, and northern England [22,37,48], and magnetically characterised RD from Lancaster and Birmingham [11]. The present work builds on this prior RD characterisation and explores the effects of UF-RDPs (which include

UF magnetic particles) at the cellular level in order to investigate the possible consequences of the presence of such particles in human organs including the brain and heart [22,44].

Table 1. Road-deposited dust sample collection sites. * The Cable Street monitoring station in Lancaster started measuring $PM_{2.5}$ from October 2020. An approximate average for Lancaster $PM_{2.5}$ is 8 μg/m^3 based on data available at http://www.ukairquality.net/ (accessed on 5 November 2021). PM data obtained from [62–64], population data from [65–67].

City	Population Size (2018)	Site	Traffic (Vehicles/Day)	Date Collected	Avg. Annual PM_{10} (μg/m^3) 2018	Avg. Annual $PM_{2.5}$ (μg/m^3) 2018
Lancaster	144,426	(A6) Cable Street	~12,000	18/10/18	22	No data *
Birmingham	1,141,400	(A38) Bristol Road Observation Site	~32,000	20/09/19	18	12
Mexico City	8,781,300	Constitución de la República Avenue	~19,200	06/03/17	47	22

2.2. *Ultrafine Particle Extraction from Road-Deposited Dust*

The bulk dust samples were dispersed (via sonication) in triple-filtered 100% ethanol and filtered through multiple (at least 3) 0.22 μm polyether sulphone (PES) filters based on a protocol as per [51]. A filter pore size of 0.22 μm was used in order to provide sufficient material for repeat cellular analyses. After ethanol evaporation, the concentrated particles (<220 nm, hereafter 'UF-RDPs') were weighed in a room with controlled temperature (20 °C) and humidity (50%), with a Mettler AT250 balance (accuracy 0.00001 g). To form stock solutions, the UF-RDPs were sonicated and re-suspended in 0.5% triple filtered bovine serum albumin (BSA) in dH_2O [68]. Each filtered sample originated from a single original bulk RD sample.

2.3. *Inductively Coupled Plasma (ICP) Mass Spectrometry (MS) and Optical Emission Spectroscopy (OES)*

The metal content of the UF-RDP samples was quantified by ICP analyses, after acid digestion, at the University of Edinburgh. A subsample of filtered UF-RDPs was taken from each stock and the ethanol fully evaporated. Dried samples were weighed in savillex Teflon vessels and digested overnight (100 °C) in digestion mixture (3 mL HNO_3, 2 mL HCL, 0.5 mL HF, all double distilled). After complete digestion and evaporation, samples were acidified in 2–5% ultrapure HNO_3 then analysed for metals and elements of interest using an Agilent Varian Vista Pro (ICP-OES) or Attom Nu (ICP-MS) with the following settings: analysis mode = deflector jump, dwell time/peak = 1 millisecond, number of sweeps = 500, number of cycles = 3, resolution = 300.

2.4. *Superconducting Quantum Interference Device (SQUID) Magnetometry*

Magnetic methods are non-destructive analyses (see SI for more detailed explanation) that have been used to identify combustion- and friction-derived magnetic nanoparticles in human brain [22,37] and heart tissue [44], as well as to characterise RD and brake wear particles [11]. The magnetic content of the bulk (unfiltered) RDPs was measured with a 2G RAPID cryogenic magnetometer (at the Centre for Environmental Magnetism & Palaeomagnetism, Lancaster University, Lancaster, UK) by imparting an isothermal remanent magnetisation (IRM) at 1 Tesla (T) (using a Newport Instruments electromagnet) at room temperature (Supplementary Figure S1). To identify the presence of ultrafine (~20–30 nm) magnetic particles, low-temperature magnetic measurements were made on the extracted UF-RDPs, using an MPMS3 SQUID magnetometer (Quantum Design, San Diego, CA, USA) at the Department of Physics, University of Cambridge. For UF-RDPs, IRM was imparted at 1T and 300 K, and was measured upon cooling to 10 K (at average rate 5 K/min) at the University of Cambridge. Then, IRM was imparted again (at 1 T and

10 K) and measured upon heating to 300 K. To increase signal-to-noise ratio, 10 DC scans were used for IRM at 300 K and 10 K (Supplementary Figure S2).

2.5. Endotoxin Quantification

Endotoxin concentrations of stock solutions were determined via a quantitative kinetic limulus amoebocyte lysate (LAL) assay kit using the manufacturer's protocol (Thermo Scientific™, Loughborough, UK).

2.6. Cell Culture

All cell culture reagents were purchased from Lonza Ltd. (Basel, Switzerland) unless stated otherwise. Human lung epithelial cells (Calu-3, ATCC HTB-55™) were selected due to their previous characterisation and study as targets of airborne particulate matter [51,53,69–71]. The immortalised cell line originated from a 25-year-old white male with lung adenocarcinoma. Calu-3 cells were maintained (until passage 20) in Eagle's minimum essential medium (EMEM) supplemented with 10% (*v/v*) filter sterilised foetal bovine serum (FBS) (Gibco™, Thermofisher), 1% (*v/v*) non-essential amino acids, penicillin (50 units/mL), and streptomycin (50 µg/mL) at 37 °C, 5% CO_2. Fluorescent and absorbance readings were conducted using a Tecan Infinite M200 pro spectrophotometer.

2.7. MTS Assay

Calu-3 cells were seeded at 40,000 cells/well in a 96-well plate and left overnight to adhere. UF-RDPs at concentrations of 0–300 µg/mL (equivalent to 0.94–94 µg/cm^2) were prepared via sonication in UltraMem supplemented with penicillin (50 units/mL) and streptomycin (50 µg/mL). Dose and exposure times were based on a previous study with Calu-3 cells and brake wear particles [51]. BSA, used here as a stabilising agent [68], was present at equal concentrations across each set of test conditions; the observed biological responses are values normalised to BSA-exposed controls. Following a 24 h exposure, cells were rinsed with Dulbecco's phosphate buffered saline (DPBS) and subjected to a 3-(4,5-dimethylthiazol-2-yl)-5-(3-carboxymethoxyphenyl)-2-(4-sulfophenyl)-2H-tetrazolium (MTS) assay, with the protocol based on a WST-1 assay as per [72].

2.8. Reactive Oxygen Species (ROS) Assay

Calu-3 cells, seeded as above, were incubated with 25 µM of the fluorescent cellular probe chloromethyl derivative of 2′,7′-dichlorodihydrofluorescein diacetate (CMD-CFH2DA) (Invitrogen™, Waltham, MA, USA) in DPBS for 45 min, prior to particle exposure, at 37 °C, 5% CO_2 [51]. Then, 100 µM tert-butyl hydroperoxide (TBHP) was used as a positive control. Fluorescence (excitation 495 nm, emission 529 nm) was measured at 0.5, 1, 2, 3, and 4 h. The 4 h timepoint was chosen to reflect the rapid clearance of UFPs from the lungs, from as little as 4 h [36]. Background controls (n = 3) consisting of UF-RDPs or TBHP in UltraMem were measured alongside the cell treatments for intrinsic fluorescence and subtracted from the experimental cell fluorescence readings.

2.9. Cytokine ELISAs

Following 24 h exposure to UF-RDPs (as above), the media samples were collected, centrifuged (15000 RCF, 15 min, 4 °C), and analysed for IL-6 and IL-8 concentrations via ELISAs conducted according to the manufacturer's protocol (IL-6 BioLegend, London, UK, IL-8 Invitrogen™, Waltham, MA, USA). Cytokine concentrations were calculated from standards fitted using a four-parameter logistic curve-fit with program MyAssays (http://www.myassays.com/, accessed on 13 August 2021) (see SI, supplementary methods). UFPs have a large surface area for cytokine adsorption [73] which can cause interference with ELISA results. Here, known cytokine standards were spiked with UF-RDPs and measured via ELISA. Values were within 5–10% of unspiked samples measured in parallel (data not shown), suggesting any adsorption has limited impact on the assay. Data were normalised to the control, however, the raw values (ng/mL) can be seen in Supplementary Figures S3 and S4.

To account for potential changes in cell number, the ELISA data were adjusted using Equation (1):

$$\left(\frac{IL\text{-}6\ fold\text{-}change\ relative\ to\ control}{MTS\ fold\text{-}change\ relative\ to\ control}\right) \times 100. \quad (1)$$

2.10. Statistical Analysis

All experimental results represent 3–4 individual experiments. Each set of UF-RDPs was tested independently under the same experimental conditions using identical control conditions in each case. Data were normalised to the control from the independent experiment to allow for comparison across experiments. Data are presented as mean ± standard error of the mean (SEM). Statistical analysis was conducted using SPSS 24 (IBM). Normality tests were performed using the Shapiro–Wilk test. A one-way analysis of variance (ANOVA) (with Dunnett's post hoc) was performed to compare particle treatments with unexposed control. Comparison of location and concentration was assessed by a two-way ANOVA (with Bonferroni correction post hoc). Statistical significance levels used are: *, $p \leq 0.05$; **, $p \leq 0.01$; ***, $p \leq 0.001$; ****, $p \leq 0.0001$ where * may be substituted for ^, # or $_0$ depending on the city or comparison being made.

3. Results

In terms of metal compositions of the three sets of UF-RDPs, it is notable that nearly all metals analysed are most abundant in the Mexico City UF-RDPs (18 out of 24), compared to the UK samples (Figure 2; Supplementary Table S2). For example, mass concentrations of Cu and Fe are 67.2 ppm and 77.1 ppm for Mexico City, followed by 29.1 ppm and 55.9 ppm for Lancaster and 19.7 ppm and 14.7 ppm for Birmingham. Conversely, Na is most abundant in the Lancaster UF-RDPs (105,489 ppm), followed by Mexico City (42,163 ppm) and Birmingham (1003 ppm). Pb concentrations are 8.0 ppm for Lancaster, 1.2 ppm for Mexico City, and 0.7 ppm for Birmingham. Cd was the only analysed element occurring in the greatest concentrations in the Birmingham UF-RDPs (6.1 ppm), followed by Lancaster (1.1 ppm) and Mexico City (below detection limit).

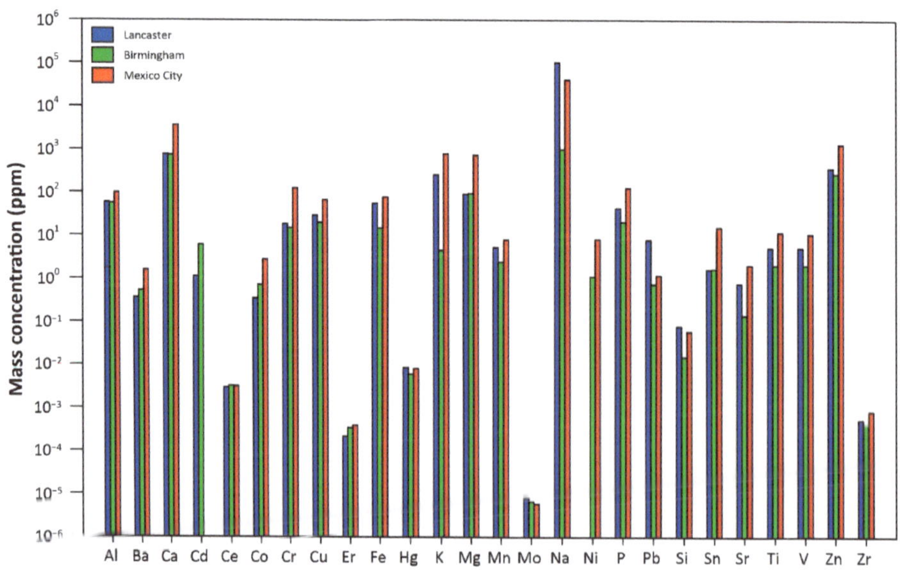

Figure 2. Elemental composition of the UF-RDPs from Lancaster, Birmingham, and Mexico City.

In terms of their magnetic content, the measured mass concentration of magnetite in the Mexico City UF-RDPs was ~0.24–0.79 wt.%. The presence of a broad Verwey transition identifies the presence specifically of magnetite (a mixed Fe^{2+}/Fe^{3+} iron oxide) with ultrafine (~20–30 nm) magnetite particles evident from the large (~43%) increase in remanence at low temperature (10 K) compared to that at 300 K (Supplementary Figures S1 and S2) [74,75]. Due to the low sample mass extracted, the magnetic content of the Lancaster and Birmingham UF-RDPs was unmeasurable; however, IRM data for the bulk samples can be seen in Supplementary Figure S1.

Detectable but minor levels of endotoxin were present in all three UF-RDP samples (8.75–9.25 EU/mg), as assessed by LAL assay (Supplementary Table S3).

Calu-3 cells were exposed to UF-RDP doses of between 0 and 300 μg/mL (0.94–94 μg/cm^2). It is noteworthy that significant cellular responses were elicited even at low and intermediate UF-RDP doses, especially for the Mexico City UF-RDPs. Although the maximum dose is larger than any single typical environmental exposure, such high doses may be indicative of responses elicited due to the chronic, repeated, and accumulated exposures to which urban dwellers are subjected in their life-course.

Following 24 h exposure to UF-RDPs (0–300 μg/mL), Calu-3 cell viability varied significantly by sample location. For the Lancaster UF-RDPs, cell viability increased, by 25–35% (Figure 3), in a dose-dependent manner (15 μg/mL upwards), similar to our previous studies with synthetic magnetite nanoparticles (unpublished data). Treatment with the Birmingham UF-RDPs caused a significant decrease in cell viability but only at the highest exposure dose (300 μg/mL, 65% decrease). In contrast, Calu-3 cells were most sensitive to the Mexico City UF-RDPs where a dose-dependent decrease in viability was seen, even at the lowest dose (3 μg/mL, 12% decline), up to a 30% decline at 300 μg/mL.

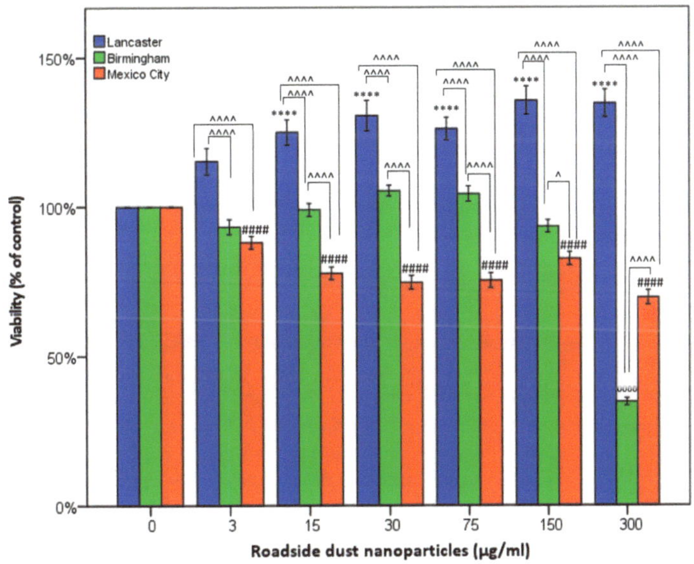

Figure 3. Cytotoxicity of <220 nm-sized road-deposited dust particles (UF-RDPs) on Calu-3 cells. Calu-3 cells were exposed to UF-RDPs (0–300 μg/mL) from Lancaster and Birmingham (UK) and Mexico City (Mexico) for 24 h and subjected to an MTS assay, which reflects cell viability. A one-way ANOVA with Dunnett's post hoc was conducted, comparing treated cells to the untreated control (* Lancaster, $_0$ Birmingham, # Mexico City) and two-way ANOVA with Bonferroni correction to compare the impacts of the UF-RDPs from the three different locations (^). Statistical significance levels used are: *, $p \leq 0.05$; ****, $p \leq 0.0001$ where * may be substituted for ^, # or $_0$ depending on the city or comparison being made.

Elevated ROS was observed from the 30 min time point for all three cities (Supplementary Figures S4 and S5A). At the 4 h timepoint, Mexico City UF-RDPs were the most potent, stimulating increased ROS from 75 µg/mL. Birmingham UF-RDPs induced little increase in ROS generation except at the maximum dose (300 µg/mL, 120%), when cell viability also showed maximum decline (Figure 4). Lancaster UF-RDPs were least potent in terms of ROS generation.

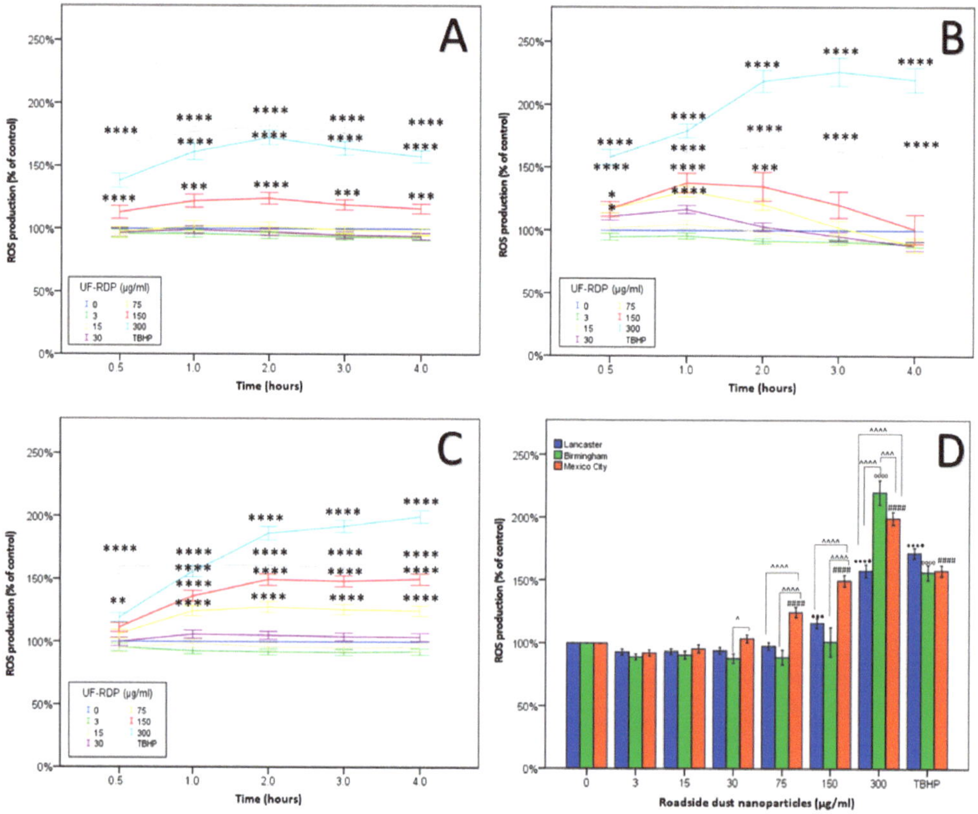

Figure 4. Oxidative stress in Calu-3 cells induced by <220 nm-sized road-deposited dust particles (UF-RDPs). Calu-3 cells were loaded with the ROS probe CM-DCFH-DA and exposed to UF-RDPs (0–300 µg/mL) from Lancaster (**A**) and Birmingham (**B**) (UK) and Mexico City (Mexico) (**C**). Generation of ROS was measured after 0.5 h, 1 h, 2 h, 3 h, and 4 h (**D**) exposure. Tert-butyl hydroperoxide (TBHP) was used at 100 µM as a positive control. A one-way ANOVA with Dunnett's post hoc was conducted, comparing treated cells to the untreated control (* for A–C and for D: • Lancaster, $_{0}$ Birmingham, # Mexico City) and two-way ANOVA with Bonferroni correction to compare the impacts of the UF-RDPs from the three different locations (^). Statistical significance levels used are: *, $p \leq 0.05$; **, $p \leq 0.01$; ***, $p \leq 0.001$; ****, $p \leq 0.0001$ where * may be substituted for ^, #, • or $_{0}$ depending on the city or comparison being made.

The release of IL-6 and IL-8 cytokines following 24 h exposure to UF-RDPs was quantified by ELISA (Figure 5). To account for potential changes in cell number, the ELISA results (Figure 5) were adjusted using the MTS data (Figure 3). Unadjusted data and absolute values in ng/mL are given in the SI (Supplementary Figures S3, S4 and S6) and generally show similar trends to unadjusted data at non-lethal doses. Following exposure to the extracted UF-RDPs, a dose-dependent increase in IL-6 release was observed in response to doses of 75 µg/mL or above from all three cities, with the greatest increase in release

(1648%, corresponding to an increase of 8.3 ng/mL) in response to 300 µg/mL Birmingham UF-RDPs (Figure 5A). An overall dose-dependent increase in IL-8 was observed for Mexico City and Birmingham UF-RDPs, and to a lesser extent for Lancaster UF-RDPs (Figure 5B). For Mexico City UF-RDPs, IL-8 release peaked at 75 µg/mL and then declined at the higher doses, possibly due to high cytotoxicity at these doses. The largest increase in both IL-6 and IL-8 was stimulated by the Birmingham UF-RDPs (adjusted data, 1648%/increase of 8.3 ng/mL and 408%/increase of 12.3 ng/mL, respectively). Compared with IL-6, IL-8 release displayed significantly greater variation with city source.

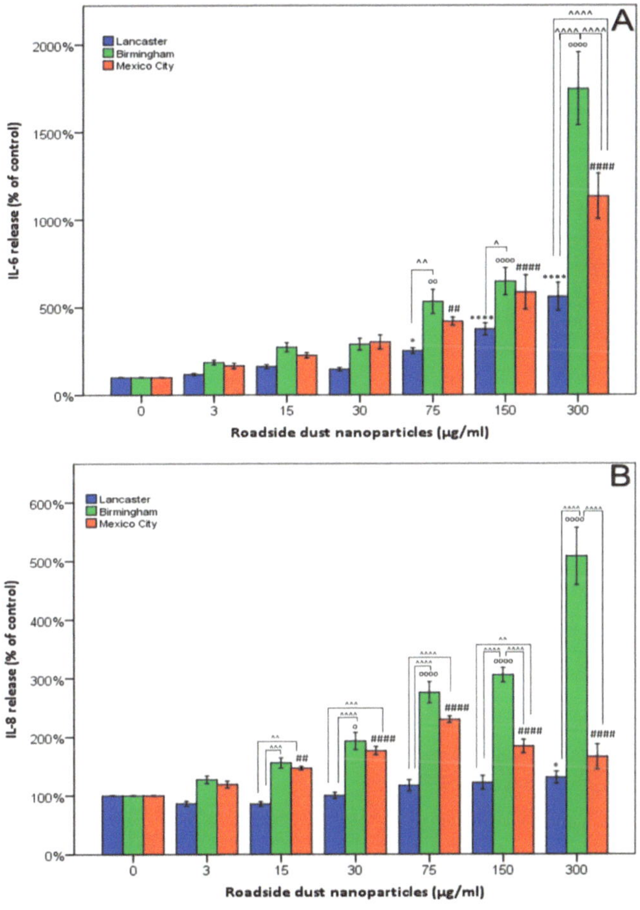

Figure 5. Release of pro-inflammatory cytokines in Calu-3 cells treated with <220 nm-sized road-deposited dust particles (UF-RDPs). Calu-3 cells were exposed (0–300 µg/mL) to UF-RDPs from Lancaster and Birmingham (UK) and Mexico City (Mexico) for 24 h. IL-6 (**A**) and IL-8 (**B**) concentrations in the media were quantified via ELISA. Data were adjusted using MTS (cell viability) data from Calu-3 cells treated in the same manner. A one-way ANOVA with Dunnett's post hoc was conducted, comparing treated cells to the untreated control. (* Lancaster, $_0$ Birmingham, # Mexico City) and two-way ANOVA with Bonferroni correction to compare the impacts of the UF-RDPs from the three different locations (^). Statistical significance levels used are: *, $p \leq 0.05$; **, $p \leq 0.01$; ***, $p \leq 0.001$; ****, $p \leq 0.0001$ where * may be substituted for ^, # or $_0$ depending on the city or comparison being made.

4. Discussion

The increased viability resulting from the Lancaster UF-RDPs may represent an increase in cell number (proliferation) or metabolic activity (Figure 3). The Pb content of the Lancaster UF-RDPs—sampled close to the city bus station—is notably high, ~8 x higher compared with those from Birmingham and Mexico City. Cell proliferation might thus reflect the tumorigenic effect of Pb [76]. Pb was reportedly 4 x higher in settled bus dust relative to background soil [77]. Proliferation might also reflect replenishment of damaged epithelial cells, and/or airway remodelling [52]; i.e., changes in the composition, structure, or thickness of (structural) elements of the airway. In response to high PM exposure, lungs from female life-long residents of Mexico City displayed extensive airway remodelling including formation of fibrous tissue (pulmonary fibrosis) [78]. Conversely, lead oxide nanoparticles have been implicated in the induction of apoptosis following mitochondrial damage [79]. Lead nanoparticles, however, did not induce apoptosis in A549 lung cells [80], so the apoptotic effect may be specific to the type of lead nanoparticle and/or cell type, and in the case of PM is also likely influenced by other compounds and elements present in the heterogenous mixture. It was not possible to assess cell proliferation due to the limited sample material available, but future work would usefully include the assessment of cell proliferation, for example, via trypan blue.

In contrast, the Mexico City UF-RDPs induced a dose-dependent decrease in cell viability, reflecting reduced metabolic activity or reduced proliferation, and/or cell death. This decreased viability likely reflects the abundance of metals in these UFPs, including Fe, Zn, Mn, Pb, Cu, Cr, and Ni [12,13]. Transition metals can catalyse ROS production in situ via the Fenton reaction, leading to oxidative damage to lipids, DNA, and proteins, and eventually cell death [81]. The majority of the analysed metals occur in greatest concentrations in the Mexico City UF-RDPs (Figure 2; Supplementary Table S2). Of these metals, Ba, Co, and Ni may have the strongest influence on the observed cytotoxic response. This is because the other metals (Cr, Cu, V, and Zn) which are most abundant in the Mexico City UF-RDPs are present at higher concentrations in Lancaster UF-RDPs than Birmingham UF-RDPs, yet there was no decline in cell viability in response to Lancaster UF-RDPs compared to Birmingham (Figures 2 and 3; Supplementary Table S2). Ni may be of particular importance, as it is present in the Birmingham and Mexico City UF-RDPs but not in the Lancaster particles (Figure 2; Supplementary Table S2); Ni in RD from South Korea was correlated with cytotoxicity [56]. Birmingham UF-RDPs only decreased Calu-3 viability at the highest dose (300 μg/mL), when antioxidant defences likely were overwhelmed.

At the maximum dose (300 μg/mL), intracellular ROS levels were elevated within 30 min of exposure to the UF-RDPs from all three cities. Lancaster UF-RDPs were least potent in terms of ROS generation; in contrast, higher ROS concentrations were observed following Birmingham and Mexico City UF-RDP exposures (Figure 4). Together, these data suggest that the lower ROS levels induced by Lancaster UF-RDPs could stimulate proliferation (increased cell viability), whereas the higher ROS levels induced by Mexico City (and Birmingham) UF-RDPs result in cell death via oxidative damage. Some similar responses have been reported for RD samples $\leq$2.5 μm and $\leq$10 μm. Up to 180% increased ROS production was observed in human corneal epithelial cells after 24 h exposure to RD from residential areas of the city of Gangdong-Gu, Korea [59]. Re-aerosolised $RD_{2.5}$ from 10 Chinese cities displayed correlation between cellular ROS production and heavy metal concentrations (Cr, Mn, Zn, Ni, Pb) [55].

In our UF-RDPs, ultrafine magnetite particles (~20–30 nm) were abundant in the Mexican sample, and we have previously detected magnetite/maghemite in bulk RD at the same sampling sites in Lancaster and Birmingham [11]. Given the catalytic role of Fe (especially Fe^{2+}) in the Fenton reaction, ultrafine magnetite may play a particular role in the dose-dependent increases in ROS generation seen here [25].

Dose-dependent increases in IL-6 and IL-8 were observed in response to UF-RDPs from all three cities (Figure 5). Toxic concentrations of UF-RDPs may result in cell death, which has been associated with an increase in IL-6 secretion [82], so examination of sublethal concentrations (i.e., viability <80% [58]) is important. Using this criterion, none of the Lancaster UF-RDP doses were lethal, whereas for Mexico City UF-RDPs, 15, 30, 75, and 300 μg/mL doses (corresponding to viability of 78, 75, 75, and 70%, respectively) were lethal. For the Birmingham UF-RDPs, only the maximum dose (300 μg/mL) was lethal. Excluding these data and focusing on sublethal doses, elevated IL-6 and IL-8 release was observed in response to UF-RDPs from all three cities from a relatively low dose (e.g., 15 μg/mL for IL-8). The largest increases in cytokine release were in response to the Birmingham UF-RDPs (Figure 5). Interestingly, Puisney et al. (using the same cell line and methods) report no change in IL-8 secretion in response to 0–300 μg/mL doses of UF brake wear particles [51] but an increase (up to ~350%) in IL-6 secretion. These differences in observed biological response may thus reflect differences in the PM samples tested—UF, dynamometer-derived brake wear [51]—compared to our UF-RDPs. The elements most abundant in the dynamometer-derived brake wear particles were Fe and Cu, followed by Si, Al, and Zn, whereas our UF-RDPs are dominated by Na and Ca, followed by Zn, K, and Mg. Roadside PM contains not only traffic- and industry-derived compounds, but also naturally derived elements/metals, including Al, Ca, Fe, K, Mg, or Na [83]. Some of the naturally (soil-)derived compounds might also be involved in the Calu-3 biological responses. Alternatively, it is possible that transition metals increase ROS/oxidative damage, decreasing cell viability, and also trigger an inflammatory response in the form of increased IL-6 and IL-8 secretion (Figure 6) [84,85].

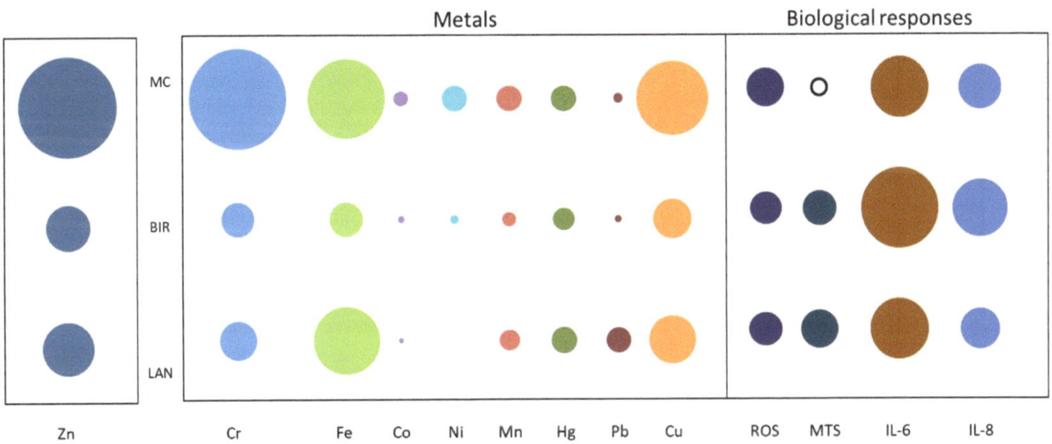

Figure 6. Summary of biological responses and heavy metal concentrations of UF-RDPs from Lancaster, Birmingham, and Mexico City in Calu-3 cells. The bubble plot depicts the relative abundance of selected metals (ppm) in UF-RDPs from Mexico City (MC), Birmingham (BIR), and Lancaster (LAN). The concentration of Zn is ~10-fold greater than the other metals so has been scaled separately. The biological responses to 75 μg/mL UF-RDPs are represented as percentages of the untreated control, measured at 24 h (MTS, IL-6, and IL-8) or 4 h (ROS). Solid black outline indicates a decrease, whilst filled circles indicate increases.

RD is, of course, a heterogenous mixture, and other components (not measured here) such as the organic fraction (e.g., carbonaceous compounds, PAHs) are also known to be cytotoxic [19], induce ROS generation [55], and alter inflammatory responses [86]. It is improbable, however, that (semi-)volatile compounds are responsible for the cellular responses observed here since most of these would have been lost in the particle extraction procedure (filtration in ethanol and evaporation). Endotoxins are also present in RD. Endotoxins—pyrogenic molecules shed from Gram-negative bacteria which are ubiquitous in the environment—are known to induce ROS production, stimulate inflammatory responses in humans, and are toxic at high levels [87]. Endotoxins were present albeit at low levels (8.75–9.80 EU/mg) in all three UF-RDP samples (Supplementary Table S3). Environmental endotoxins are primarily associated with the coarse (2.5–10 μm) fraction of PM, and have been implicated in ROS generation [88]. It is possible the endotoxins present in our UF-RDPs contribute to ROS generation and pro-inflammatory responses; nanoscale endotoxin can penetrate to the alveoli and poses a higher risk to human health than its larger counterparts [89]. However, exposing Calu-3 cells to 10 μg/mL LPS (5000 EU/mL) for 4 h showed no change in ROS generation (data not shown), suggesting that the presence of environmental endotoxins at the concentrations measured did not contribute to the ROS generation observed at this timepoint in these cells.

The different biological responses seen here in response to UF-RDPs from the three cities likely derive from differences in UF-RDP physicochemical compositions, and which reflect the biogenic and anthropogenic activities in the surrounding area [13], as well as current air pollution regulations. ROS generation is generally higher in response to PM in the developing world; for example, PM from Beirut (more permissive regulations) was more redox-active than PM from Los Angeles, and Ni and V (tracers of fuel oil combustion) were enriched 5–6 × in $PM_{0.25}$ from Beirut compared to Los Angeles [40]. Here, we see ~8 × higher Ni and 2–5 × higher V in Mexico City compared to the two UK cities. All three samples were collected from heavily trafficked road sites (Table 1), where many metals are emitted from traffic-related sources. Ba, Cu, and Fe (most abundant in Mexico City UF-RDPs) appear co-associated, and likely originate from the same source, i.e., from brake wear (e.g., [90]). Zn, also most abundant in Mexico City, has often been reported in tyre wear PM emissions [9,91]. A recent study in Toronto, Canada, observed similar co-associations between traffic-derived metals in RD, including Zn, Mo, Cr, Sn, Pb, and Ba [20]. All these metals (along with the remaining transition metals) likely contributed to the strong oxidative response observed in the lung cells exposed to UF-RDPs from the three cities studied here (Figure 6).

One limitation of this study is the difference in sampling collection time (year and season) (Table 1). Variable biological responses may reflect the season of sampling, associated with varying ambient PM sources [14]. Finally, because of the low mass of extractable UFPs, particle number concentrations were not analysed for the three samples; nor was the magnetic content of the UFPs measurable for the UK samples. To enhance further our understanding of the impacts of RD from different locations, with different physicochemical properties, future work would usefully identify the particle size distribution, morphology, and any possible particle changes within the biological media used.

Due to its abundance, reactivity, and pervasiveness at/near roadsides, growing attention is focused on the health impacts of ultrafine PM and RD. The differential oxidative stress, inflammatory, and cytotoxic responses to UF-RDPs we have observed in human lung epithelial cells represent processes which may be pathogenic if exposures to UF-RDPs are prolonged and/or chronic. We have examined lung cells but the effects of exposure to UF-RDPs extend well beyond the lung (Figure 1). For example, ultrafine magnetite nanoparticles, like those present in our Mexico City UF-RDPs, have been found in the human brain [22]. Although a direct inhalation route via the olfactory bulb is possible and likely, our measured brain and road-deposited magnetite concentrations [11,48], coupled with modelled olfactory deposition rates [92], indicate that inhalation and circulatory transport must dominate CNS translocation of such particles. Similarly, exogenous metal-rich nanoparticles have been found in the heart [25,44] and the placenta [93]. Thus, any consequences of exposure to UF-RDPs (including iron-rich, magnetic nanoparticles) are most likely to be systemic.

5. Conclusions

The first in vitro exposure of human lung epithelial (Calu-3) cells to the ultrafine fraction of road dust from three contrasting cities (Lancaster and Birmingham, UK, and Mexico City, Mexico) resulted in differential oxidative, cytotoxic, and inflammatory responses. Even at low and intermediate UF-RDP doses, significant cellular responses were elicited, especially by the Mexico City UF-RDPs. Given that any (semi-)volatile components are likely to have been lost during the particle extraction process, variations in the solid, metal content of the three sets of UF-RDPs are likely to be linked with the different observed cell responses.

Airborne PM is subject to regulations which rely currently on mass-based metrics such as PM_{10} and $PM_{2.5}$, rather than particle number/composition, while RD is not currently limited/regulated by any legislation. Such mass-based metrics cannot capture the differential biological impacts induced by roadside particles as observed here. We observed a stronger cytotoxic response from Mexico City UF-RDPs compared to the UK cities, and a stronger inflammatory response from Birmingham UF-RDPs. It is therefore illogical to apply the same PM limits across different cities, where exposure to a mass- or even number-based PM limit in one city may have far worse consequences for health than exposure to the same limit in a different city. A combination of new, biologically relevant metrics, identification of specific components (e.g., metals) of RD that cause toxic effects, and localised regulations each appear critical to mitigating the global pandemic of the health impacts of particulate air pollution and road-deposited dust. A metric that provides an intermediate measurement, such as the lung-deposited surface area (LDSA) which utilises particle size distribution to estimate the surface area concentration of particles that deposit in the lung alveolar region, might be more appropriate [94]. Not only particle number but also physicochemical composition, and/or specific elemental components (e.g., transition metals), likely require regulation in order to achieve substantive mitigation of the human health impacts of exposure to RD and airborne PM.

Supplementary Materials: The following supporting information can be downloaded at: https://www.mdpi.com/article/10.3390/antiox11091814/s1, Magnetic properties of the total and ultrafine (<220 nm) road-deposited dust particles, Table S1: Overview of cytotoxicity studies of road-deposited dust and roadside airborne PM in human and rodent models, Table S2: Elemental composition of the analysed UF-RDPs from Lancaster, Birmingham, and Mexico City, Table S3: Endotoxin content of ultrafine road-deposited dust particles, Figure S1: Isothermal remanence magnetisation (IRM) for the bulk (unfiltered) PM from the three studied sampling sites, Figure S2: Zero-field changes in IRM during cooling of the Mexican UF-RDPs, after acquisition of IRM (at 1T) at 300 K, Figure S3: Release of pro-inflammatory cytokines (ng/mL) in Calu-3 cells treated with ultrafine road-deposited dust particles, Figure S4: MTS-adjusted release of pro-inflammatory cytokines (ng/mL) in Calu-3 cells treated with ultrafine road-deposited dust particles, Figure S5: Oxidative stress in Calu-3 cells induced by ultrafine road-deposited dust particles, Figure S6: Release of pro-inflammatory cytokines in Calu-3 cells treated with ultrafine road-deposited dust particles.

Author Contributions: J.H.: conceptualisation, methodology, validation, formal analysis, investigation, writing—original draft, writing—review and editing, visualisation, B.A.M.: conceptualisation, writing—review and editing, supervision, T.G.: methodology, investigation, writing—review and editing, F.B.: resources, writing—review and editing, D.A.: supervision, investigation. All authors have read and agreed to the published version of the manuscript.

Funding: J.H. received a funded studentship from the Sir John Fisher Foundation, and T.G. received a funded studentship from Jaguar Land Rover. We would like to thank Defying Dementia (Lancaster University) for funding part of this work. F.B. received financial support from the DGAPA UNAM project IN208621.

Institutional Review Board Statement: Not applicable.

Informed Consent Statement: Not applicable.

Data Availability Statement: All of the data is contained within the article and the supplementary materials.

Acknowledgments: We would like to thank Laetitia Pichevin, Edinburgh University for performing the sample digestions and ICP-MS and OES analyses, and Hassan A. Sheikh, University of Cambridge for performing MPMS analyses. F.B. appreciates the technical support of Rubén Cejudo.

Conflicts of Interest: The authors declare no conflict of interest. The funders had no role in the design of the study; in the collection, analyses, or interpretation of data; in the writing of the manuscript; or in the decision to publish the results.

References

1. Southerland, V.A.; Brauer, M.; Mohegh, A.; Hammer, M.S.; van Donkelaar, A.; Martin, R.V.; Apte, J.S.; Anenberg, S.C. Global urban temporal trends in fine particulate matter ($PM_{2.5}$) and attributable health burdens: Estimates from global datasets. *Lancet Planet. Health* **2022**, *6*, e139–e146. [CrossRef]
2. WHO. Ambient (Outdoor) Air Pollution. Available online: https://www.who.int/news-room/fact-sheets/detail/ambient-(outdoor)-air-quality-and-health (accessed on 28 September 2021).
3. Zhang, Z.; Zhu, D.; Cui, B.; Ding, R.; Shi, X.; He, P. Association between particulate matter air pollution and lung cancer. *Thorax* **2020**, *75*, 85. [CrossRef] [PubMed]
4. Kim, O.-J.; Lee, S.H.; Kang, S.-H.; Kim, S.-Y. Incident cardiovascular disease and particulate matter air pollution in South Korea using a population-based and nationwide cohort of 0.2 million adults. *Environ. Health* **2020**, *19*, 113. [CrossRef]
5. Weichenthal, S.; Olaniyan, T.; Christidis, T.; Lavigne, E.; Hatzopoulou, M.; Van Ryswyk, K.; Tjepkema, M.; Burnett, R. Within-city Spatial Variations in Ambient Ultrafine Particle Concentrations and Incident Brain Tumors in Adults. *Epidemiology* **2020**, *31*, 177–183. [CrossRef]
6. Chen, H.; Kwong, J.C.; Copes, R.; Tu, K.; Villeneuve, P.J.; van Donkelaar, A.; Hystad, P.; Martin, R.V.; Murray, B.J.; Jessiman, B.; et al. Living near major roads and the incidence of dementia, Parkinson's disease, and multiple sclerosis: A population-based cohort study. *Lancet* **2017**, *389*, 718–726. [CrossRef]
7. Mortamais, M.; Gutierrez, L.-A.; de Hoogh, K.; Chen, J.; Vienneau, D.; Carrière, I.; Letellier, N.; Helmer, C.; Gabelle, A.; Mura, T.; et al. Long-term exposure to ambient air pollution and risk of dementia: Results of the prospective Three-City Study. *Environ. Int.* **2021**, *148*, 106376. [CrossRef]
8. Heal, M.R.; Hammonds, M.D. Insights into the composition and sources of rural, urban and roadside carbonaceous PM_{10}. *Environ. Sci. Technol.* **2014**, *48*, 8995–9003. [CrossRef]
9. Hong, N.; Guan, Y.; Yang, B.; Zhong, J.; Zhu, P.; Ok, Y.S.; Hou, D.; Tsang, D.C.W.; Guan, Y.; Liu, A. Quantitative source tracking of heavy metals contained in urban road deposited sediments. *J. Hazard. Mater.* **2020**, *393*, 122362. [CrossRef]
10. Mon, E.E.; Phay, N.; Agusa, T.; Bach, L.T.; Yeh, H.-M.; Huang, C.-H.; Nakata, H. Polycyclic Aromatic Hydrocarbons (PAHs) in Road Dust Collected from Myanmar, Japan, Taiwan, and Vietnam. *Arch. Environ. Contam. Toxicol.* **2020**, *78*, 34–45. [CrossRef]
11. Gonet, T.; Maher, B.A.; Kukutschová, J. Source apportionment of magnetite particles in roadside airborne particulate matter. *Sci. Total Environ.* **2021**, *752*, 141828. [CrossRef] [PubMed]
12. Yang, Y.; Vance, M.; Tou, F.; Tiwari, A.; Liu, M.; Hochella, M.F. Nanoparticles in road dust from impervious urban surfaces: Distribution, identification, and environmental implications. *Environ. Sci. Nano* **2016**, *3*, 534–544. [CrossRef]
13. Pant, P.; Baker, S.J.; Shukla, A.; Maikawa, C.; Godri Pollitt, K.J.; Harrison, R.M. The PM10 fraction of road dust in the UK and India: Characterization, source profiles and oxidative potential. *Sci. Total Environ.* **2015**, *530–531*, 445–452. [CrossRef] [PubMed]
14. Gualtieri, M.; Mantecca, P.; Corvaja, V.; Longhin, E.; Perrone, M.G.; Bolzacchini, E.; Camatini, M. Winter fine particulate matter from Milan induces morphological and functional alterations in human pulmonary epithelial cells (A549). *Toxicol. Lett.* **2009**, *188*, 52–62. [CrossRef]
15. Harrison, R.M.; Jones, A.M.; Gietl, J.; Yin, J.; Green, D.C. Estimation of the Contributions of Brake Dust, Tire Wear, and Resuspension to Nonexhaust Traffic Particles Derived from Atmospheric Measurements. *Environ. Sci. Technol.* **2012**, *46*, 6523–6529. [CrossRef]
16. Al-Kattan, A.; Wichser, A.; Vonbank, R.; Brunner, S.; Ulrich, A.; Zuin, S.; Nowack, B. Release of TiO2 from paints containing pigment-TiO2 or nano-TiO2 by weathering. *Environ. Sci. Process. Impacts* **2013**, *15*, 2186–2193. [CrossRef] [PubMed]
17. Prichard, H.M.; Fisher, P.C. Identification of Platinum and Palladium Particles Emitted from Vehicles and Dispersed into the Surface Environment. *Environ. Sci. Technol.* **2012**, *46*, 3149–3154. [CrossRef]
18. Kim, E.-A.; Koh, B. Utilization of road dust chemical profiles for source identification and human health impact assessment. *Sci. Rep.* **2020**, *10*, 14259. [CrossRef]
19. Liao, C.-M.; Chiang, K.-C. Probabilistic risk assessment for personal exposure to carcinogenic polycyclic aromatic hydrocarbons in Taiwanese temples. *Chemosphere* **2006**, *63*, 1610–1619. [CrossRef]
20. Wiseman, C.L.S.; Levesque, C.; Rasmussen, P.E. Characterizing the sources, concentrations and resuspension potential of metals and metalloids in the thoracic fraction of urban road dust. *Sci. Total Environ.* **2021**, *786*, 147467. [CrossRef]
21. Halsall, C.J.; Maher, B.A.; Karloukovski, V.V.; Shah, P.; Watkins, S.J. A novel approach to investigating indoor/outdoor pollution links: Combined magnetic and PAH measurements. *Atmos. Environ.* **2008**, *42*, 8902–8909. [CrossRef]
22. Maher, B.A.; Ahmed, I.A.M.; Karloukovski, V.; MacLaren, D.A.; Foulds, P.G.; Allsop, D.; Mann, D.M.A.; Torres-Jardón, R.; Calderón-Garciduenas, L. Magnetite pollution nanoparticles in the human brain. *Proc. Natl. Acad. Sci. USA* **2016**, *113*, 10797–10801. [CrossRef] [PubMed]
23. Dadras, A.; Riazi, G.H.; Afrasiabi, A.; Naghshineh, A.; Ghalandari, B.; Mokhtari, F. In vitro study on the alterations of brain tubulin structure and assembly affected by magnetite nanoparticles. *JBIC J. Biol. Inorg. Chem.* **2013**, *18*, 357–369. [CrossRef] [PubMed]

24. Plascencia-Villa, G.; Ponce, A.; Collingwood, J.F.; Arellano-Jiménez, M.J.; Zhu, X.; Rogers, J.T.; Betancourt, I.; José-Yacamán, M.; Perry, G. High-resolution analytical imaging and electron holography of magnetite particles in amyloid cores of Alzheimer's disease. *Sci. Rep.* **2016**, *6*, 24873. [CrossRef] [PubMed]
25. Maher, B.A. Airborne Magnetite- and Iron-Rich Pollution Nanoparticles: Potential Neurotoxicants and Environmental Risk Factors for Neurodegenerative Disease, Including Alzheimer's Disease. *J. Alzheimer's Dis.* **2019**, *71*, 361–375. [CrossRef]
26. Ramesh, V.; Ravichandran, P.; Copeland, C.L.; Gopikrishnan, R.; Biradar, S.; Goornavar, V.; Ramesh, G.T.; Hall, J.C. Magnetite induces oxidative stress and apoptosis in lung epithelial cells. *Mol. Cell. Biochem.* **2012**, *363*, 225–234. [CrossRef]
27. Pereira, G.M.; Teinilä, K.; Custódio, D.; Gomes Santos, A.; Xian, H.; Hillamo, R.; Alves, C.A.; Bittencourt de Andrade, J.; Olímpio da Rocha, G.; Kumar, P.; et al. Particulate pollutants in the Brazilian city of São Paulo: 1-year investigation for the chemical composition and source apportionment. Atmos. *Chem. Phys.* **2017**, *17*, 11943–11969. [CrossRef]
28. Sahu, S.K.; Beig, G.; Parkhi, N.S. Emissions inventory of anthropogenic PM2.5 and PM10 in Delhi during Commonwealth Games 2010. *Atmos. Environ.* **2011**, *45*, 6180–6190. [CrossRef]
29. Chen, S.; Zhang, X.; Lin, J.; Huang, J.; Zhao, D.; Yuan, T.; Huang, K.; Luo, Y.; Jia, Z.; Zang, Z.; et al. Fugitive Road Dust PM2.5 Emissions and Their Potential Health Impacts. *Environ. Sci. Technol.* **2019**, *53*, 8455–8465. [CrossRef]
30. Chen, R.; Hu, B.; Liu, Y.; Xu, J.; Yang, G.; Xu, D.; Chen, C. Beyond PM2.5: The role of ultrafine particles on adverse health effects of air pollution. *Biochim. Biophys. Acta* **2016**, *1860*, 2844–2855. [CrossRef]
31. DEFRA. *Non-Exhaust Emissions from Road Traffic*; Department for Environment, Food and Rural Affairs: London, UK, 2019; p. 89.
32. Oberdörster, G.; Ferin, J.; Lehnert, B.E. Correlation between particle size, in vivo particle persistence, and lung injury. *Environ. Health Perspect.* **1994**, *102* (Suppl. S5), 173–179.
33. Wang, B.; Yin, J.-J.; Zhou, X.; Kurash, I.; Chai, Z.; Zhao, Y.; Feng, W. Physicochemical Origin for Free Radical Generation of Iron Oxide Nanoparticles in Biomicroenvironment: Catalytic Activities Mediated by Surface Chemical States. *J. Phys. Chem. C* **2013**, *117*, 383–392. [CrossRef]
34. Braakhuis, H.M.; Park, M.V.; Gosens, I.; De Jong, W.H., Cassee, F.R. Physicochemical characteristics of nanomaterials that affect pulmonary inflammation. *Part. Fibre Toxicol.* **2014**, *11*, 18. [CrossRef] [PubMed]
35. Donaldson, K.; Stone, V.; Seaton, A.; MacNee, W. Ambient particle inhalation and the cardiovascular system: Potential mechanisms. *Environ. Health Perspect.* **2001**, *109* (Suppl. S4), 523–527. [PubMed]
36. Oberdörster, G.; Sharp, Z.; Atudorei, V.; Elder, A.; Gelein, R.; Kreyling, W.; Cox, C. Translocation of inhaled ultrafine particles to the brain. *Inhal. Toxicol.* **2004**, *16*, 437–445. [CrossRef]
37. Calderón-Garcidueñas, L.; González-Maciel, A.; Reynoso-Robles, R.; Hammond, J.; Kulesza, R.; Lachmann, I.; Torres-Jardón, R.; Mukherjee, P.S.; Maher, B.A. Quadruple abnormal protein aggregates in brainstem pathology and exogenous metal-rich magnetic nanoparticles. The substantia nigrae is a very early target in young urbanites and the gastrointestinal tract likely a key brainstem portal. *Environ. Res.* **2020**, 110139. [CrossRef] [PubMed]
38. Jin, S.P.; Li, Z.; Choi, E.K.; Lee, S.; Kim, Y.K.; Seo, E.Y.; Chung, J.H.; Cho, S. Urban particulate matter in air pollution penetrates into the barrier-disrupted skin and produces ROS-dependent cutaneous inflammatory response in vivo. *J. Dermatol. Sci.* **2018**, *91*, 175–183. [CrossRef]
39. Hiemstra, P.S.; McCray, P.B., Jr.; Bals, R. The innate immune function of airway epithelial cells in inflammatory lung disease. *Eur. Respir. J.* **2015**, *45*, 1150–1162. [CrossRef]
40. Saffari, A.; Daher, N.; Shafer, M.M.; Schauer, J.J.; Sioutas, C. Global Perspective on the Oxidative Potential of Airborne Particulate Matter: A Synthesis of Research Findings. *Environ. Sci. Technol.* **2014**, *48*, 7576–7583. [CrossRef]
41. Yu, S.-H.; Tang, D.-W.; Hsieh, H.-Y.; Wu, W.-S.; Lin, B.-X.; Chuang, E.-Y.; Sung, H.-W.; Mi, F.-L. Nanoparticle-induced tight-junction opening for the transport of an anti-angiogenic sulfated polysaccharide across Caco-2 cell monolayers. *Acta Biomater.* **2013**, *9*, 7449–7459. [CrossRef]
42. Geiser, M. Update on macrophage clearance of inhaled micro- and nanoparticles. *J. Aerosol Med. Pulm. Drug Deliv.* **2010**, *23*, 207–217. [CrossRef]
43. Snow, S.J.; Henriquez, A.R.; Costa, D.L.; Kodavanti, U.P. Neuroendocrine Regulation of Air Pollution Health Effects: Emerging Insights. *Toxicol. Sci.* **2018**, *164*, 9–20. [CrossRef] [PubMed]
44. Calderón-Garcidueñas, L.; González-Maciel, A.; Mukherjee, P.S.; Reynoso-Robles, R.; Pérez-Guillé, B.; Gayosso-Chávez, C.; Torres-Jardón, R.; Cross, J.V.; Ahmed, I.A.M.; Karloukovski, V.V.; et al. Combustion- and friction-derived magnetic air pollution nanoparticles in human hearts. *Environ. Res.* **2019**, *176*, 108567. [CrossRef] [PubMed]
45. Maher, B.A.; González-Maciel, A.; Reynoso-Robles, R.; Torres-Jardón, R.; Calderón-Garcidueñas, L. Iron-rich air pollution nanoparticles: An unrecognised environmental risk factor for myocardial mitochondrial dysfunction and cardiac oxidative stress. *Environ. Res.* **2020**, *188*, 109816. [CrossRef]
46. Chen, R.; Li, H.; Cai, J.; Wang, C.; Lin, Z.; Liu, C.; Niu, Y.; Zhao, Z.; Li, W.; Kan, H. Fine Particulate Air Pollution and the Expression of microRNAs and Circulating Cytokines Relevant to Inflammation, Coagulation, and Vasoconstriction. *Environ. Health Perspect.* **2018**, *126*, 017007. [CrossRef] [PubMed]
47. Liao, Y.H.; Chen, W.L.; Wang, C.C.; Lai, C.H. Associations between Personal Exposure to Metals in Fine Particulate Matter and Autonomic Nervous System Dysfunction among Healthy Adults. *Aerosol Air Qual. Res.* **2020**, *20*, 1842–1849. [CrossRef]
48. Hammond, J.; Maher, B.A.; Ahmed, I.A.M.; Allsop, D. Variation in the concentration and regional distribution of magnetic nanoparticles in human brains, with and without Alzheimer's disease, from the UK. *Sci. Rep.* **2021**, *11*, 9363. [CrossRef] [PubMed]

49. Tsai, D.-H.; Amyai, N.; Marques-Vidal, P.; Wang, J.-L.; Riediker, M.; Mooser, V.; Paccaud, F.; Waeber, G.; Vollenweider, P.; Bochud, M. Effects of particulate matter on inflammatory markers in the general adult population. *Part. Fibre Toxicol.* **2012**, *9*, 24. [CrossRef] [PubMed]
50. Andreau, K.; Leroux, M.; Bouharrour, A. Health and cellular impacts of air pollutants: From cytoprotection to cytotoxicity. *Biochem. Res. Int.* **2012**, *2012*, 493894. [CrossRef]
51. Puisney, C.; Oikonomou, E.K.; Nowak, S.; Chevillot, A.; Casale, S.; Baeza-Squiban, A.; Berret, J.-F. Brake wear (nano)particle characterization and toxicity on airway epithelial cells in vitro. *Environ. Sci. Nano* **2018**, *5*, 1036–1044. [CrossRef]
52. Abbas, I.; Garçon, G.; Saint-Georges, F.; Billet, S.; Verdin, A.; Gosset, P.; Mulliez, P.; Shirali, P. Occurrence of molecular abnormalities of cell cycle in L132 cells after in vitro short-term exposure to air pollution PM2.5. *Chem.-Biol. Interact.* **2010**, *188*, 558–565. [CrossRef]
53. Alfaro-Moreno, E.; Torres, V.; Miranda, J.; Martínez, L.; García-Cuellar, C.; Nawrot, T.S.; Vanaudenaerde, B.; Hoet, P.; Ramírez-López, P.; Rosas, I.; et al. Induction of IL-6 and inhibition of IL-8 secretion in the human airway cell line Calu-3 by urban particulate matter collected with a modified method of PM sampling. *Environ. Res.* **2009**, *109*, 528–535. [CrossRef] [PubMed]
54. Huang, M.; Kang, Y.; Wang, W.; Chan, C.Y.; Wang, X.; Wong, M.H. Potential cytotoxicity of water-soluble fraction of dust and particulate matters and relation to metal(loid)s based on three human cell lines. *Chemosphere* **2015**, *135*, 61–66. [CrossRef] [PubMed]
55. Sun, J.; Yu, J.; Shen, Z.; Niu, X.; Wang, D.; Wang, X.; Xu, H.; Chuang, H.-C.; Cao, J.; Ho, K.-F. Oxidative stress–inducing effects of various urban PM2.5 road dust on human lung epithelial cells among 10 Chinese megacities. *Ecotoxicol. Environ. Saf.* **2021**, *224*, 112680. [CrossRef]
56. Koh, B.; Kim, E.-A. Comparative analysis of urban road dust compositions in relation to their potential human health impacts. *Environ. Pollut.* **2019**, *255*, 113156. [CrossRef] [PubMed]
57. Park, M.; Joo, H.S.; Lee, K.; Jang, M.; Kim, S.D.; Kim, I.; Borlaza, L.J.S.; Lim, H.; Shin, H.; Chung, K.H.; et al. Differential toxicities of fine particulate matters from various sources. *Sci. Rep.* **2018**, *8*, 17007. [CrossRef]
58. Tung, N.T.; Ho, K.-F.; Niu, X.; Sun, J.; Shen, Z.; Wu, F.; Cao, J.; Dung, H.B.; Thuy, T.P.C.; Hsiao, T.-C.; et al. Loss of E-cadherin due to road dust PM2.5 activates the EGFR in human pharyngeal epithelial cells. *Environ. Sci. Pollut. Res.* **2021**, *28*, 53872–53887. [CrossRef] [PubMed]
59. Yoon, S.; Han, S.; Jeon, K.J.; Kwon, S. Effects of collected road dusts on cell viability, inflammatory response, and oxidative stress in cultured human corneal epithelial cells. *Toxicol. Lett.* **2018**, *284*, 152–160. [CrossRef]
60. Morgan, T.E.; Davis, D.A.; Iwata, N.; Tanner, J.A.; Snyder, D.; Ning, Z.; Kam, W.; Hsu, Y.-T.; Winkler, J.W.; Chen, J.-C.; et al. Glutamatergic neurons in rodent models respond to nanoscale particulate urban air pollutants in vivo and in vitro. *Environ. Health Perspect.* **2011**, *119*, 1003–1009. [CrossRef]
61. Woodward, N.C.; Levine, M.C.; Haghani, A.; Shirmohammadi, F.; Saffari, A.; Sioutas, C.; Morgan, T.E.; Finch, C.E. Toll-like receptor 4 in glial inflammatory responses to air pollution in vitro and in vivo. *J. Neuroinflamm.* **2017**, *14*, 84. [CrossRef]
62. Birmingham City Council. 2019 Air Quality Annual Status Report (ASR). 2019. Available online: http://62.65.40.208/birmingham//Reports/2019_Birmingham_City_Council_ASR.pdf (accessed on 26 August 2021).
63. Lancaster City Council. 2019 Air Quality Annual Status Report (ASR) for Lancaster City Council. 2020. Available online: https://www.lancaster.gov.uk/assets/attach/6104/Air%20Quality%20Annual%20Status%20Report%20-%20Lancaster%202020.pdf (accessed on 26 August 2021).
64. SEDEMA. Mexico City Air Quality Monitoring System Public Database. Available online: http://www.aire.cdmx.gob.mx/default.php (accessed on 26 August 2021).
65. ONS. UK Population Pyramid Interactive. 2019. Available online: https://www.ons.gov.uk/peoplepopulationandcommunity/populationandmigration/populationestimates/articles/ukpopulationpyramidinteractive/2020-01-08 (accessed on 26 August 2021).
66. Birmingham City Council. Birmingham Population. 2018. Available online: https://www.birmingham.gov.uk/info/20057/about_birmingham/1294/population_and_census/2 (accessed on 26 August 2021).
67. Romero, T. Total Population in Mexico City between 2008 and 2018. 2021. Available online: https://www.statista.com/statistics/1038080/mexico-city-total-population/ (accessed on 26 August 2021).
68. Tantra, R.; Tompkins, J.; Quincey, P. Characterisation of the de-agglomeration effects of bovine serum albumin on nanoparticles in aqueous suspension. *Colloids Surf. B* **2010**, *75*, 275–281. [CrossRef]
69. Gorr, M.W.; Youtz, D.J.; Eichenseer, C.M.; Smith, K.E.; Nelin, T.D.; Cormet-Boyaka, E.; Wold, L.E. In vitro particulate matter exposure causes direct and lung-mediated indirect effects on cardiomyocyte function. *Am. J. Physiol.-Heart Circ. Physiol.* **2015**, *309*, H53–H62. [CrossRef] [PubMed]
70. He, R.-W.; Gerlofs-Nijland, M.E.; Boere, J.; Fokkens, P.; Leseman, D.; Janssen, N.A.H.; Cassee, F.R. Comparative toxicity of ultrafine particles around a major airport in human bronchial epithelial (Calu-3) cell model at the air–liquid interface. *Toxicol. In Vitro* **2020**, *68*, 104950. [CrossRef] [PubMed]
71. Cooney, D.J.; Hickey, A.J. Cellular response to the deposition of diesel exhaust particle aerosols onto human lung cells grown at the air–liquid interface by inertial impaction. *Toxicol. In Vitro* **2011**, *25*, 1953–1965. [CrossRef] [PubMed]
72. Vietti, G.; Ibouraadaten, S.; Palmai-Pallag, M.; Yakoub, Y.; Bailly, C.; Fenoglio, I.; Marbaix, E.; Lison, D.; van den Brule, S. Towards predicting the lung fibrogenic activity of nanomaterials: Experimental validation of an in vitro fibroblast proliferation assay. *Part. Fibre Toxicol.* **2013**, *10*, 52. [CrossRef]

73. Kocbach, A.; Totlandsdal, A.I.; Låg, M.; Refsnes, M.; Schwarze, P.E. Differential binding of cytokines to environmentally relevant particles: A possible source for misinterpretation of in vitro results? *Toxicol. Lett.* **2008**, *176*, 131–137. [CrossRef]
74. Özdemir, Ö.; Dunlop, D.J. Hallmarks of maghemitization in low-temperature remanence cycling of partially oxidized magnetite nanoparticles. *J. Geophys. Res. Solid Earth* **2010**, 115. [CrossRef]
75. Verwey, E.J.W. Electronic Conduction of Magnetite (Fe3O4) and its Transition Point at Low Temperatures. *Nature* **1939**, *144*, 327–328. [CrossRef]
76. Rousseau, M.-C.; Straif, K.; Siemiatycki, J. IARC carcinogen update. *Environ. Health Perspect.* **2005**, *113*, A580–A581. [CrossRef]
77. Botsou, F.; Moutafis, I.; Dalaina, S.; Kelepertzis, E. Settled bus dust as a proxy of traffic-related emissions and health implications of exposures to potentially harmful elements. *Atmos. Pollut. Res.* **2020**, *11*, 1776–1784. [CrossRef]
78. Churg, A.; Brauer, M.; del Carmen Avila-Casado, M.; Fortoul, T.I.; Wright, J.L. Chronic exposure to high levels of particulate air pollution and small airway remodeling. *Environ. Health Perspect.* **2003**, *111*, 714–718. [CrossRef]
79. Dumková, J.; Smutná, T.; Vrlíková, L.; Le Coustumer, P.; Večeřa, Z.; Dočekal, B.; Mikuška, P.; Čapka, L.; Fictum, P.; Hampl, A.; et al. Sub-chronic inhalation of lead oxide nanoparticles revealed their broad distribution and tissue-specific subcellular localization in target organs. *Part. Fibre Toxicol.* **2017**, *14*, 55. [CrossRef] [PubMed]
80. Shimizu, K.; Horie, M.; Tabei, Y.; Kashiwada, S. Proinflammatory response caused by lead nanoparticles triggered by engulfed nanoparticles. *Environ. Toxicol.* **2021**, *36*, 2040–2050. [CrossRef] [PubMed]
81. Mazuryk, O.; Stochel, G.; Brindell, M. Variations in Reactive Oxygen Species Generation by Urban Airborne Particulate Matter in Lung Epithelial Cells—Impact of Inorganic Fraction. *Front. Chem.* **2020**, *8*, 581752. [CrossRef] [PubMed]
82. Vanden Berghe, T.; Kalai, M.; Denecker, G.; Meeus, A.; Saelens, X.; Vandenabeele, P. Necrosis is associated with IL-6 production but apoptosis is not. *Cell. Signal.* **2006**, *18*, 328–335. [CrossRef]
83. de Caritat, P.; Reimann, C. Comparing results from two continental geochemical surveys to world soil composition and deriving Predicted Empirical Global Soil (PEGS2) reference values. *Earth Planet. Sci. Lett.* **2012**, *319–320*, 269–276. [CrossRef]
84. Wang, G.; Zhang, G.; Gao, X.; Zhang, Y.; Fan, W.; Jiang, J.; An, Z.; Li, J.; Song, J.; Wu, W. Oxidative stress-mediated epidermal growth factor receptor activation regulates PM(2.5)-induced over-secretion of pro-inflammatory mediators from human bronchial epithelial cells. *Biochim. Biophys. Acta Gen. Subj.* **2020**, *1864*, 129672. [CrossRef]
85. Shukla, A.; Timblin, C.; BeruBe, K.; Gordon, T.; McKinney, W.; Driscoll, K.; Vacek, P.; Mossman, B.T. Inhaled particulate matter causes expression of nuclear factor (NF)-kappaB-related genes and oxidant-dependent NF-kappaB activation in vitro. *Am. J. Respir. Cell Mol. Biol.* **2000**, *23*, 182–187. [CrossRef]
86. Veranth, J.M.; Moss, T.A.; Chow, J.C.; Labban, R.; Nichols, W.K.; Walton, J.C.; Watson, J.G.; Yost, G.S. Correlation of in vitro cytokine responses with the chemical composition of soil-derived particulate matter. *Environ. Health Perspect.* **2006**, *114*, 341–349. [CrossRef]
87. Thorn, J. The inflammatory response in humans after inhalation of bacterial endotoxin: A review. *Inflamm. Res.* **2001**, *50*, 254–261. [CrossRef]
88. Michael, S.; Montag, M.; Dott, W. Pro-inflammatory effects and oxidative stress in lung macrophages and epithelial cells induced by ambient particulate matter. *Environ. Pollut.* **2013**, *183*, 19–29. [CrossRef]
89. Yue, Y.; Chen, H.; Setyan, A.; Elser, M.; Dietrich, M.; Li, J.; Zhang, T.; Zhang, X.; Zheng, Y.; Wang, J.; et al. Size-Resolved Endotoxin and Oxidative Potential of Ambient Particles in Beijing and Zürich. *Environ. Sci. Technol.* **2018**, *52*, 6816–6824. [CrossRef] [PubMed]
90. Gonet, T.; Maher, B.A.; Nyirő-Kósa, I.; Pósfai, M.; Vaculík, M.; Kukutschová, J. Size-resolved, quantitative evaluation of the magnetic mineralogy of airborne brake-wear particulate emissions. *Environ. Pollut.* **2021**, *288*, 117808. [CrossRef] [PubMed]
91. Gustafsson, M.; Blomqvist, G.; Järlskog, I.; Lundberg, J.; Janhäll, S.; Elmgren, M.; Johansson, C.; Norman, M.; Silvergren, S. Road dust load dynamics and influencing factors for six winter seasons in Stockholm, Sweden. *Atmos. Environ. X* **2019**, *2*, 100014. [CrossRef]
92. Garcia, G.J.M.; Schroeter, J.D.; Kimbell, J.S. Olfactory deposition of inhaled nanoparticles in humans. *Inhal. Toxicol* **2015**, *27*, 394–403. [CrossRef]
93. Liu, N.M.; Miyashita, L.; Maher, B.A.; McPhail, G.; Jones, C.J.P.; Barratt, B.; Thangaratinam, S.; Karloukovski, V.; Ahmed, I.A.; Aslam, Z.; et al. Evidence for the presence of air pollution nanoparticles in placental tissue cells. *Sci. Total Environ.* **2021**, *751*, 142235. [CrossRef] [PubMed]
94. Salo, L.; Hyvärinen, A.; Jalava, P.; Teinilä, K.; Hooda, R.K.; Datta, A.; Saarikoski, S.; Lintusaari, H.; Lepistö, T.; Martikainen, S.; et al. The characteristics and size of lung-depositing particles vary significantly between high and low pollution traffic environments. *Atmos. Environ.* **2021**, *255*, 118421. [CrossRef]

antioxidants

Article

Codium fragile Suppressed Chronic $PM_{2.5}$-Exposed Pulmonary Dysfunction via TLR/TGF-β Pathway in BALB/c Mice

Tae Yoon Kim [1,†], Jong Min Kim [1,†], Hyo Lim Lee [1], Min Ji Go [1], Seung Gyum Joo [1], Ju Hui Kim [1], Han Su Lee [1], Won Min Jeong [2], Dong Yeol Lee [2], Hyun-Jin Kim [1] and Ho Jin Heo [1,*]

1 Division of Applied Life Science (BK21), Institute of Agriculture and Life Science, Gyeonsang National University, Jinju 52828, Republic of Korea; kty8747@naver.com (T.Y.K.); myrock201@gnu.ac.kr (J.M.K.); gyfla059@gnu.ac.kr (H.L.L.); alswl9245@gnu.ac.kr (M.J.G.); s716g@naver.com (S.G.J.); zkfkapflove@nate.com (J.H.K.); ns3005@naver.com (H.S.L.); hyunjkim@gnu.ac.kr (H.-J.K.)

2 Research & Development Team, Gyeongnam Anti-Aging Research Institute, Sancheong 52215, Republic of Korea; jwm5618@gari.or.kr (W.M.J.); dylee1984@gari.or.kr (D.Y.L.)

* Correspondence: hjher@gnu.ac.kr; Tel.: +82-55772190

† These authors contributed equally to this work.

Abstract: This study investigated the ameliorating effect of the aqueous extract of *Codium fragile* on $PM_{2.5}$-induced pulmonary dysfunction. The major compounds of *Codium fragile* were identified as palmitic acid, stearic acid, and oleamide using GC/MS^2 and hexadecanamide, oleamide, and 13-docosenamide using UPLC-Q-TOF/MS^E. *Codium fragile* improved pulmonary antioxidant system deficit by regulating SOD activities and reducing GSH levels and MDA contents. It suppressed pulmonary mitochondrial dysfunction by regulating ROS contents and mitochondrial membrane potential levels. It regulated the inflammatory protein levels of TLR4, MyD88, p-JNK, p-NF-κB, iNOS, Caspase-1, TNF-α, and IL-1β. In addition, it improved the apoptotic protein expression of BCl-2, BAX, and Caspase-3 and attenuated the fibrous protein expression of TGF-β1, p-Smad-2, p-Smad-3, MMP-1, and MMP-2. In conclusion, this study suggests that *Codium fragile* might be a potential material for functional food or pharmaceuticals to improve lung damage by regulating oxidative stress inflammation, cytotoxicity, and fibrosis via the TLR/TGF-β1 signaling pathway.

Keywords: *Codium fragile*; particulate matter; pulmonary inflammation fibrosis; toll-like receptor

Citation: Kim, T.Y.; Kim, J.M.; Lee, H.L.; Go, M.J.; Joo, S.G.; Kim, J.H.; Lee, H.S.; Jeong, W.M.; Lee, D.Y.; Kim, H.-J.; et al. *Codium fragile* Suppressed Chronic $PM_{2.5}$-Exposed Pulmonary Dysfunction via TLR/TGF-β Pathway in BALB/c Mice. *Antioxidants* **2023**, *12*, 1743. https://doi.org/10.3390/antiox12091743

Academic Editors: Aron B. Fisher and Yasuhiro Yoshida

Received: 2 August 2023
Revised: 4 September 2023
Accepted: 8 September 2023
Published: 10 September 2023

1. Introduction

Fine dust contains several substances, including organic carbon, carbonaceous aerosols, metals and metalloids (MMs), and inorganic ions, that can negatively affect respiratory health [1]. In particular, fine dust particles, a major component of particulate matter (PM), are inhaled into the respiratory tract and interact with lung tissue [2]. These interactions promote the production of transforming growth factor-β1 (TGF-β1), a tissue growth factor, causing inflammatory responses and fibrotic processes [3]. TGF-β1 secreted from pulmonary cells affects surrounding tissues, and these effects are related to inflammation and fibrosis caused by PM [4]. When exposure to fine dust occurs, fibroblasts in lung tissues abnormally increase, and excessive accumulation of fibrous proteins such as collagen causes excessive fibrosis of tissues [5]. Lung tissue damage caused by fine dust continuously activates TGF-β1, which promotes fibroblast differentiation and collagen synthesis, thereby promoting fibrosis [6]. When $PM_{2.5}$, which classified as a size smaller than 2.5 μm, particles are inhaled into the respiratory tract, certain cells in lung tissue, such as lung epithelial and immune cells, interact with the $PM_{2.5}$ particles [5]. $PM_{2.5}$ particles adsorb various elements on their surface, and these elements activate Toll-like receptors (TLR), which play a major role in the host's immune system [1]. Activated TLRs initiate signal transduction pathways inside cells, thereby producing various inflammatory response-inducing factors [7]. Activation of the TLR pathway in this process causes continuous secretion of TGF-β1, and the interaction between $PM_{2.5}$ and TLRs leads to pulmonary fibrosis and inflammatory responses such as nuclear factor kappa-light-chain-enhancer of the activated B cell (NF-κB) pathway [8]. This inflammatory response induces oxidative stress and the production of proinflammatory mediators, leading to various diseases such as asthma, chronic obstructive pulmonary disease (COPD), and cancer [6]. To reduce damage to lung tissue from $PM_{2.5}$, it is most important to avoid direct exposure [4]. However, since PM inhalation cannot be completely inhibited, it is important to reduce the toxicity of PM in advance with the consumption of or treatment with natural materials [9].

Codium fragile, a marine green alga belonging to the family *Codiaceae*, is cultivated and used for food in various countries such as the Republic of Korea, China, and Japan, and in North America [10]. *Codium fragile* contains various bioactive phenolic compounds, such as salicylic acid, p-coumaric acid, tamarixetin, morin, naringenin, kaempferol, and quercetin, and polysaccharides such as sulfated galactan [11]. It also has been reported that *Codium fragile* shows bioactive activities such as antioxidant activity, anti-diabetes, osteoarthritis inhibitory effect, hepato-protective effect, and anti-inflammatory effect [12,13]. However, there are few studies related to the protective effect of *Codium fragile* on $PM_{2.5}$-induced lung disease. In a previous study, the aqueous extract of *Codium fragile* significantly protected against $PM_{2.5}$-induced cytotoxicity in pulmonary A549 cells and nasal RPMI2650 cells [14]. Thus, this study was conducted to estimate the ameliorating effect of the aqueous extract of *Codium fragile* against $PM_{2.5}$-induced pulmonary damage.

2. Materials and Methods

2.1. Sample Preparation

Codium fragile used in this study was obtained from Yeosu (Republic of Korea) in February 2018. The samples were washed to remove salt and impurities until the salt concentration reached 0%. The desalted samples were lyophilized using a vacuum drier (Operon, Gimpo, Republic of Korea) and extracted with 50-fold distilled water at 40 °C. The extracted samples were filtered with a No. 2 filter (Whatman Inc., Kent, UK) and concentrated using an evaporator with a vacuum (N-N series, Eyela Co., Tokyo, Japan). The re-lyophilized samples were stored at −20 °C until used.

2.2. Gas Chromatography–Tandem Mass Spectrometry (GC/MS^2)

The dried aqueous extract of *Codium fragile* was extracted with 80% methanol using a bullet blender (Next Advance Inc., Averill Park, NY, USA) and sonicated for 20 min. The extracted sample was reacted with methoxyamin at 37 °C for 90 min, and the mixture was incubated with BSTFA 70 °C for 30 min. The mixture was centrifuged at 14,000× *g* for 10 min, and the supernatant filtered by a 0.45 µm membrane filter was used for identification. The contents of physiological compounds were identified using GC/MS^2 (GC, Agilent 7890A; MS^2, Agilent 5975C, Agilent, Santa Clara, CA, USA) on a capillary column (30 m × 250 µm, 0.25 µm, DB-5MS DB-5MS Inert Column, Agilent). Initially, samples (1.0 µL) were injected in splitless mode (50:1). The injection temperature was preserved at 260 °C, and the protocol was conducted with helium gas (1.0 mL/min). Using a temperature control system, the initial column oven temperature was set at 40 °C for 5 min, then 10 °C/min to 120 °C for 8 min, then 60 °C/min to 300 °C for 3 min, and then 300 °C for 0.5 min (total run time: 16.5 min). The MS conditions used were as follows: acquisition mode, scan (40 to 250 m/z); quadrupole temperature, 150 °C. The GC/MS^2 system was analyzed using library software (NIST MS search 2.2, National Institute of Standards and Technology, Gaithersburg, MD, USA).

2.3. Ultra-Performance Liquid Chromatography–Quadrupole Time-of-Flight Mass Spectrometry (UPLC-Q-TOF/MS^E)

The extracted sample was fractionated using n-hexane, chloroform, and ethyl acetate to remove the impurities, and a fraction from distilled water of *Codium fragile*, which contained the highest content of oleamide, was used for identification. The dried sample of fraction from distilled water dissolved in methanol filtered through a 0.45 μm membrane filter and analyzed using UPLC-Q-TOF/MS^E (Vion, Waters Corp., Milford, MA, USA) with a BEH C_{18} column (100 × 2.1 mm, 1.7 μm; Waters Corp.). The mobile phases consisted of solvent A (0.1% formic acid in distilled water) and solvent B (0.1% formic acid in acetonitrile) during analysis, and gradient conditions were as follows: 1% B at 0–1 min; 1–100% B at 1–8 min; 100% B at 8–9 min; 100–1% B at 9–9.5 min; and 1% B at 9.5–12 min.

2.4. High-Performance Liquid Chromatography (HPLC)

The dried sample dissolved in methanol was filtered through a 0.45 μm membrane filter and analyzed using HPLC (Ultimate 3000 series, Dionex, Sunnyvale, CA, USA). HPLC separation of oleamide was conducted using a C_{18} column (250 × 4.6 mm, 5.0 μm, YMC-Triart, YMC, Kyoto, Japan) with a flow rate of 1.0 mL/min. The mobile phases consisted of solvent A (distilled water) and solvent B (acetonitrile), and the analysis conditions were as follows: a gradient elution of 50% A and 50% B at 0–0.1 min; 50–0% A and 50–100% B at 0.1–20 min; and 0% A and 100% B at 20–30 min. The injected volume was 20 μL, and the wavelength of the UV detector was analyzed using a diode array detector at 203 nm at 40 °C to measure the oleamide. The detected wavelength was compared to a standard compound to determine similarity.

2.5. Animal Experimental Design

BALB/c mice (6 week olds, male) were obtained from Samtako (Osan, Republic of Korea). The animals were divided into 3 or 4 per cage and controlled in standard laboratory conditions (12 h light/dark cycle; 55% humidity; 22 ± 2 °C). Experimental groups were divided into 6 groups (n = 10; 5 for ex vivo tests; 5 for mitochondrial tests) as a sham control (sham) group (chamber exposure-, sample intake-), normal control (NC) group (a clean air-exposure+, sample intake-), normal sample (NS) group (a clean air-exposure+, sample intake+; 40 mg/kg of body weight), PM group ($PM_{2.5}$ air-exposure+, sample intake-), and the *Codium fragile* groups ($PM_{2.5}$ air-exposure+, sample intake+; 50 and 100 mg/kg of body weight; CF50 and CF100, respectively). The *Codium fragile* was dissolved in pure water and orally fed using a stomach tube once a day for 12 weeks. $PM_{2.5}$ (mean diameter: 1.06 μm) was obtained from Power Technology INC. (Arizona Test Dust, Arden Hills, MN, USA). The animal was exposed to $PM_{2.5}$ (500 μg/m^3) in the chamber for 5 h/day for 12 weeks. All procedures were conducted in accordance with the Institutional Animal Care and Use Committee of Gyeongsang National University (certificate: GNU-210803-M0069, approved on 3 August 2021) and the Policy of the Ethical Committee of Ministry of Health and Welfare, Republic of Korea. The experimental design was presented as follows (Figure 1).

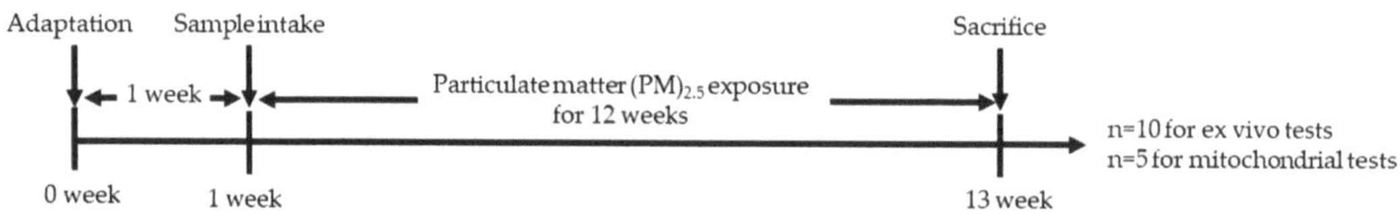

Figure 1. Experimental design of the particulate matter (PM)$_{2.5}$ exposure model and ex vivo tests for $PM_{2.5}$-induced mice.

2.6. Antioxidant System Activity

2.6.1. Preparation of Lung Tissues

After all mice were fasted for 12 h, experimental animals were sacrificed by CO_2 inhalation. Lung tissues were immediately isolated for ex vivo tests, and tissues were homogenized with 10 times the volume of phosphate-buffered saline (PBS, pH 7.4) or phosphate buffer (pH 7.8). The protein concentration of the obtained sample was evaluated according to the previous study [15].

2.6.2. Superoxide Dismutase (SOD) Activity

To assess the SOD activities, the lung tissues homogenized in PBS buffer were centrifuged at 400× *g* for 10 min at 4 °C, and the obtained pellet was extracted using 1 × cell extraction buffer with 20% (*v*/*v*) Triton X-100 and 200 mM phenylmethylsulfonyl fluoride. The mixtures incubated for 30 min on ice were centrifuged at 10,000× *g* for 10 min at 4 °C. The obtained supernatants were used for SOD activities using a SOD commercial kit (Sigma-Aldrich Chemical Co., St. Louis, MO, USA).

2.6.3. Reduced Glutathione (GSH) Contents

To assess the reduced GSH contents, the lung tissues homogenized in phosphate buffer (pH 7.8) were mixed with 200 μM phosphate buffer (pH 6–7) and centrifuged at 10,000× *g* for 15 min at 4 °C. The protein concentration of the supernatant was quantified according to the Bradford method [15]. Then, the supernatants and 5% metaphosphoric acid were mixed and centrifuged at 2000× *g* for 2 min at 4 °C. The supernatant was incubated with 0.26 M Tris-HCl (pH 7.5), 0.65 N NaOH, and 1 mg/mL o-phthaldialdehyde. The fluorescence of reactants was measured at 360 nm (excitation wavelength) and 430 nm (emission wavelength) using a fluorometer (Infinite F200, Tecan, Mannedorf, Switzerland) [16].

2.6.4. Malondialdehyde (MDA) Contents

To assess the MDA contents, the lung tissues homogenized in PBS buffer were centrifuged at 2350× *g* for 10 min at 4 °C. The supernatants were mixed with 1% phosphoric acid and 0.67% thiobarbituric acid, and these mixtures were incubated at 95 °C for 1 h. After that, the MDA contents were measured at 532 nm using a microplate reader (Epoch2, BioTek, Winooski, VT, USA) [16].

2.7. Mitochondrial Function Activity

2.7.1. Isolation of Mitochondria

Lung tissues were homogenized in mitochondrial isolation buffer (215 mM mannitol, 75 mM sucrose, 0.1% bovine serum albumin, and 20 mM HEPES sodium salt (pH 7.2) and 1 mM EGTA). The homogenized tissues were centrifuged at 1300× *g* for 5 min at 4 °C to obtain the supernatant. The supernatants were re-centrifuged at 13,000× *g* for 10 min at 4 °C, and the pellets were mixed with 0.1% digitonin and mitochondrial isolation buffer containing 1 mM EGTA for 5 min. The mixtures were centrifuged at 13,000× *g*, for 15 min at 4 °C. The re-obtained pellets mixed with mitochondrial isolation buffer were used for the evaluation of mitochondrial functions.

2.7.2. Mitochondrial Reactive Oxygen Species (ROS) Content

To assess the mitochondrial ROS levels, mitochondrial extracts were reacted to a respiration buffer containing 125 mM KCl, 2 mM KH_2SO_4, 2.5 mM malate, 20 mM HEPES, 1 mM $MgCl_2$, 5 mM pyruvate, 500 μM EGTA, and 25 μM DCF-DA. These reactants were incubated in a dark room for 20 min, and fluorescence was measured at 485 nm (excitation wavelength) and 535 nm (emission wavelength) [17].

2.7.3. Mitochondrial Membrane Potential (MMP) Levels

To assess the MMP level, mitochondrial extracts were reacted to a mitochondrial isolation buffer with 5 mM pyruvate, 5 mM malate, and 1 μM JC-1. The reactants were incubated in a dark room for 20 min, and fluorescence was measured at 530 nm (excitation wavelength) and 590 nm (emission wavelength) [17].

2.8. Western Blot

Lung tissues were homogenized using ProtinEx™ Animal cell/tissue (Gene All Biotechnology, Seoul, Republic of Korea) with 1% protease inhibitor and phosphatase inhibitor. The homogenized tissues were separated by sodium dodecyl sulfate–polyacrylamide gel electrophoresis (SDS-PAGE) and transferred to a PVDF membrane (Millipore, Billerica,

MA, USA). The transferred membranes were treated with 5% skim milk for 1 h and washed using 1 × tris-buffer saline with 0.1% Tween® 20 (TBST) buffer 3 times for 10 min. The blocked membranes were incubated in a primary antibody solution for 12 h at 4 °C. The incubated membranes were reacted with a secondary antibody (1:2500) for 1 h. The chemiluminescence of each protein was detected using an image analyzer (iBright™ CL1500 instrument, Invitrogen, Carlsbad, CA, USA). Antibody information is presented in Table 1.

Table 1. List of antibody information.

Antibody	Catalog	Concentration	Manufacturer
TLR4	sc-52962	1:1000	Santa Cruz Biotech (Dallas, TX, USA)
MyD88	sc-74532	1:1000	Santa Cruz Biotech (Dallas, TX, USA)
p-JNK	sc-6254	1:1000	Santa Cruz Biotech (Dallas, TX, USA)
p-NF-κB	3033	1:1000	Cell Signaling Tech (Danvers, MA, USA)
iNOS	sc-7271	1:1000	Santa Cruz Biotech (Dallas, TX, USA)
Caspase-1	sc-392736	1:1000	Santa Cruz Biotech (Dallas, TX, USA)
TNF-α	sc-393887	1:1000	Santa Cruz Biotech (Dallas, TX, USA)
IL-1β	sc-4592	1:1000	Santa Cruz Biotech (Dallas, TX, USA)
BCl-2	sc-509	1:1000	Santa Cruz Biotech (Dallas, TX, USA)
BAX	sc-7480	1:1000	Santa Cruz Biotech (Dallas, TX, USA)
Caspase-3	CSB-PA05689A0Rb	1:1000	Cusabio (Hubei, China)
TFG-β1	sc-130348	1:1000	Santa Cruz Biotech (Dallas, TX, USA)
p-Smad-2	3108	1:1000	Cell Signaling Tech (Danvers, MA, USA)
p-Smad-3	sc-517575	1:1000	Santa Cruz Biotech (Dallas, TX, USA)
MMP-1	sc-21731	1:1000	Santa Cruz Biotech (Dallas, TX, USA)
MMP-2	sc-13595	1:1000	Santa Cruz Biotech (Dallas, TX, USA)
β-actin	66009-1-Ig	1:1000	Proteintech (Rosemont, IL, USA)

2.9. *Statistical Analysis*

All experimental results were expressed as mean ± standard deviation (SD) and assessed by one-way analysis of variance (ANOVA) to analyze significant differences. Each set of data was evaluated using Duncan's new multiple range test ($p < 0.05$) with the statistical program (SAS version 9.4, SAS Institute Inc., Cary, NC, USA). Significant differences were statistically presented as different small letters.

3. Results

3.1. *Physiological Compounds in Codium fragile*

The compounds of the aqueous extract of *Codium fragile* were qualitatively identified using GC/MS^2 (Figure 2a and Table 2) and UPLC-Q-TOF/MS^E (Figure 2b and Table 3) and quantitatively identified using HPLC (Figure 2c). The MS^2 spectra of GC/MS^2 were analyzed as compound 1: 328 m/z (retention time (RT): 43.92 min); compound 2: 356 m/z (RT: 48.58 min); and compound 3: 330 m/z (RT: 51.82 min). These compounds were tentatively identified as palmitic acid (PubChem CID:985, compound 1), stearic acid (PubChem CID:5281, compound 2), and oleamide (PubChem CID: 5283387, compound 3). The MS^2 spectra of UPLC-Q-TOF/MS^E were analyzed as compound 1: 256 m/z (RT: 8.35 min); compound 2: 282 m/z (RT: 8.44 min); and compound 3: 338 m/z (RT: 8.73 min). These detected compounds were tentatively identified as hexadecanamide (PubChem CID: 69,421, compound 1), oleamide (PubChem CID: 5283387, compound 2), and 13-docosenamide (PubChem CID: 5,365,371, compound 3). In the result of HPLC, oleamide contents were 7.74 mg/g dried weight compared to the retention time and UV-VIS spectrum of a standard material.

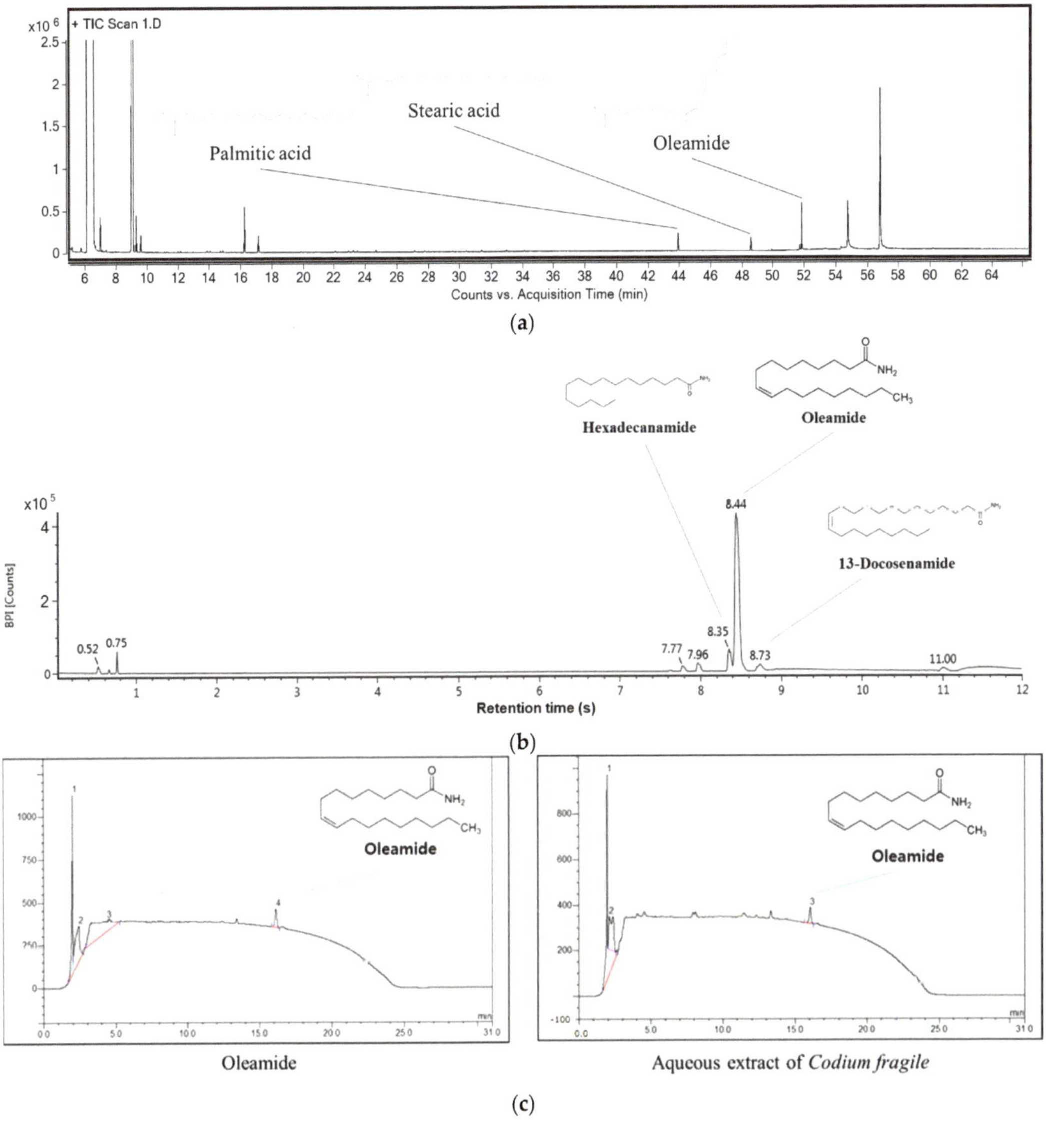

Figure 2. UPLC Q–TOF/MS^E chromatography in negative ion mode of *Codium fragile*. (**a**) MS^2 spectra of gas chromatography (GC)/MS^2; (**b**) Ultra-Performance Liquid Chromatography–Quadrupole Time-of-Flight Mass Spectrometry (UPLC-Q-TOF/MS^E); (**c**) High-Performance Liquid Chromatography (HPLC).

Table 2. Identification of main compounds of the aqueous extract of *Codium fragile* using GC/MS[2] chromatography.

No.	RT (min) [1]	Parent Ion	Fragment (*m/z*)	Compound
1	43.92	328	313, 269, 201, 117, 73, 43	Palmitic acid
2	48.58	356	341, 309, 241, 201, 117	Stearic acid
3	51.82	330	282, 249, 167, 149, 122	Oleamide

[1] RT: retention time.

Table 3. Identification of main compounds of the aqueous extract of *Codium fragile* using UPLC-Q-TOF/MS[E] chromatography.

No.	RT (min) [1]	*m/z* [M + H]+	Fragment (*m/z*)	Compound
1	8.35	256	80, 88, 184, 201	Hexadecanamide
2	8.44	282	135, 149, 247, 265	Oleamide
3	8.73	338	80, 106, 309	13-docosenamide

[1] RT: retention time.

3.2. Effect of *Codium fragile* on Antioxidant System Biomarkers

As a result of analyzing SOD activities, there were no significant differences between the sham group (7.39 unit/mg of protein), NC group (8.44 unit/mg of protein), and NS group (7.65 unit/mg of protein) (Figure 3a). The SOD activities of the PM group (4.71 unit/mg of protein) significantly decreased compared to the NC groups. However, the CF50 and CF100 groups (6.40 unit/mg of protein and 7.31 unit/mg of protein, respectively) were significantly ameliorated compared to the PM group. As a result of analyzing reduced GSH levels, there were no significant differences between the sham group (119.04% of control), the NC group (100% of control), and the NS group (97.83% of control) (Figure 3b). The GSH contents of the PM group (49.59% of control) were significantly decreased compared to the NC group. However, the CF50 and CF100 groups (82.24% of control and 115.99% of control, respectively) were significantly ameliorated compared to the PM group. As a result of analyzing MDA contents, there were no significant differences between the sham group (55.70 pmole/mg of protein), the NC group (57.30 pmole/mg of protein), and the NS group (60.90 pmole/mg of protein) (Figure 3c). The MDA levels of the PM group (84.60 pmole/mg of protein) were significantly increased compared to the NC group. However, the CF50 and CF100 groups (68.30 pmole/mg of protein and 68.50 pmole/mg of protein, respectively) were significantly ameliorated compared to the PM group.

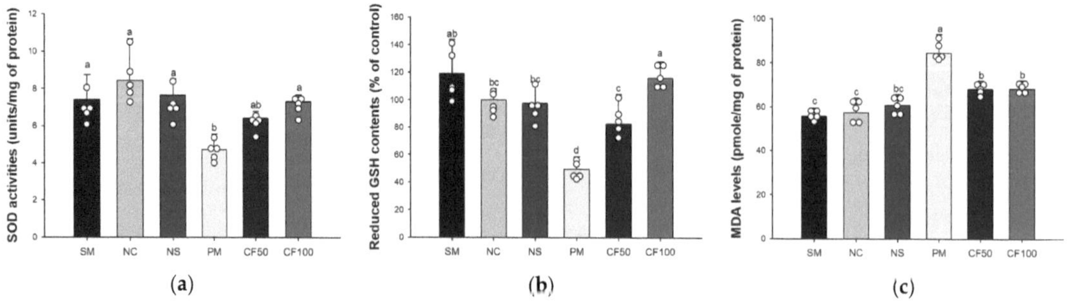

Figure 3. Protective effect of the aqueous extract of *Codium fragile* on $PM_{2.5}$-induced biochemical changes related to antioxidant system. (**a**) Superoxide dismutase (SOD) contents; (**b**) reduced glutathione (GSH) level; (**c**) malondialdehyde (MDA) contents. The results shown are mean ± SD (n = 5). Data were statistically considered at $p < 0.05$, and different small letters represent statistical differences.

3.3. *Effect of Codium fragile on Mitochondrial Activity*

As a result of analyzing mitochondrial ROS levels, there were no significant differences between the sham group (104.31% of control), the NC group (100% of control), and the NS group (87.06% of control). The ROS levels of the PM group (179.70% of control) were significantly increased compared to the NC group (Figure 4a). However, the CF50 and CF100 groups (87.19% of control and 90.52% of control, respectively) were significantly ameliorated compared to the PM group. As a result of analyzing MMP levels, there were no significant differences between the Sham group (103.20% of control), the NC group (100% of control), and the NS group (96.96% of control). The MMP levels of the PM group (71.70%) were significantly decreased compared to the NC group (Figure 4b). However, the CF50 and CF100 groups (120.35% and 85.40%, respectively) were significantly ameliorated compared to the PM group.

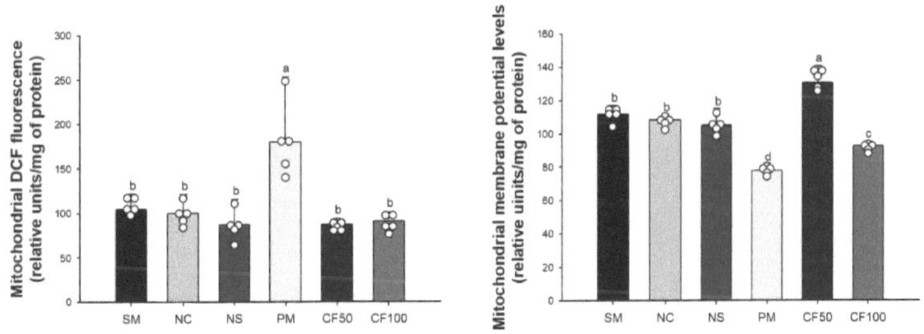

(a) (b)

Figure 4. Protective effect of the aqueous extract of *Codium fragile* on $PM_{2.5}$-induced mitochondrial dysfunction. (**a**) Reactive oxygen species (ROS) contents; (**b**) mitochondrial membrane potential (MMP) levels. The results shown are mean ± SD (n = 5). Data were statistically considered at $p < 0.05$, and different small letters represent statistical differences by one-way analysis of variance (ANOVA) and Duncan's new multiple range test.

3.4. *Effect of Codium fragile on $PM_{2.5}$-Induced Pulmonary Inflammatory-Related Factors*

The expression levels of TLR4 (152.14%), myeloid differentiation primary response 88 (MyD88) (266.74%), phosphorylated c-Jun N-terminal kinase (p-JNK) (158.73%), p-NF-κB (189.59%), inducible nitric oxide synthase (iNOS) (145.44%), Caspase-1 (215.34%), tumor necrosis factor-α (TNF-α) (307.61%), and interleukin-1β (IL-1β) (230.52%) of the PM group were significantly increased compared to NC group (Figure 5). However, the CF100 group (81.19%, 109.14%, 102.43%, 100.17%, 96.45%, 121.76%, 126.83%, and 107.73%, respectively) was ameliorated compared to the NC group.

3.5. *Effect of Codium fragile on $PM_{2.5}$-Induced Pulmonary Apoptosis-Related Factors*

The expression level of B-cell lymphoma 2 (BCl-2) (46.53%) of the PM group was significantly decreased compared to the NC group (Figure 6). However, the CF100 group (82.50%) was ameliorated compared to the NC group. The expression levels of BCl-2 associated X (BAX) (167.78%), BAX/BCl-2 ratio (276.85%), and Caspase-3 (250.18%) of the PM group were significantly increased compared to the NC group. However, the CF100 group (87.52%, 87.44%, and 82.81%, respectively) was ameliorated compared to the NC group.

3.6. *Effect of Codium fragile on $PM_{2.5}$-Induced Pulmonary Fibrosis-Related Factors*

The expression levels of TGF-β1 (151.86%), phosphorylated small mothers against decapentaplegic (p-Smad)-2 (187.56%), p-Smad-3 (268.35%), matrix metalloproteinase-1 (MMP-1) (151.59%), and matrix metalloproteinase-2 (MMP-2) (254.89%) of the PM group were significantly increased compared to the NC group (Figure 7). However, the CF100 group (90.44%, 75.66%, 89.33%, 74.45%, and 81.74%, respectively) was ameliorated compared to the NC group.

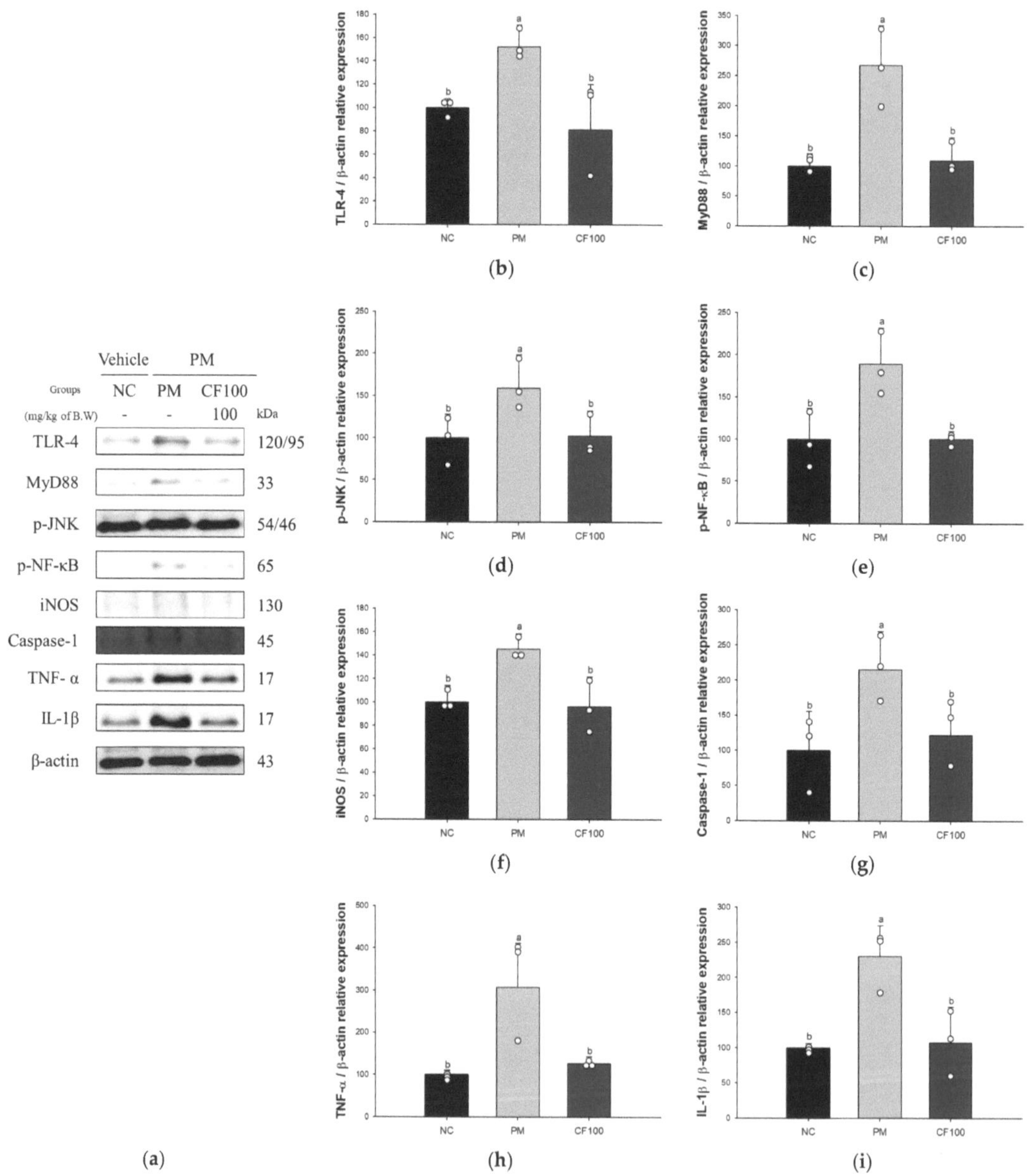

Figure 5. Regulatory effect of the aqueous extract of *Codium fragile* on $PM_{2.5}$-induced inflammatory related protein expression in Western blot. (**a**) Western blot band image; protein expression levels of (**b**) Toll-like receptors 4 (TLR4); (**c**) myeloid differentiation primary response 88 (MyD88); (**d**) phosphorylated c-Jun N-terminal kinase (p-JNK); (**e**) phosphorylated nuclear factor kappa-light-chain-enhancer of the activated B cell (p-NF-κB); (**f**) inducible nitric oxide synthase (iNOS); (**g**) Caspase 1; (**h**) tumor necrosis factor-α (TNF-α); (**i**) interleukin-1β (IL-1β). The results shown are mean ± SD (n = 3). Data were statistically considered at $p < 0.05$, and different small letters represent statistical differences by one-way analysis of variance (ANOVA) and Duncan's new multiple range test.

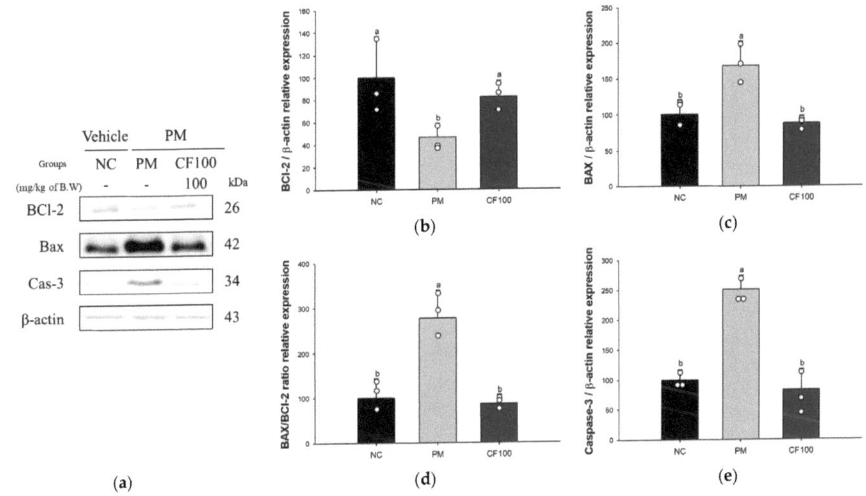

Figure 6. Regulatory effect of the aqueous extract of *Codium fragile* on $PM_{2.5}$-induced apoptosis-related protein expression in Western blot. (**a**) Western blot band image; protein expression levels of (**b**) B-cell lymphoma 2 (BCl-2); (**c**) BCl-2 associated X (BAX); (**d**) BAX/BCl-2 ratio; (**e**) Caspase-3. The results shown are mean ± SD (n = 3). Data were statistically considered at $p < 0.05$, and different small letters represent statistical differences by one-way analysis of variance (ANOVA) and Duncan's new multiple range test.

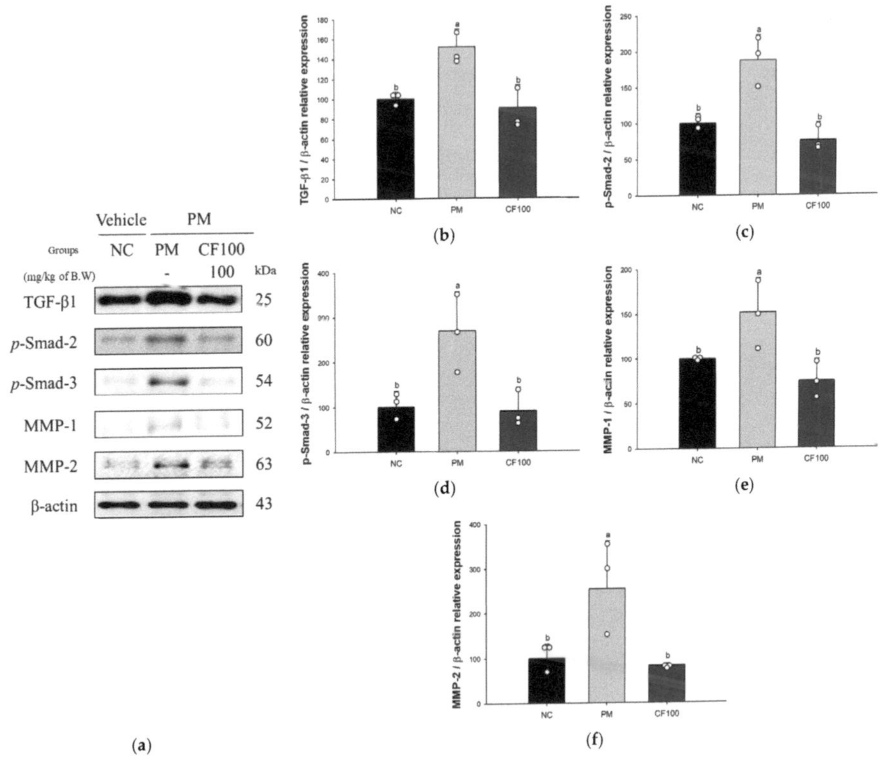

Figure 7. Regulatory effect of the aqueous extract of *Codium fragile* on $PM_{2.5}$-induced pulmonary fibrosis related protein expression in Western blot. (**a**) Western blot band image; protein expression

levels of (**b**) transforming growth factor-β1 (TGF-β1); (**c**) phosphorylated small mothers against decapentaplegic (p-Smsd)-2; (**d**) p-Smad-3; (**e**) matrix metalloproteinase-1 (MMP-1); (**f**) matrix metalloproteinase-2 (MMP-2). The results shown are mean ± SD (n = 3). Data were statistically considered at $p < 0.05$, and different small letters represent statistical differences by one-way analysis of variance (ANOVA) and Duncan's new multiple range test.

4. Discussion

Environmental air pollutant $PM_{2.5}$ causes various health problems including respiratory disease, asthma, COPD, and lung cancer [6]. Chronic exposure to $PM_{2.5}$ causes continuous inflammation, loss of oxidative stress scavenging function, and pulmonary fibrosis, which ultimately cause death of lung tissue and various lung diseases [4]. In particular, depending on MMS and other compositions, fine dust exhibits various toxicities and easily damages lung tissue [1]. Therefore, it is most important to block fine dust from the air, but since this is not easy, it is important to increase antioxidant activity in the body. In addition, since there are few treatments for pulmonary fibrosis, it is important to prevent it in advance through the consumption of antioxidants or natural products with strong physiological activity [9]. On the other hand, *Codium fragile* has a considerable possibility of use as an excellent natural product with various physiological activities [12]. On the other hand, studies of *Codium fragile* on fine dust toxicity related to protective effects or specific mechanisms are not clear. Thus, this study was estimated to assess the pulmonary ameliorating effect of the aqueous extract of *Codium fragile* against chronic $PM_{2.5}$-inhalated inflammation and fibrosis in BALB/c mice.

$PM_{2.5}$ contains various materials such as heavy metals and organic compounds that cause oxidative stress [1]. $PM_{2.5}$, which is composed of these reactive compounds, reaches the lung tissues and causes lipid peroxidation, DNA denaturation, structural degeneration, and mitochondrial dysfunction through damage to the antioxidant system [18]. In particular, lung tissues are in direct contact with $PM_{2.5}$ and have a structure, that is vulnerable to PM, and damage to the alveoli causes inflammatory responses by producing cytokines due to structural dysfunction. In addition, when $PM_{2.5}$ is deposited in lung tissues, it is difficult to remove and continuously leads to damage to the antioxidant system [19]. Oxidative stress derived from $PM_{2.5}$ increases the production of various radicals, and superoxide resulting from this process ultimately leads to the depletion of antioxidant substances such as SOD, GSH, catalase, and glutathione peroxidase (GPx) and lung damage [20]. Therefore, to estimate the protective effects of *Codium fragile* against $PM_{2.5}$-induced oxidative stress, SOD activities, reduced GSH contents, and MDA levels were estimated in lung tissues (Figure 3). According to a previous study, *Codium fragile* contains various flavonoid compounds such as naringenin, kaempferol, catechin, and epicatechin with considerable antioxidant activities [11]. Administration of these flavonoids had significant protective effects against oxidative stress-induced cytotoxicity in the systemic antioxidant system [21]. Considerable amounts of δ-tocopherol and α-tocopherol were found in *Codium fragile*, and these tocopherols showed significant antioxidant activity against cytotoxicity [22]. Oleamide inhibited lipid peroxidation and the imbalance in the ratio of reduced/oxidized GSH in 3-nitropropionic acid-induced antioxidant damage [23]. In addition, a sulfated polysaccharide from *Codium fragile* attenuated the disruption of antioxidant enzymes such as SOD, catalase, and GPx, and inhibited cytotoxicity indicators such as aspartate aminotransferase (AST), alanine aminotransferase (ALT), and lactate dehydrogenase (LDH) in high fat diet-induced liver and kidney cytotoxicity [10,24]. Based on these results, the aqueous extract of *Codium fragile* containing various compounds with physiological activities might have an ameliorating effect against $PM_{2.5}$-induced oxidative stress and cytotoxicity in lung tissues.

Inflammatory response and radicals generated by $PM_{2.5}$ lead to mitochondrial dysfunction in lung tissue [25]. Repeated and chronic exposure to $PM_{2.5}$ causes reduced mitochondrial fusion and impaired dynamics with the reduction of optic atrophy 1 (OPA1) and mitofusin2 (MFN2) [26]. $PM_{2.5}$ causes mitochondrial morphology damage with mitochondrial swelling, vacuole formation, and crystal destruction [27]. In addition, in damaged mitochondria, PTEN-induced kinase 1 (PINK1), a mitophagy regulatory protein, accumulates in the mitochondrial outer membrane, thereby inducing an abnormal mitophagy process resulting in a decrease in mitochondrial volume, inhibition of mitochondrial respiratory function, and an increase in mitochondrial ROS levels [28]. This mitochondrial deficit induces abnormal energy metabolism in lung tissues by reducing the MMP level and damaging the sodium–potassium pump and calcium pump [26]. Thus, in this study, the protective effects of the aqueous extract of *Codium fragile* against $PM_{2.5}$-induced mitochondrial dysfunction were confirmed, and mitochondrial ROS contents and MMP levels were ameliorated (Figure 4). Catechins, the bio-active compounds in *Codium fragile*, had a considerable protective effect against mitochondrial damage by regulating the mitochondrial complex and MMP levels [21]. Oleamide, a major compound in the aqueous extract of *Codium fragile*, regulated the inhibition of neuronal excitability with the activation of cannabinoid receptors against quinolinic acid-induced mitochondrial and synaptic dysfunction [29]. Oleamide also regulated mitochondrial dysfunction and death in 3-nitropropionic acid-induced mitochondrial deficit with the regulation of mitochondrial complex and cannabinoid receptors [23]. The administration of *Codium fragile* containing lysophosphatidyl choline and canthaxanthin promoted mitochondrial biogenesis with the regulation of peroxisome proliferator-activated receptor-gamma coactivator (PGC)-1α-related pathway [30]. Furthermore, the intake of *Codium fragile* regulated intestinal microbiota involved in mitochondrial energy metabolism such as pyruvate fermentation and glycolysis [31]. Thus, various phenolic compounds and unsaturated fatty acids in *Codium fragile* might help maintain lung health by suppressing $PM_{2.5}$-induced mitochondrial damage related to energy metabolism.

Absorbed $PM_{2.5}$ induces the inflammatory response in various tissues by binding into TLRs and stimulates MyD88 resulting in extensive inflammatory damage [32]. Stimulated TLRs continuously activate mitogen-activated protein kinase (MAPK), including extracellular signal-regulated kinase (ERK)1/2, p38 kinase, and JNK, and NF-κB pathway stimulating the secretion of cytokines and chemokines such as 1L-1β, interleukin-12, TNF-α, and monocyte chemoattractant protein-1. In addition, because $PM_{2.5}$ is composed of complex components including organic carbon, radicals, carbonaceous aerosols, inorganic ions, heavy metals, and polycyclic aromatic hydrocarbons, it activates various receptors such as aryl hydrocarbon receptors, hormone receptors, angiotensin type 1 receptors as well as TLRs [32]. The activated receptors stimulate inflammatory response, hormonal imbalance, and apoptotic signal through MAPK/NF-κB/phosphoinositide 3-kinase (PI3K)/protein kinase B (Akt) pathways. These activated signals increase the gene expression of cyclooxigenase-2 (COX-2) and iNOS-producing inflammatory cytokines [32]. Therefore, to evaluate the anti-inflammatory effect of the aqueous extract of *Codium fragile*, inflammatory protein expression levels in lung tissues were confirmed, and the consumption of this extract significantly suppressed inflammation in lung tissues (Figure 5). Similar to this study, sulfated polysaccharides isolated from *Codium fragile* significantly down-regulated inflammatory indicators such as prostaglandin E2, nitric oxide, and TNF-α in RAW 264.7 cells [33]. In addition, baicalin as one of the bio-active compounds of *Codium fragile* suppressed inflammation via PI3K/sirtuin 1 (SIRT1)/MAPK/NF-κB pathway [8]. The aqueous extract from *Codium fragile* decreased nitrite production, protein expression of iNOS, matrix metalloproteinase-13, a disintegrin and metalloproteinase with thrombospondin motifs (ADAMTS)-4, and ADAMTS-5 against IL-1β-induced osteoarthritis with the regulation of the MAPK/NF-κB signal [13]. *Codium fragile* also inhibited inflammatory cytokines such as TNF-α, IL-1β, and IL-6 and nuclear translocation of NF-κB by suppressing the phosphorylation and degradation of IκB-α [34], and suppressed inflammatory indicators such as COX-2, iNOS, prostaglandin E2, and release of nitric oxide (NO) [35]. In addition, treatment of kaempferol, as one of the flavonoid compounds, significantly inhibited IgE and lipopolysaccharide-induced inflammation via nuclear factor erythroid

2-related factor 2 (Nrf2)/SHIP1 on bone marrow-derived mast cells (BMMCs) [36]. In various previous studies, *Codium fragile* might help significantly inhibit the inflammatory reaction caused by $PM_{2.5}$, and in particular, it is judged to be able to suppress inflammation through the NF-kB pathway.

$PM_{2.5}$ promotes the initiation of an inflammatory response as well as the generation of oxidative stress, resulting in cytotoxicity [4]. Oxidative stress induced by fine dust causes damage to the antioxidant system, dysfunction of mitochondria, and damage to lung cell membranes [16]. Damage to pulmonary cells increases the level of intracellular Ca^{2+} and results in the release of cytochrome c from inside the mitochondria [37]. This process acts as a signal for apoptosis and induces a caspase cascade by causing an imbalance of mitochondria-related proteins such as BAX, BCl-2, and BCl-Xl/BCl-2-associated death promoter homolog (Bad) [37]. Therefore, continuous and chronic exposure to $PM_{2.5}$ stimulates intracellular apoptosis, which ultimately leads to cell death, which causes lung tissue dysfunction [38]. Therefore, to estimate the ameliorating effect of the aqueous extract of *Codium fragile*, apoptotic expression levels in lung tissues were evaluated, and the administration of *Codium fragile* significantly down-regulated pulmonary apoptosis (Figure 6). Oleamide significantly suppressed the nuclear condensation and activation of Caspase-3 in cerebellar granule neurons induced by K^+ deprivation. However, oleamide isomers without the Δ9-cis double bond, such as elaidic acid or stearic acid, did not affect cell death [39]. In addition, rutin, one of the flavonoids of *Codium fragile*, inhibited apoptosis by regulating the expression of BCl-2/BAX ratio, Caspase-9, and cleaved poly ADP-ribose polymerase (PARP) in endometriosis development in a rat model [37]. The treatment of p-coumaric acid suppressed apoptosis signaling in ethanol-induced hepatotoxicity by attenuating the expression of BAX, caspases, AST, and LDH via the PI3K/Akt pathway [40]. In conclusion, the aqueous extract of *Codium fragile* with physiological activities significantly suppressed apoptosis and might be used as a material to protect against $PM_{2.5}$-induced cytotoxicity. However, studies on factors related to other caspase cascades, including activated caspase-3, caspase-9, and cleaved PARP, need to be investigated in further experiments.

Fine dust in the air is absorbed into lung tissue and increases the expression of TGF-β1, which plays an important role in damage and repair signaling [4]. TGF-β1 secreted from fibroblasts and myofibroblasts activates the TGF-β1 receptor to phosphorylate Smad2/3 and stimulates the expression of MMPs [41]. Through this process, sub-signals such as type I collagen (Col1) and α-smooth muscle actin (α-SMA) related to collagen accumulation are continuously stimulated and continue to cause fibrosis and cancer [5]. In particular, heavy metals in $PM_{2.5}$ can easily accumulate in lung tissue and continuously stimulate the TGF-β pathway, causing damage to lung tissue [41]. Therefore, to evaluate the protective effect of the aqueous extract of *Codium fragile*, the pulmonary fibrous protein expression levels were confirmed. The administration of *Codium fragile* significantly attenuated pulmonary fibrosis (Figure 7). Tamarixetin, a quercetin derivative of *Codium fragile*, inhibited cardiac fibrosis by regulating the protein expression of TGF-β1, collagen I, collagen III, and matrix metalloproteinase-9 [42]. The consumption of gallic acid regulated hepatic fibrosis by regulating the mRNA levels of MMP-2 and tissue inhibitor of MMP-1 in carbon tetrachloride-induced liver injury mice [43]. In addition, morin as a flavonoid ameliorated allergic airway fibrosis by regulating the expression of matrix metalloproteinase-9 and cytokine levels of IgE and Th2 in bronchoalveolar lavage fluid in ovalbumin-induced mice [44]. Baicalein, one of the flavones in *Codium fragile*, inhibited pulmonary fibrosis by reducing microRNA-21 levels, which play an important role in the pathogenesis of pulmonary fibrosis, and by suppressing the up regulated expression levels of TGF-β1 and p-Smad 2/3 in bleomycin-treated rats [45]. Based on these results, the aqueous extract of *Codium fragile* with phenolic compounds significantly suppressed $PM_{2.5}$-induced lung fibrosis via the TGF-β1/matrix metalloproteinase/Smad pathway. Furthermore, it has been reported that lung fibrosis is sensitively affected by lipid changes [46]. Impairments and changes in fatty acid metabolism are associated with the pathogenesis of pulmonary fibrosis, and changes in the profile and metabolome of fatty acids are associated with disease progression and outcome [47]. The accumulation of triglyceride in the form of lipid droplets in alveolar epithelial cells induces endoplasmic reticulum (ER) stress and induces apoptosis of these cells through the expression of TGF-β1 [48]. On the other hand, stearic acid inhibited liver fibrosis by reducing α-SMA, collagen I expression, and ROS production in TGF-β1-induced fibroblasts [49]. In particular, ω-3 fatty acids have been reported to have excellent activity to improve pulmonary fibrosis [46]. In conclusion, *Codium fragile*, containing a large amount of unsaturated fatty acids and stearic acid, is judged to have an activity to improve pulmonary fibrosis by regulating the TGF-β1 pathway and changes in the profile and metabolome of fatty acids.

5. Conclusions

In conclusion, exposure to $PM_{2.5}$ damaged the pulmonary antioxidant system and mitochondrial function and caused an inflammatory response, apoptosis, and fibrosis. However, the aqueous extract of *Codium fragile* had a protective effect against $PM_{2.5}$-induced pulmonary cytotoxicity by regulating the TLR/TGF-β1/NF-κB pathway in BALB/c mice. This study proved that *Codium fragile*, a marine green alga, has a considerable and significant therapeutic effect against $PM_{2.5}$-induced pulmonary damage, and might be a potential and beneficial resource for functional food to improve lung health (Figure 8). However, as discussed above, $PM_{2.5}$ contains unspecific compounds such as heavy metals, VOCs, PAHs, and other organic compounds. Since the individual effect of these toxic materials is not clear, additional studies are needed. Moreover, additional studies related to genetic changes, nutritional studies, epigenetics, and effects on physiological activity of extracts of *Codium fragile* from exposure to $PM_{2.5}$ due to environmental factors should be conducted.

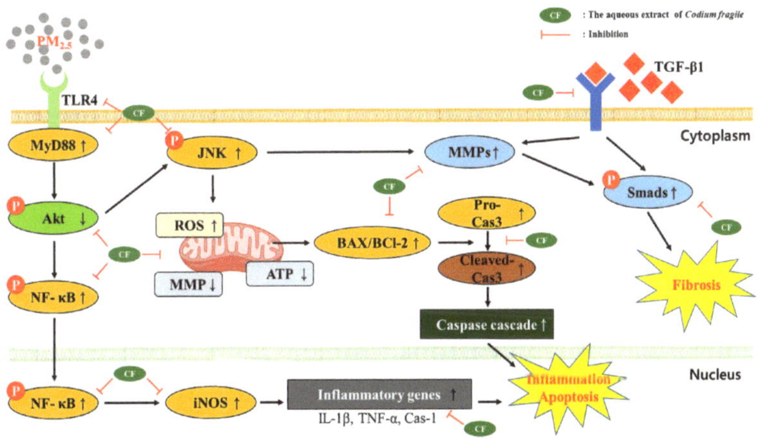

Figure 8. A schematic illustration presents the ameliorating effect of the aqueous extract of *Codium fragile* against particulate matter $(PM)_{2.5}$-exposed pulmonary damage and fibrosis via TLR/TGF-β1 and NF-κB pathways. (↑) upregularation; (↓) downregulateion in image.

Author Contributions: Conceptualization, T.Y.K., J.M.K. and H.J.H.; methodology, T.Y.K. and H.L.L.; software, H.-J.K.; validation, W.M.J. and D.Y.L.; investigation, T.Y.K., J.M.K., H.L.L., M.J.G., S.G.J., J.H.K. and H.S.L.; resources, H.J.H.; data curation, T.Y.K. and M.J.G.; writing—original draft preparation, J.M.K.; writing—review and editing, H.J.H.; visualization, M.J.G.; supervision, H.J.H.; project administration, H.J.H.; funding acquisition, H.J.H. All authors have read and agreed to the published version of the manuscript.

Funding: This study was supported by the Basic Science Research Program through the National Research Foundation (NRF) of Korea (NRF 2018R1D1A3B07043398) funded by the Ministry of Education, Republic of Korea, and Korean Institute of Marine Science & Technology Promotion (KIMST), funded by the Ministry of Oceans and Fisheries (2017029713), Republic of Korea.

Institutional Review Board Statement: All procedures were conducted in accordance with the Institutional Animal Care and Use Committee of Gyeongsang National University (certificate: GNU-210803-M0069, approved on 3 August 2021) and the policy of the Ethical Committee of the Ministry of Health and Welfare, Republic of Korea.

Informed Consent Statement: Not applicable.

Data Availability Statement: The data underlying this article are shared upon reasonable request to the corresponding author.

Conflicts of Interest: The authors declare no conflict of interest.

Abbreviations

The following abbreviations are used in this manuscript.

BAX	BCl-2 associated X
BCl-2	B-cell lymphoma 2
GSH	Glutathione
IL-1β	interleukin-1β
iNOS	inducible nitric oxide synthase
MDA	Malondialdehyde
MMP-1	matrix metalloproteinase-1
MMP-2	matrix metalloproteinase-2
MMP	mitochondrial membrane potential
MyD88	myeloid differentiation primary response 88
NF-κB	nuclear factor kappa-light-chain-enhancer of the activated B cell
p-JNK	phosphorylated c-Jun N-terminal kinase
p-Smad	phosphorylated small mothers against decapentaplegic
PM	particulate matter
$PM_{2.5}$	particulate matter, which classified as a size smaller than 2.5 μm
ROS	reactive oxygen species
SOD	superoxide dismutase
TGF-β1	transforming growth factor-β1
TLR	Toll-like receptors
TNF-α	tumor necrosis factor-α

References

1. Wei, T.; Tang, M. Biological effects of airborne fine particulate matter ($PM_{2.5}$) exposure on pulmonary immune system. *Environ. Toxicol. Pharmacol.* **2018**, *60*, 195–201. [CrossRef]
2. Wang, F.; Liu, J.; Zeng, H. Interactions of particulate matter and pulmonary surfactant: Implications for human health. *Adv. Colloid Interface Sci.* **2020**, *284*, 102244. [CrossRef]
3. Wang, S.; Chen, Y.; Hong, W.; Li, B.; Zhou, Y.; Ran, P. Chronic exposure to biomass ambient particulate matter triggers alveolar macrophage polarization and activation in the rat lung. *J. Cell Mol. Med.* **2022**, *26*, 1156–1168. [CrossRef]
4. Wang, W.; Zheng, F.; Zhang, A. Arsenic-induced lung inflammation and fibrosis in a rat model: Contribution of the HMGB1/RAGE, PI3K/AKT, and TGF-β1/SMAD pathways. *Toxicol. Appl. Pharmacol.* **2021**, *432*, 115757. [CrossRef] [PubMed]
5. Cutroneo, K.R.; White, S.L.; Phan, S.H.; Ehrlich, H.P. Therapies for bleomycin induced lung fibrosis through regulation of TGF-β1 induced collagen gene expression. *J. Cell Physiol.* **2007**, *211*, 585–589. [CrossRef] [PubMed]
6. Durham, A.L.; Adcock, I.M. The relationship between COPD and lung cancer. *Lung Cancer* **2015**, *90*, 121–127. [CrossRef] [PubMed]
7. Vijay, K. Toll like receptors in immunity and inflammatory diseases: Past, present, and future. *Int. Immunopharmacol.* **2018**, *59*, 391–412. [CrossRef] [PubMed]
8. Li, Y.; Song, K.; Zhang, H.; Yuan, M.; An, N.; Wei, Y.; Wang, L.; Sun, Y.; Xing, Y.; Gao, Y. Anti-inflammatory and immunomodulatory effects of baicalin in cerebrovascular and neurological disorders. *Brain Res. Bull.* **2020**, *164*, 314–324. [CrossRef]
9. Sun, J.; Zheng, X. A review of oil-suspended particulate matter aggregation—A natural process of cleansing spilled oil in the aquatic environment. *J. Environ. Monit.* **2009**, *11*, 1801–1809. [CrossRef]

10. Kolsi, R.B.A.; Jardak, N.; Hajkacem, F.; Chaaben, R.; El Feki, A.; Rebai, T.; Moussi, K.; Fki, L.; Belghith, H. Anti-obesity effect and protection of liver-kidney functions by *Codium fragile* sulphated polysaccharide on high fat diet induced obese rats. *Int. J. Biol. Macromol.* **2017**, *102*, 119–129. [CrossRef]
11. Keskinkaya, H.B.; Deveci, E.; Güneş, E.; Okudan, E.Ş.; Akköz, C.; Gümüş, N.E.; Karakurt, S. Chemical composition, in vitro antimicrobial and antioxidant activities of marine macroalgae *Codium fragile* (Suringar) Hariot. *Commagene J. Biol.* **2022**, *6*, 94–104. [CrossRef]
12. Magwaza, S.T.N.; Islam, M.S. Roles of marine macroalgae or seaweeds and their bioactive compounds in combating overweight, obesity and diabetes: A comprehensive review. *Mar. Drugs* **2023**, *21*, 258. [CrossRef]
13. Moon, S.M.; Lee, S.A.; Han, S.H.; Park, B.R.; Choi, M.S.; Kim, J.S.; Kim, S.G.; Kim, H.J.; Chun, H.S.; Kim, D.K.; et al. Aqueous extract of *Codium fragile* alleviates osteoarthritis through the MAPK/NF-κB pathways in IL-1β-induced rat primary chondrocytes and a rat osteoarthritis model. *Biomed. Pharmacother.* **2018**, *97*, 264–270. [CrossRef] [PubMed]
14. Kim, G.H.; Park, S.K.; Kang, J.Y.; Kim, J.M.; Shin, E.J.; Moon, J.H.; Kim, M.J.; Lee, H.L.; Jeong, H.R.; Heo, H.J. Protective effect of *Codium fragile* extract on fine dust ($PM_{2.5}$)-induced toxicity in nasal cavity, lung, and brain cells. *Korean J. Food Sci. Technol.* **2021**, *53*, 223–229.
15. Bradford, M.M. A rapid and sensitive method for the quantitation of microgram quantities of protein utilizing the principle of protein-dye binding. *Anal. Biochem.* **1976**, *72*, 248–254. [CrossRef]
16. Kim, J.M.; Kang, J.Y.; Park, S.K.; Moon, J.H.; Kim, M.J.; Lee, H.L.; Jeong, H.R.; Kim, J.C.; Heo, H.J. Powdered green tea (matcha) attenuates the cognitive dysfunction via the regulation of systemic inflammation in chronic $PM_{2.5}$-exposed BALB/c mice. *Antioxidants* **2021**, *10*, 1932. [CrossRef] [PubMed]
17. Kim, J.M.; Lee, U.; Kang, J.Y.; Park, S.K.; Kim, J.C.; Heo, H.J. Matcha improves metabolic imbalance-induced cognitive dysfunction. *Oxidative Med. Cell. Longev.* **2020**, *2020*, 8882763. [CrossRef]
18. Wei, M.; Bao, G.; Li, S.; Yang, Z.; Cheng, S.; Le, W. $PM_{2.5}$ exposure triggers cell death through lysosomal membrane permeabilization and leads to ferroptosis insensitivity via the autophagy dysfunction/p62-KEAP1-NRF2 activation in neuronal cells. *Ecotoxicol. Environ. Saf.* **2022**, *248*, 114333. [CrossRef]
19. Liu, K.; Hua, S.; Song, L. $PM_{2.5}$ exposure and asthma development: The key role of oxidative stress. *Oxidative Med. Cell. Longev.* **2022**, *2022*, 3618806. [CrossRef]
20. Liu, X.; Meng, Z. Effects of airborne fine particulate matter on antioxidant capacity and lipid peroxidation in multiple organs of rats. *Inhal. Toxicol.* **2005**, *17*, 467–473. [CrossRef]
21. Kim, J.M.; Heo, H.J. The roles of catechins in regulation of systemic inflammation. *Food Sci. Biotechnol.* **2022**, *31*, 957–970. [CrossRef] [PubMed]
22. Di Mascio, P.; Murphy, M.E.; Sies, H. Antioxidant defense systems: The role of carotenoids, tocopherols, and thiols. *Am. J. Clin. Nutr.* **1991**, *53*, 194–200. [CrossRef]
23. Reyes-Soto, C.Y.; Villaseca-Flores, M.; Ovalle-Noguez, E.A.; Nava-Osorio, J.; Galván-Arzate, S.; Rangel-López, E.; Maya-López, M.; Retana-Márquez, S.; Túnez, I.; Tinkov, A.A.; et al. Oleamide reduces mitochondrial dysfunction and toxicity in rat cortical slices through the combined action of cannabinoid receptors activation and induction of antioxidant activity. *Neurotox. Res.* **2022**, *140*, 2167–2178. [CrossRef] [PubMed]
24. Kalita, P.; Ahmed, A.B.; Sen, S.; Chakraborty, R. A comprehensive review on polysaccharides with hypolipidemic activity: Occurrence, chemistry and molecular mechanism. *Int. J. Biol. Macromol.* **2022**, *206*, 681–698. [CrossRef]
25. Li, R.; Kou, X.; Geng, H.; Xie, J.; Tian, J.; Cai, Z.; Dong, C. Mitochondrial damage: An important mechanism of ambient $PM_{2.5}$ exposure-induced acute heart injury in rats. *J. Hazard. Mater.* **2015**, *287*, 392–401. [CrossRef]
26. Li, R.; Kou, X.; Geng, H.; Xie, J.; Yang, Z.; Zhang, Y.; Cai, Z.; Dong, C. Effect of ambient $PM_{2.5}$ on lung mitochondrial damage and fusion/fission gene expression in rats. *Chem. Res. Toxicol.* **2015**, *28*, 408–418. [CrossRef]
27. Dornhof, R.; Maschowski, C.; Osipova, A.; Giere, R.; Seidl, M.; Merfort, I.; Humar, M. Stress fibers, autophagy and necrosis by persistent exposure to $PM_{2.5}$ from biomass combustion. *PLoS ONE* **2017**, *12*, 0180291. [CrossRef]
28. Lin, Q.; Zhang, C.F.; Guo, J.L.; Su, J.L.; Guo, Z.K.; Li, H.Y. Involvement of NEAT1/PINK1-mediated mitophagy in chronic obstructive pulmonary disease induced by cigarette smoke or $PM_{2.5}$. *J. Transl. Med.* **2022**, *10*, 277. [CrossRef]
29. Maya-López, M.; Rubio-López, L.C.; Rodríguez-Alvarez, I.V.; Orduño-Piceno, J.; Flores-Valdivia, Y.; Colonnello, A.; López, E.R.; Túnez, I.; Prospéro-García, O.; Santamaría, A. A cannabinoid receptor-mediated mechanism participates in the neuroprotective effects of oleamide against excitotoxic damage in rat brain synaptosomes and cortical slices. *Neurotox. Res.* **2020**, *37*, 126–135. [CrossRef]
30. Ahn, J.; Kim, M.J.; Ahn, J.; Ha, T.Y.; Jung, C.H.; Seo, H.D.; Jang, Y.J. Identifying *Codium fragile* extract components and their effects on muscle weight and exercise endurance. *Food Chem.* **2021**, *353*, 129463. [CrossRef]
31. Meinita, M.D.N.; Harwanto, D.; Choi, J.S. Seaweed exhibits therapeutic properties against chronic diseases: An overview. *Appl. Sci.* **2022**, *12*, 2638. [CrossRef]
32. Zhu, P.; Zhang, W.; Feng, F.; Qin, L.; Ji, W.; Li, D.; Liang, R.; Zhang, Y.; Wang, Y.; Li, M.; et al. Role of angiotensin-converting enzyme 2 in fine particulate matter-induced acute lung injury. *Sci. Total Environ.* **2022**, *825*, 153964. [CrossRef]
33. Lee, S.A.; Moon, S.M.; Choi, Y.H.; Han, S.H.; Park, B.R.; Choi, M.S.; Kim, J.S.; Kim, Y.H.; Kim, D.K.; Kim, C.S. Aqueous extract of *Codium fragile* suppressed inflammatory responses in lipopolysaccharide-stimulated RAW264.7 cells and carrageenan-induced rats. *Biomed. Pharmacother.* **2017**, *93*, 1055–1064. [CrossRef]

34. Patel, S. Therapeutic importance of sulfated polysaccharides from seaweeds: Updating the recent findings. *3 Biotech.* **2012**, *2*, 171–185. [CrossRef]
35. Michalak, I.; Tiwari, R.; Dhawan, M.; Alagawany, M.; Farag, M.R.; Sharun, K.; Emran, T.B.; Dhama, K. Antioxidant effects of seaweeds and their active compounds on animal health and production—A review. *Vet. Q.* **2022**, *42*, 48–67. [CrossRef]
36. Nagata, K.; Araumi, S.; Ando, D.; Ito, N.; Ando, M.; Ikeda, Y.; Takahashi, M.; Noguchi, S.; Yasuda, Y.; Nakano, N.; et al. Kaempferol Suppresses the Activation of Mast Cells by Modulating the Expression of FcεRI and SHIP1. *Int. J. Mol. Sci.* **2023**, *24*, 5997. [CrossRef]
37. Talebi, H.; Farahpour, M.R.; Hamishehkar, H. The effectiveness of Rutin for prevention of surgical induced endometriosis development in a rat model. *Sci. Rep.* **2021**, *11*, 7180. [CrossRef] [PubMed]
38. Li, Y.; Batibawa, J.W.; Du, Z.; Liang, S.; Duan, J.; Sun, Z. Acute exposure to PM2.5 triggers lung inflammatory response and apoptosis in rat. *Ecotoxicol. Environ. Saf.* **2021**, *222*, 112526. [CrossRef]
39. Yang, J.Y.; Abe, K.; Xu, N.J.; Matsuki, N.; Wu, C.F. Oleamide attenuates apoptotic death in cultured rat cerebellar granule neurons. *Neurosci. Lett.* **2002**, *328*, 165–169. [CrossRef]
40. Sabitha, R.; Nishi, K.; Gunasekaran, V.P.; Agilan, B.; David, E.; Annamalai, G.; Vinothkumar, R.; Perumal, M.; Subbiah, L.; Ganeshan, M. p-Coumaric acid attenuates alcohol exposed hepatic injury through MAPKs, apoptosis and Nrf2 signaling in experimental models. *Chem. Biol. Interact.* **2020**, *321*, 109044. [CrossRef] [PubMed]
41. Liu, S.; Zhang, W.; Zhang, F.; Roepstorff, P.; Yang, F.; Lu, Z.; Ding, W. TMT-based quantitative proteomics analysis reveals airborne $PM_{2.5}$-induced pulmonary fibrosis. *Int. J. Environ. Res. Public. Health* **2019**, *16*, 98. [CrossRef] [PubMed]
42. Fan, C.; Li, Y.; Yang, H.; Cui, Y.; Wang, H.; Zhou, H.; Zhang, J.; Du, B.; Zhai, Q.; Chen, X.; et al. Tamarixetin protects against cardiac hypertrophy via inhibiting NFAT and AKT pathway. *J. Mol. Histol.* **2019**, *50*, 343–354. [CrossRef] [PubMed]
43. Wang, J.; Tang, L.; White, J.; Fang, J. Inhibitory effect of gallic acid on CCl_4-mediated liver fibrosis in mice. *Cell Biochem. Biophys.* **2014**, *69*, 21–26. [CrossRef] [PubMed]
44. Ma, Y.; Ge, A.; Zhu, W.; Liu, Y.N.; Ji, N.F.; Zha, W.J.; Zhang, J.X.; Zeng, X.N.; Huang, M. Morin attenuates ovalbumin-induced airway inflammation by modulating oxidative stress-responsive MAPK signaling. *Oxidative Med. Cell. Longev.* **2016**, *2016*, 5843672. [CrossRef]
45. Gao, Y.; Lu, J.; Zhang, Y.; Chen, Y.; Gu, Z.; Jiang, X. Baicalein attenuates bleomycin-induced pulmonary fibrosis in rats through inhibition of miR-21. *Pulm. Pharmacol. Ther.* **2013**, *26*, 649–654. [CrossRef]
46. Wygrecka, M.; Hadzic, S.; Potaczek, D.P.; Alexopoulos, I.; El Agha, E.; Schaefer, L. Decoding the role of fatty acids and their metabolites in lung fibrosis. *Pol. Arch. Intern. Med.* **2023**, *133*, 16520. [CrossRef] [PubMed]
47. Lyu, Y.; Guo, C.; Zhang, H. Fatty acid metabolism-related genes in bronchoalveolar lavage fluid unveil prognostic and immune infiltration in idiopathic pulmonary fibrosis. *Front. Endocrinol.* **2022**, *13*, 1001563. [CrossRef] [PubMed]
48. Santos, C.R.; Schulze, A. Lipid metabolism in cancer. *FEBS J.* **2012**, *279*, 2610–2623. [CrossRef] [PubMed]
49. Kim, H.S.; Yoo, H.J.; Lee, K.M.; Song, H.E.; Kim, S.J.; Lee, J.O.; Hwang, J.J.; Song, J.W. Stearic acid attenuates profibrotic signalling in idiopathic pulmonary fibrosis. *Respirology* **2021**, *26*, 255–263. [CrossRef] [PubMed]

antioxidants

Article

Effects of Particulate Matter Inhalation during Exercise on Oxidative Stress and Mitochondrial Function in Mouse Skeletal Muscle

Jinhan Park [1,†], Junho Jang [1,†], Byunghun So [1,†], Kanggyu Lee [1], Dongjin Yeom [1], Ziyi Zhang [2], Woo Shik Shin [3] and Chounghun Kang [1,4,*]

1 Graduate School of Health and Exercise Science, Inha University, Incheon 22212, Republic of Korea; sportsjinhan@gmail.com (J.P.); jangju2489@gmail.com (J.J.); sportshun@inha.ac.kr (B.S.); skyarn@nate.com (K.L.); duarkans@gmail.com (D.Y.)

2 Tianjin Key Laboratory of Exercise Physiology and Sports Medicine, Institute of Sport, Exercise & Health, Tianjin University of Sport, Tianjin 300381, China; zhangzy427@tjus.edu.cn

3 Department of Pharmaceutical Sciences, Northeast Ohio Medical University, Rootstown, OH 44272, USA; wshin@neomed.edu

4 Department of Physical Education, College of Education, Inha University, Incheon 22212, Republic of Korea

* Correspondence: ck@inha.ac.kr; Tel.: +82-32-860-8647

† These authors contributed equally to this work.

Abstract: Particulate matter (PM) has deleterious consequences not only on the respiratory system but also on essential human organs, such as the heart, blood vessels, kidneys, and liver. However, the effects of PM inhalation on skeletal muscles have yet to be sufficiently elucidated. Female C57BL/6 or mt-Keima transgenic mice were randomly assigned to one of the following four groups: control (CON), PM exposure alone (PM), treadmill exercise (EX), or PM exposure and exercise (PME). Mice in the three-treatment group were subjected to treadmill running (20 m/min, 90 min/day for 1 week) and/or exposure to PM (100 μg/m^3). The PM was found to exacerbate oxidative stress and inflammation, both at rest and during exercise, as assessed by the levels of proinflammatory cytokines, manganese-superoxide dismutase activity, and the glutathione/oxidized glutathione ratio. Furthermore, we detected significant increases in the levels of in vivo mitophagy, particularly in the PM group. Compared with the EX group, a significant reduction in the level of mitochondrial DNA was recorded in the PME group. Moreover, PM resulted in a reduction in cytochrome *c* oxidase activity and an increase in hydrogen peroxide generation. However, exposure to PM had no significant effect on mitochondrial respiration. Collectively, our findings in this study indicate that PM has adverse effects concerning both oxidative stress and inflammatory responses in skeletal muscle and mitochondria, both at rest and during exercise.

Keywords: particulate matter; oxidative stress; skeletal muscle; mitochondria; in vivo mitophagy; treadmill exercise

Citation: Park, J.; Jang, J.; So, B.; Lee, K.; Yeom, D.; Zhang, Z.; Shin, W.S.; Kang, C. Effects of Particulate Matter Inhalation during Exercise on Oxidative Stress and Mitochondrial Function in Mouse Skeletal Muscle. *Antioxidants* **2024**, *13*, 113. https://doi.org/10.3390/antiox13010113

Academic Editor: Yasuhiro Yoshida

Received: 7 December 2023
Revised: 15 January 2024
Accepted: 15 January 2024
Published: 17 January 2024

1. Introduction

Particulate matter (PM) is primarily derived from the combustion of coal, vehicle exhausts, and diverse industrial processes. It forms secondarily in the atmosphere through intricate chemical reactions involving sulfur dioxide and nitrogen oxides [1,2]. The size of PM is commonly classified as either coarse (PM_{10} < 10 μm) or fine ($PM_{2.5}$ < 2.5 μm) [3]. When inhaled, suspended PM enters the body, and while coughing and saliva can remove some of the inhaled PM_{10} particles from the oral and nasal cavities, the remaining $PM_{2.5}$ particles can readily penetrate the distal parts of the lungs, depositing in the alveoli [4–6]. As $PM_{2.5}$ particles attach to alveolar tissues and can circulate in the bloodstream, acute or chronic exposure to PM can potentially contribute to an increase in systemic inflammation and oxidative stress [7–9]. Such exposure can accordingly exacerbate any preexisting respiratory, cardiovascular, and endocrine system disorders, including hypertension, chronic obstructive pulmonary disease, asthma, and type 2 diabetes [10–12].

Skeletal muscles play essential roles not only in movement but also as sites for a range of key biochemical processes, including the production of myokines, regulation of hormones, energy metabolism via the exchange of extracellular metabolites and oxygen, and intracellular signaling [13,14]. Evidence obtained to date indicates that similar to other organs, the detrimental effects of PM on skeletal muscle are associated with the circulation of proinflammatory cytokines stimulated by the presence of PM accumulating in the lungs. However, these discoveries are primarily based on the findings of in vitro studies [15–17], making it difficult to identify any specific systemic mechanisms. Accordingly, gaining more meaningful insights into the effects of oxidative stress and inflammatory responses provoked by PM deposition in the lung on skeletal muscle function and integrity, requires overcoming the limitations of in vitro studies by developing animal models that mimic atmospheric inhalation of particulate matter.

Prolonged aerobic exercise offers numerous health benefits, including enhancement of antioxidant and immune functions, which can contribute to inhibiting excessive inflammatory responses and the generation of reactive oxygen species (ROS), particularly in skeletal muscle [18–20]. Under normal physiological conditions, ROS play a key role in various cellular activities, including cellular energy metabolism, signal transduction, and the regulation of gene expression. However, when produced in excess, they can cause damage to cellular biomolecules, including lipids, proteins, and nucleic acids, thereby promoting cellular aging and eventually cell death [21]. High-intensity acute exercise, accompanied by a rapid increase in oxygen consumption, promotes excessive ROS production, leading to an imbalance in the oxidative–antioxidative homeostasis of cells [22]. However, in skeletal muscle adapted through long-term endurance training, the fibers undergo various structural adaptations and an increase in antioxidant enzyme activity, enhancing the capacity to adapt to high-intensity acute exercises and reducing oxidative stress levels [23,24]. The plasticity of skeletal muscle to adapt to exercise-related stress is characterized by an increase in myofibrillar cross-sectional area and changes in myofibrillar composition, promoted by resistance training and associated increases in metabolic and biochemical capacity. These adaptations are accompanied by morphological and functional changes, particularly in the mitochondria [25], the primary site of ROS production in response to exercise [26]. Morphological and functional changes in mitochondria, pivotal in cellular respiration and signal transduction, often correlate with elevated levels of oxidative stress in cells, and evidence indicates that PM may also influence mitochondrial homeostasis by inducing ROS production and inflammation [27].

The beneficial effects of exercise on the human body are widely acknowledged, but the repercussions of inhaling PM during physical activity are still being assessed, with a major focus on cardiorespiratory complications [28]. In this context, in the present study we sought to examine the effects of PM inhalation during exercise on oxidative stress, inflammatory responses, and mitochondrial function in the skeletal muscle of mice.

2. Materials and Methods

2.1. Experimental Animals

To eliminate the potential effects of testosterone on the accumulation of ROS and mitochondria-mediated cell death [29,30], female C57BL/6 mice (8 weeks old) were randomly divided into one of the following four groups: control (CON), PM exposure (PM), treadmill exercise (EX), and PM exposure + exercise (PME). For the study of in vivo mitophagy, mt-Keima transgenic (heterozygous type (+/−) FVB/N) mice (kindly provided by Dr. Jeanho Yun, Dong-A University) were bred and maintained in a specific pathogen-free facility. These mice harbor a mitochondria-targeting sequence derived from COX VIII that binds to the pH-dependent fluorescent protein Keima, thereby facilitating the detection of mitophagy. This model reflects the physiological status of mitochondria based on a dual-fluorescence probe, with a green fluorescence being indicative of normal conditions (pH 8.0) and red fluorescence signifying acidic lysosomal conditions (pH 4.5) [31,32]. All mice were housed in a temperature (22 °C)- and humidity (40–60%)-controlled environment illuminated on a 12 h light/12 h dark cycle, and had free access to allergen-free food and water. The protocols used in this study were approved by the Institutional Animal Care and Use Committee (IACUC: approval number INHA 190211-616, approval date 11 February 2019; TJUS: approval number TJUS 2022-021, approval date: February 2022).

2.2. PM Chamber and Treadmill Exercise

Commercial PM samples (Urban Particulate Matter; NIST1648A, Sigma, St. Louis, MO, USA) were used in a specifically constructed PM chamber designed to maintain a PM concentration of 100 μg/m^3 during experimental assessments (Korea patent registration: 10-2529955 and 10-2529956, Republic of Korea). PM inhalation and/or exercise treatments were performed for 90 min per day for 7 days, with the incline and speed of the mouse treadmill being set at 20 m/min on a 5-degree uphill slope (Figure 1).

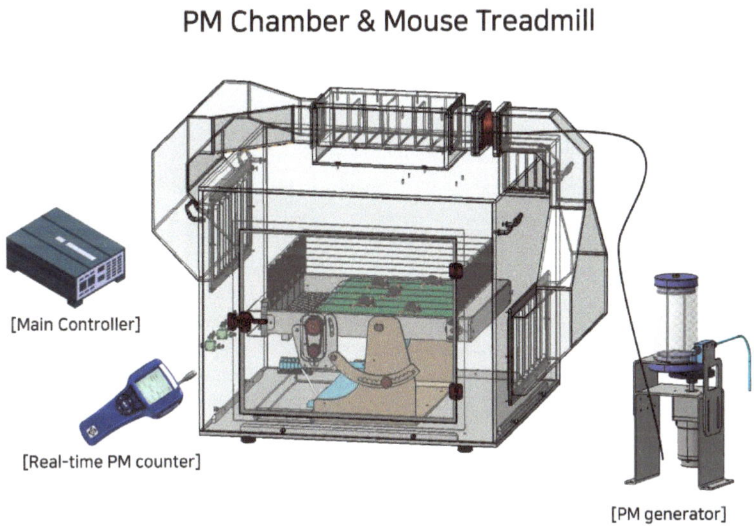

Figure 1. A schematic diagram of the particulate matter (PM) chamber used in this study. The PM generator introduces fine dust into the chamber, the concentration of which is continuously monitored in real-time using an AeroTrak Handheld Particle Counter 9303 (TSI, Shoreview, MN, USA). The PM chamber and its main controller include a mouse treadmill with adjustable speed and duration.

2.3. In Vivo Mitophagy Analysis

To evaluate in vivo mitophagy, we utilized mt-Keima transgenic mice. For in vivo observations, quadriceps muscle tissues obtained from mice were initially washed with cold phosphate-buffered saline, after which 1.0-μm-thick sections were cut using a brain slicer matrix and placed in confocal dishes (SPL). Nuclear staining was performed on ice using Hoechst 33342 and 4,6-diamidino-2-phenylindole solution (5 μg/mL) for 5 min (Thermo Fisher Scientific, Waltham, MA, USA). Fluorescence measurements were obtained using a laser confocal microscope (LSM 510 META; ZEISS) with excitation wavelengths of 488 nm (green) and 561 nm (red) and an emission wavelength of 620 nm. Changes in mitophagy were based on assessments of the red to green fluorescence ratio. Statistical analysis was conducted using ImageJ version 1.8.0 software.

2.4. Quantification of Mitochondrial DNA

The mitochondrial DNA (mtDNA) was purified from total DNA using a Nucleospin RNA Plus kit (MACHEREY-NAGEL, Düren, Nordrhein-Westfalen, Germany), with nuclear DNA (nDNA) being isolated using standard protocols. To quantify the amounts of mtDNA present per nuclear genome, we used the following primers pairs: mtDNA forward primer, 5′-CCTATCACCCTTGCCATCAT-3′ and mtDNA reverse primer, 5′-GAGGCTGTTG-CTTGTGTGAC-3′; nuclear DNA forward primer, 5′-ATGGAAAGCCTGCCATCATG-3′ and nuclear DNA reverse primer, 5′-TCCTTGTTGTTCAGCATCAC-3′. The quantification of relative copy number differences was performed using the ΔΔCt method of the difference in threshold amplification between mtDNA and nuclear DNA. The RT-PCR thermal cycling conditions were 95 °C for 15 min and 50 °C for 40 s.

2.5. Enzyme-Linked Immunosorbent Assay

Interleukin-6 (IL-6), tumor necrosis factor-alpha (TNF-α), interleukin-1β (IL-1β), superoxide dismutase (SOD), and manganese superoxide dismutase (MnSOD) were measured using a Quantikine™ ELISA kit (R&D System, Inc., NE Minneapolis, MN, USA). For each assay, gastrocnemius muscle samples were prepared according to the manufacturer's protocol.

2.6. Permeabilization of Muscle Fibers and Measurement of Respiration and H_2O_2 Generation

The permeabilization of muscle fibers and measurements of respiration and H_2O_2 production were performed using modified versions of previously described methods [33]. Samples of the red gastrocnemius muscle of mice (2–4 mg) were dissected and mechanically separated. The muscle fibers were permeabilized for 30 min using saponin (30 μg/mL) in buffer Z (pH 7.1; 30 mM KCl, 10 mM KH_2PO_4, 0.6 mg/mL BSA, 5 mM $MgCl_2$-$6H_2O$, 1 mM EGTA, 105 mM K-MES) supplemented with 1 mM EGTA (wash buffer), after which the preparations were washed three times with wash buffer. The permeabilized fiber bundles thus obtained were utilized to simultaneously measure the rates of oxygen consumption (OCR) and H_2O_2 generation using an Oroboros Oxygraph-2k device (O2k; OROBOROS Instruments, Innsbruck, Austria). OCR was assessed using an oxygen probe, while the rate of H_2O_2 production was evaluated using a green fluorescence sensor of the O2k-Fluo LED2 module. OCR measurements were standardized by incorporating antimycin A to account for non-mitochondrial oxygen consumption. The production of H_2O_2 was determined based on a standard H_2O_2 calibration curve.

2.7. Cyclooxygenase (COX) Activity Assay

COX activity was estimated using a COX activity assay (ab204699; Abcam, Cambridge, UK) according to the manufacturer's instructions. Fluorescence (λEx/Em = 535/587 nm) was measured using a microplate reader in kinetic mode, and COX activity was expressed as μU/mg.

2.8. Malondialdehyde (MDA) Measurements

For the determination of the levels of MDA in gastrocnemius muscle tissues, 20 mg samples were initially homogenized in 250 μL of 7.5% trichloroacetic acid. After centrifugation and filtration, the resulting supernatants were combined with an equal volume of a mixture containing 10% trichloroacetic acid and 0.5% TBA. The samples were then boiled in a dry thermoblock for 30 min, followed by cooling. The absorbance of the TBA–MDA complex thus obtained was measured at 532 nm and corrected for non-specific absorbance at 600 nm to account for background noise.

2.9. Glutathione/Oxidized Glutathione Assay

Total and oxidized levels of glutathione (GSH and GSSG, respectively) were determined using a Glutathione Assay Kit provided by Cayman Chemical Company. Assays were performed by initially homogenizing gastrocnemius muscles in cold buffer (50 mM MES, pH 6 to 7, 1 mM EDTA) followed by centrifugation at 10,000× *g* for 15 min at 4 °C.

The resulting supernatants were deproteinized using MPA reagent, followed by further centrifugation at 3000× g for 2 min. The resulting supernatants were treated with TEAM Reagent to quantify the total GSH present. To determine the levels of GSSG, we added 2-vinylpyridine to the supernatants and determined the level of reduced GSH by subtracting GSSG from the total GSH content.

2.10. Statistical Analysis

Data were analyzed using SPSS 22.0 software for statistics and GraphPad Prism 8 (v8.0.2, 2019) for visualization. The normality of the distribution for outcomes was assessed using the Shapiro–Wilk test and QQ plot. The differences between groups were tested via the student's t-test, by comparing the groups. The interaction between PM exposure and treadmill exercise was analyzed via a two-way analysis of variance (ANOVA). Test were two-tailed, and significance was set at $p < 0.05$. All values are presented as the means and standard error (SEM).

3. Results

To assess the generation of lipid peroxidation by-products in gastrocnemius muscles, we measured the levels of MDA. Compared with the CON and EX groups, we detected a significant increase in MDA levels in the two respective PM inhalation groups ($p < 0.001$; Figure 2A). To assess the levels of proinflammatory cytokines, we measured the production of IL-6, TNF-α, and IL-1β. Compared with the levels in mice in the CON group, the levels of IL-6 were found to be 2.75-fold higher in the PM-treated mice ($p < 0.001$; Figure 2B). Similarly, compared with the levels in the EX group, the levels of IL-6 were found to be 49% higher in the PME group mice ($p < 0.001$). Likewise, compared with the levels in the CON and EX groups, we detected 2.6- and 4.0-fold higher levels of TNF-α in the PM and PME groups, respectively ($p < 0.001$; Figure 2C). Notably, two-way ANOVA revealed a significant interaction between exercise and PM inhalation regarding TNF-α levels ($p < 0.05$). Furthermore, we detected increases of 79% and 90% in IL-1β in response to the PM and PME treatments, respectively (both $p < 0.001$; Figure 2D).

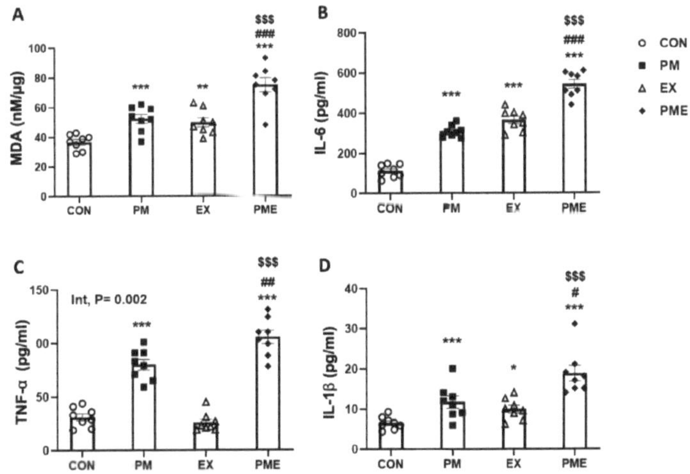

Figure 2. Inflammatory response levels in the gastrocnemius muscle of C57BL/6 mice exposed to particulate matter during exercise. (**A**) MDA (malondialdehyde), (**B**) IL-6, (**C**) TNFα, and (**D**) IL-1β. Control (CON, n = 8), particulate matter exposure (PM, n = 8), exercise (EX, n = 8), and particulate matter exposure + exercise (PME, n = 8). Data are expressed as the means ± SEM. Statistical significance is assigned as * $p < 0.05$, ** $p < 0.01$, *** $p < 0.001$ vs. CON, # $p < 0.05$, ## $p < 0.01$, ### $p < 0.001$ vs. PM and $$$ $p < 0.001$ vs. EX. Two-way ANOVA results were shown with p values. Int, interaction.

We detected no significant differences among groups regarding total SOD levels. The levels of MnSOD, the only SOD enzyme located within the mitochondrial matrix, were found to be 30% lower in the PM group compared with that in the CON group ($p < 0.01$; Figure 3B) and 40% lower in the PME group compared with that in the EX group ($p < 0.01$). We also observed a significant reduction of 7.3% in the total GSH content in response to PM treatment ($p < 0.05$), and a reduction of 17% in PME mice compared with that in the EX group ($p < 0.01$; Figure 3C). Conversely, we recorded a 29% increase in the levels of oxidized GSH (GSSG) in the PM group compared with that in the CON group ($p < 0.05$), and an increase of 19% in the PME group compared with that in the EX group ($p < 0.05$; Figure 3D). Moreover, compared with the CON and EX groups, we detected reductions of 25% and 28% in the ratio of GSH to GSSG (GSH/GSSG) in the respective groups exposed to PM ($p < 0.01$; Figure 3E).

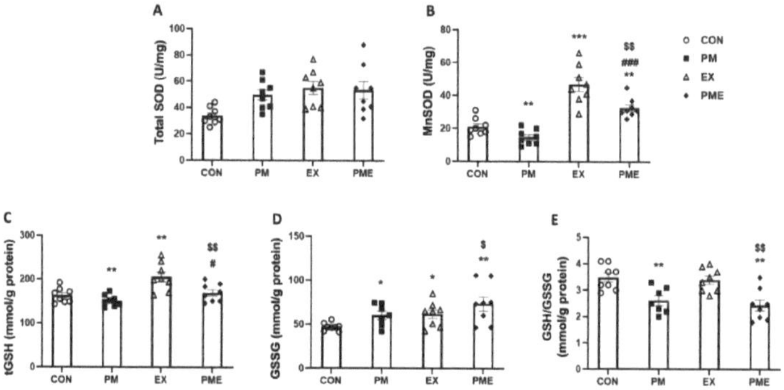

Figure 3. Redox status in the gastrocnemius skeletal muscle of C57BL/6 mice exposed to particulate matter during exercise. (**A**) Total superoxide dismutase (SOD), (**B**) manganese superoxide dismutase (MnSOD), (**C**) total glutathione (tGSH), (**D**) glutathione disulfide (oxidized glutathione, GSSG), and (**E**) GSH/GSSG ratio. Control (CON, n = 8), particulate matter exposure (PM, n = 8), exercise (EX, n = 8), and particulate matter exposure + exercise (PME, n = 8). Data are expressed as the means ± SEM. Statistical significance is defined as * $p < 0.05$, ** $p < 0.01$, *** $p < 0.001$ vs. CON, # $p < 0.05$, ### $p < 0.001$ vs. PM and \$ $p < 0.05$, \$\$ $p < 0.01$ vs. EX.

Further analysis of the levels of in vivo mitophagy, using model transgenic mice that express the pH-dependent fluorescent protein mt-Keima, revealed a significant increase in the levels of mitophagy (as indicated by an increase in the red/green fluorescence ratio) in the PM, EX, and PME groups compared with that in the CON group ($p < 0.001$). Interestingly, although we detected a significant increase in mitophagy in mice subjected to exercise ($p < 0.001$) compared with the CON group mice, this increase was significantly reversed following exposure to PM ($p < 0.01$; Figure 4A). To examine mitochondrial biogenesis, we assessed the ratio of mtDNA to nDNA and, in line with expectations, detected a significant exercise-induced increase in this ratio in the CON and PM group mice ($p < 0.001$; Figure 4B). However, compared with the CON or EX groups, we detected no significant difference in those mice exposed to PM.

To evaluate mitochondrial function, we initially analyzed the activity of mitochondrial cytochrome *c* oxidase (COX), which plays an essential role in ATP production, and accordingly detected reductions in activity of 8.5% and 32% in PM vs. CON ($p < 0.01$), and PME vs. EX ($p < 0.001$) comparisons, respectively (Figure 5A). Moreover, we detected an interactive effect between PM inhalation and exercise ($p < 0.05$). We then evaluated mitochondrial respiration and H_2O_2 generation in permeabilized red gastrocnemius muscle fiber bundles, and in line with expectations, recorded an exercised-induced increase in the rate of mitochondrial complex I + II oxygen consumption (+glutamate, malate, ADP, and succinate) ($p < 0.001$; Figure 5B). In contrast to COX activity, compared with the CON and

EX groups, we detected no significant changes in the corresponding groups in which mice had been exposed to PM, although non-significant reductions in levels were detected in the PME group compared with those in the EX group. Similarly, there were no significant exercise-associated differences between the CON, PM, EX, and PME groups concerning respiratory control ratio values ($p < 0.001$; Figure 5C). However, exercise was found to promote a significant increase in the generation of state I H_2O_2 ($p < 0.001$), with exposure PM also inducing the production of 48% and 60% higher levels of H_2O_2 compared with those of CON and EX groups, respectively ($p < 0.001$; Figure 5D).

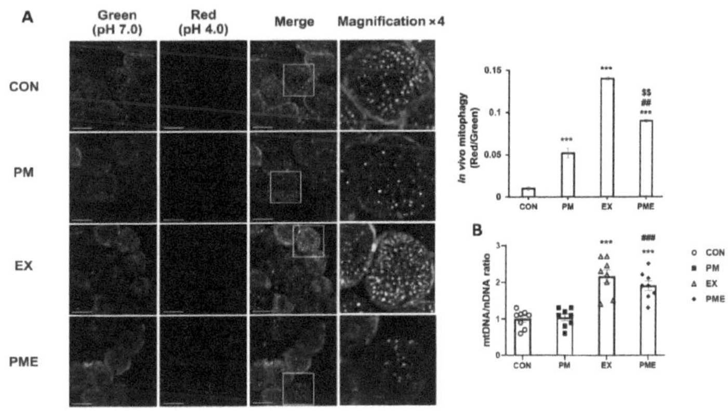

Figure 4. In vivo mitophagy levels in the skeletal muscles of mt-Keima mice were modified by particulate matter exposure and exercise. (**A**) Representative confocal images showing superimposed red/green signals in the skeletal muscle of mt-Keima mouse. The yellow signal Indicates merged red and green fluorescence. The white square indicates area shown magnified in each inset (4x magnification). Control (CON, n = 4), particulate matter exposure (PM, n = 4), exercise (EX, n = 4), and particulate matter exposure + exercise (PME, n = 4). (**B**) Level of mitochondrial biogenesis (mtDNA to nDNA ratio). Control (CON, n = 8), particulate matter exposure (PM, n = 8), exercise (EX, n = 8), and particulate matter exposure + exercise (PME, n = 8). Data are presented as the mean ± SEM. Statistical significance is assigned as *** $p < 0.001$ vs. CON, ## $p < 0.01$ and ### $p < 0.001$ vs. PM, and $$ $p < 0.01$ vs. EX. Scale bar, 50 μm.

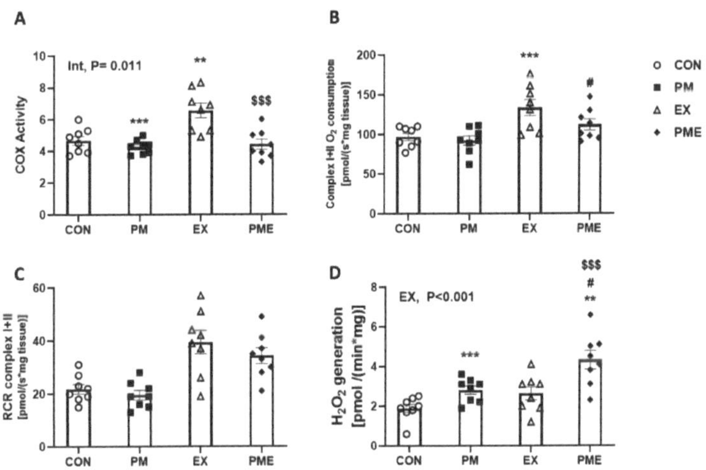

Figure 5. Mitochondrial respiratory function in the muscles of C57BL/6 mice exposed to particulate matter during exercise. (**A**) COX activity assay, (**B**) O_2 consumption, (**C**) respiratory control ratio (RCR),

(**D**) basal rate of hydrogen peroxide (H_2O_2) generation. Control (CON, n = 8), particulate matter exposure (PM, n = 8), exercise (EX, n = 8), and particulate matter exposure + exercise (PME, n = 8). Data are presented as the means ± SEM. Statistical significance is assigned as ** $p < 0.01$, *** $p < 0.001$ vs. CON, # $p < 0.05$ vs. PM and $$$ $p < 0.05$ vs. EX. Two-way ANOVA results were shown with *p* values. Int, interaction; EX, exercise.

4. Discussion

It is assumed that PM has adverse effects on skeletal muscle integrity, either through direct deposition in skeletal muscle tissues or by inducing the release of proinflammatory cytokines in the lungs. In this study, we accordingly examined the effects of PM inhalation on the redox status, inflammatory responses, and mitochondrial function of skeletal muscle. To date, experimental animals have been injected with or exposed to different concentrations of PM to gain insights into the mechanisms by which these particles influence the function and integrity of different body tissues [34]. In the present study, we developed a novel experimental model based on a PM generation chamber containing a mouse treadmill designed to mimic actual human PM inhalation conditions (Figure 1), using which, we examined the effects of PM inhalation during exercise on selected dependent variables. The findings of several studies conducted to date have indicated that exposure to PM increases oxidative stress in the pulmonary and cardiovascular systems, leading to lipid peroxidation and subsequent downstream inflammatory signals [9,35]. It has also been demonstrated that in addition to the lungs, deposits of PM can accumulate in a number of other organs, including the brain, liver, heart, and peripheral blood vessels [13,36]. These findings accordingly indicate that by triggering local and/or systemic inflammation and oxidative stress, prolonged exposure to PM may heighten the risk of direct damage to multiple vital organs [37]. In the present study, we also detected elevated levels of lipid oxidation (as evaluated by the oxidative stress-related production of MDA) in the gastrocnemius muscle of mice exposed to PM, signifying that PM may also have a detrimental effect on skeletal muscles.

Biological ROS and reactive nitrogen species can be generated in different cellular compartments, including the mitochondria, peroxisomes, endoplasmic reticulum, and phagocytes. An excessive production of ROS can disrupt cellular homeostasis and compromise the immune system [38], and multiple studies have provided evidence of elevated levels of oxidative stress in skeletal muscle following exercise, associated with the production of diverse ROS free radicals, including superoxide anion radicals ($O_2^{\bullet-}$), hydroxyl radicals (·OH), hydroperoxyl radicals (HOO.), singlet oxygen ($1O_2$), and free nitrogen radicals [39]. Consistently, in the present study we established the inhalation of PM exacerbates oxidative stress, as evidenced by a reduction in MnSOD activity and GSH/GSSG levels. However, heightened oxidative stress in skeletal muscles, resulting from either exercise or PM exposure, can engender varying degrees of adaptability. For example, an acute spike in exercise-induced oxidative stress is unlikely to culminate in chronic problems. However, due to the establishment of higher antioxidant capacities promoted by regular exercise [40,41], this may not hold true for PM exposure.

Concerning cytokine production, a previous study on humans, in which the authors measured the levels of IL-6 and TNF-α in the exhaled breath condensate of individuals with and without asthma, revealed a strong correlation between the concentration of PM and elevated levels of proinflammatory cytokines in people with asthma [42]. Consistently, other studies have established a positive association between short-term exposure to PM_{10} and elevated levels of circulating IL-1β, IL-6, and TNF-α in the general adult population. These positive correlations accordingly indicate an association between air pollution and heightened cardiovascular risk [43]. Furthermore, the findings of other studies have revealed elevated levels of proinflammatory cytokines in different organs, including the brain, kidneys, and lungs, of animals exposed to different concentrations of PM, either via inhalation or injection [37,44–46]. In the present study, we similarly found that respiratory inhalation of PM during exercise induced an increase in proinflammatory cytokines in the skeletal muscles of mice. Rapid fluctuations in the levels of IL-6, IL-1β, and TNF-α within skeletal muscle imply their role in regulating muscle cell degradation and apoptosis, as well as muscle fiber atrophy and hypertrophy [47]. These responses would thus tend to indicate that prolonged exposure to heightened levels of proinflammatory cytokines is associated with unfavorable physiological outcomes [48]. Notably, we detected a correlation between PM inhalation and exercise regarding their effects on TNF-α, indicating the necessity for further mechanistic studies to gain a more comprehensive understanding of the specific mechanisms by which PM inhalation stimulates the release of TNF-α during exercise. Interestingly, in addition to modifying the inflammatory cytokine profile, we also established that PM induces changes in the activity of MnSOD, an enzyme found exclusively in the mitochondrial matrix. Similar to the findings of previous studies indicating that PM also inhibits MnSOD activity in various different tissues, in addition to its effect on muscle tissue, we found that PM significantly reversed the exercise-induced increase in MnSOD activity. Additionally, it has previously been observed that the lung tissues of mice exposed to high concentrations of $PM_{2.5}$ for three months were characterized by a significantly lower GSH/GSSG ratio, while the findings of another study on $PM_{2.5}$-induced lung fibrosis revealed a reduction in the GSH/GSSG ratio and MnSOD activity [49,50]. It is worth noting that the gastrocnemius muscle of both the control and exercise groups of mice examined in the present study were characterized by reductions in the GSH/GSSG ratio, which we speculate could be attributable to the systemic dissemination of inflammatory factors induced by PM in the lungs and airways, as stated in the introduction. However, it is also plausible that PM with extremely small particle size can be deposited directly in muscle tissue. Accordingly, further research is required to determine the precise contributory mechanisms.

Muscle loss and weakness associated with reduced physical activity and exercise are common features of a range of disorders, including diabetes, cancer, kidney failure, and heart failure, and occur as part of the general aging process, a condition referred to as sarcopenia [51]. This catabolic state is associated with marked changes in mitochondrial content, morphology, and function. A deterioration in skeletal muscle functional capacity may involve a reduction in oxidative capacity and resistance to fatigue [52], which may occur because of mitochondrial dysfunction. Mitochondria are the primary energy-producing organelles in cells that support a diverse range of biological processes associated with metabolism, growth, and the regeneration of skeletal muscle [53]. The maladaptive responses linked to malfunctioning mitochondria are attributed to changes in mitochondrial quality control, which includes mitochondrial synthesis (biogenesis), remodeling (dynamics), and degradation (mitophagy) [54,55]. Physical activity contributes to enhancing mitochondrial function by activating mitochondrial biogenesis and mitophagy, which may underlie the beneficial effects of physical activity in the context of several diseases [56]. As anticipated, we discovered that exercise was associated with a significant increase in the mtDNA/nDNA ratio of skeletal muscle. Nevertheless, we found no evidence to indicate that PM inhalation influences mitochondrial biogenesis, which thereby tends to indicate that mitochondrial quality control, which reflects mitochondrial homeostasis, is activated by other factors. Mitophagy has frequently been evaluated by quantifying proteins associated with this process. For example, it has been established that in skeletal muscle, mitophagy is mediated via the activation of Unc-51-like autophagy-activating kinase (ULK1), resulting in the formation of autophagosomes. Additionally, the expression of BNIP3/NIX has been found to trigger the initiation of mitophagy. Subsequently, light chain 3-I (LC3-I) undergoes conversion to a lipid-modified form, light chain 3-II (LC3-II, the phosphatidylethanolamine conjugated form of LC3-I). Additionally, the autophagosomal structural protein, p62/SQSTM1, recruits damaged mitochondria, which leads to the clearance of damaged mitochondria via the degradative activity of LC3. Finally, lysosomal degradative enzymes degrade the mitochondria after several additional processes [57,58]. However, although multiple studies have quantitatively analyzed these proteins to assess mitophagy, it is notably more difficult to characterize the dynamics of mitophagy in vivo by scrutinizing these proteins as a static representation. Consequently, in this study we employed an in vivo model of mitophagy, using mt-Keima mice, which express a pH-dependent fluorescent protein, mt-Keima, thereby enabling a more intuitive and accurate assessment of mitophagy [32]. Using this model, we established that exercise was associated with an increase in in vivo mitophagy, as previously observed in other studies. Moreover, the level was significantly increased by the PM, with values 4–6 times higher than those observed in the EX and PME groups. Accordingly, this in vivo PM-induced increase in mitochondrial removal signaling within skeletal muscles can be considered a novel finding of the present study. Nevertheless, the systemic effects of PM remain unclear, and it is yet to be established whether localized inflammatory and oxidative stresses are promoted directly by the deposition of PM in tissues. In addition, further studies are needed to determine whether PM molecules mediate regulation of the specific proteins involved in mitophagy.

Using permeabilized red gastrocnemius muscle fibers to assess mitochondrial function, we detected increases in both O_2 consumption and respiratory control ratio for co mplex I + II in response to exercise, whereas levels tended to decline in mice exposed to PM treatment, albeit non-significantly. Contrastingly, exposure to PM induced a significant increase in the production of H_2O_2, with the most pronounced effects being detected in response to the inhalation of PM during exercise. Collectively, these findings indicate that although PM triggers a marked increase in mitochondrial ROS production, this does not appear to significantly impair mitochondrial respiration. This outcome would imply that the implicated reactions are controlled dose-dependently by pollutants, such as PM, which should be verified by further experiments assessing different PM exposure concentrations, and durations and intensities of exercise. As an initial investigation into the impact of PM hyper-inhalation on skeletal muscles during exercise, the findings of this study certainly merit additional validation regarding diseases affecting muscle. Models of sarcopenia, an age-related degeneration of skeletal muscle, can also be used. Furthermore, conducting additional studies to examine the effects of different types and intensities of exercise could contribute to the development of appropriate exercise programs to counteract the potentially detrimental effects of exposure to ambient air pollution, particularly that attributable to PM. Additionally, it remains unclear whether intramuscular adhesion of PM or circulatory effects of inflammatory responses in the respiratory system, such as the lungs, are responsible. Any recommendations in this regard should, nevertheless, consider the tradeoff between the benefits of physical activity and the adverse effects of PM inhalation.

5. Conclusions

PM adheres to the lung bronchi and alveoli, causing localized inflammation. Given its small particle sizes, $PM_{2.5}$ can circulate systemically, and has potentially adverse effects on the heart, blood vessels, liver, and kidneys. Our findings in this study reveal that inhaling PM also exacerbates inflammation and oxidative stress in skeletal muscle. We also observed mitochondrial oxidative stress, which is comparable to excessive inhalation of PM during exercise (Figure 6). Further studies should focus on analyzing the effects of exposure to different concentrations of PM, in conjunction with different intensities and durations of exercise, which could contribute to guiding exercise schedules for PM-polluted environments.

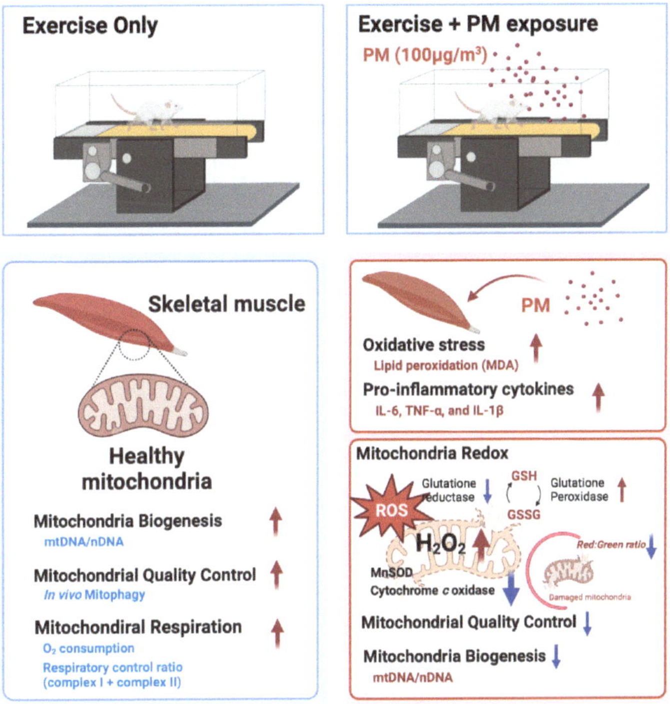

Figure 6. A graphical summary depicting the impact of particulate matter (PM) and exercise on mouse skeletal muscle and mitochondrial function. The exercise group exhibited an increase in mitochondrial biogenesis (mtDNA/nDNA), mitophagy, and mitochondrial respiration. PM inhalation during treadmill exercise leads to increased oxidative stress and elevated levels of proinflammatory cytokines, as evaluated based on analyses of malondialdehyde (MDA) and interleukin-6 (IL-6), tumor necrosis factor-alpha (TNF-α), and interleukin-1 beta (IL-1β). The ratio of reduced (GSH) to oxidized (GSSG) glutathione levels is reduced in response to PM exposure, and mitochondrial biogenesis and mitophagy are reduced in the muscles of mice exposed to PM during exercise.

Author Contributions: Conceptualization, C.K.; methodology, C.K.; software, Z.Z. and W.S.S.; validation, Z.Z., W.S.S. and C.K.; formal analysis, Z.Z. and W.S.S.; investigation, Z.Z., W.S.S., J.P. and C.K.; resources, C.K.; data curation, Z.Z. and W.S.S.; writing—original draft preparation, C.K.; writing—review and editing, J.P., J.J., B.S., K.L., D.Y. and C.K.; visualization, J.P., J.J. and B.S.; supervision, C.K.; project administration, C.K.; funding acquisition, C.K. All authors have read and agreed to the published version of the manuscript.

Funding: This research was funded by an INHA University Research Grant to C.K.

Institutional Review Board Statement: The animal study protocol was approved by the Institutional Animal Care and Use Committee of Inha University and Tianjin University of Sport (certificate: INHA 190211-616, approved 11 February 2019; TJUS 2022-021, approved February 2022).

Informed Consent Statement: Not applicable.

Data Availability Statement: The data obtained in this study are available upon reasonable request to the corresponding author.

Acknowledgments: The authors thank Jin-Soo Kim at the Exercise Medicine Research Institute, Edith Cowan University, Australia, for his assistance in graphical representation.

Conflicts of Interest: The authors declare that they have no conflicts of interest.

References

1. Shah, A.S.; Langrish, J.P.; Nair, H.; McAllister, D.A.; Hunter, A.L.; Donaldson, K.; Newby, D.E.; Mills, N.L. Global association of air pollution and heart failure: A systematic review and meta-analysis. *Lancet* **2013**, *382*, 1039–1048. [CrossRef] [PubMed]
2. Tian, Y.; Harrison, R.M.; Feng, Y.; Shi, Z.; Liang, Y.; Li, Y.; Xue, Q.; Xu, J. Size-resolved source apportionment of particulate matter from a megacity in northern China based on one-year measurement of inorganic and organic components. *Environ. Pollut.* **2021**, *289*, 117932. [CrossRef] [PubMed]
3. Agency, U.E.P. National ambient air quality standards for particulate matter. *Fed. Regist.* **1997**, *62*, 138.
4. Dockery, D.W.; Pope, C.A., 3rd. Acute respiratory effects of particulate air pollution. *Annu. Rev. Public Health* **1994**, *15*, 107–132. [CrossRef] [PubMed]
5. Prahalad, A.K.; Inmon, J.; Dailey, L.A.; Madden, M.C.; Ghio, A.J.; Gallagher, J.E. Air pollution particles mediated oxidative DNA base damage in a cell free system and in human airway epithelial cells in relation to particulate metal content and bioreactivity. *Chem. Res. Toxicol.* **2001**, *14*, 879–887. [CrossRef] [PubMed]
6. Wang, F.; Liu, J.; Zeng, H. Interactions of particulate matter and pulmonary surfactant: Implications for human health. *Adv. Colloid. Interface Sci.* **2020**, *284*, 102244. [CrossRef] [PubMed]
7. Rajagopalan, S.; Brook, R.D. Air pollution and type 2 diabetes: Mechanistic insights. *Diabetes* **2012**, *61*, 3037–3045. [CrossRef]
8. Pope, C.A.; Bhatnagar, A.; McCracken, J.P.; Abplanalp, W.; Conklin, D.J.; O'Toole, T. Exposure to Fine Particulate Air Pollution Is Associated with Endothelial Injury and Systemic Inflammation. *Circ. Res.* **2016**, *119*, 1204–1214. [CrossRef]
9. Haberzettl, P.; Bhatnagar, A.; Conklin, D.J. Particulate Matter and Oxidative Stress—Pulmonary and Cardiovascular Targets and Consequences. In *Systems Biology of Free Radicals and Antioxidants*; Laher, I., Ed.; Springer: Berlin/Heidelberg, Germany, 2014; pp. 1557–1586.
10. Schraufnagel, D.E.; Balmes, J.R.; Cowl, C.T.; De Matteis, S.; Jung, S.H.; Mortimer, K.; Perez-Padilla, R.; Rice, M.B.; Riojas-Rodriguez, H.; Sood, A.; et al. Air Pollution and Noncommunicable Diseases: A Review by the Forum of International Respiratory Societies' Environmental Committee, Part 2: Air Pollution and Organ Systems. *Chest* **2019**, *155*, 417–426. [CrossRef]
11. Yang, Z.; Mahendran, R.; Yu, P.; Xu, R.; Yu, W.; Godellawattage, S.; Li, S.; Guo, Y. Health Effects of Long-Term Exposure to Ambient PM(2.5) in Asia-Pacific: A Systematic Review of Cohort Studies. *Curr. Environ. Health Rep.* **2022**, *9*, 130–151. [CrossRef]
12. To, T.; Zhu, J.; Villeneuve, P.J.; Simatovic, J.; Feldman, L.; Gao, C.; Williams, D.; Chen, H.; Weichenthal, S.; Wall, C.; et al. Chronic disease prevalence in women and air pollution—A 30-year longitudinal cohort study. *Environ. Int.* **2015**, *80*, 26–32. [CrossRef]
13. Feng, S.; Huang, F.; Zhang, Y.; Feng, Y.; Zhang, Y.; Cao, Y.; Wang, X. The pathophysiological and molecular mechanisms of atmospheric PM2.5 affecting cardiovascular health: A review. *Ecotoxicol. Environ. Saf.* **2023**, *249*, 114444. [CrossRef]
14. Iizuka, K.; Machida, T.; Hirafuji, M. Skeletal muscle is an endocrine organ. *J. Pharmacol. Sci.* **2014**, *125*, 125–131. [CrossRef]
15. Huang, Y.C. The role of in vitro gene expression profiling in particulate matter health research. *J. Toxicol. Environ. Health B Crit. Rev.* **2013**, *16*, 381–394. [CrossRef]
16. Smyth, T.; Georas, S.N. Effects of ozone and particulate matter on airway epithelial barrier structure and function: A review of in vitro and in vivo studies. *Inhal. Toxicol.* **2021**, *33*, 177–192. [CrossRef] [PubMed]
17. Valderrama, A.; Ortiz-Hernández, P.; Agraz-Cibrián, J.M.; Tabares-Guevara, J.H.; Gómez, D.M.; Zambrano-Zaragoza, J.F.; Taborda, N.A.; Hernandez, J.C. Particulate matter (PM(10)) induces in vitro activation of human neutrophils, and lung histopathological alterations in a mouse model. *Sci. Rep.* **2022**, *12*, 7581. [CrossRef]
18. Garber, C.E.; Blissmer, B.; Deschenes, M.R.; Franklin, B.A.; Lamonte, M.J.; Lee, I.M.; Nieman, D.C.; Swain, D.P. American College of Sports Medicine position stand. Quantity and quality of exercise for developing and maintaining cardiorespiratory, musculoskeletal, and neuromotor fitness in apparently healthy adults: Guidance for prescribing exercise. *Med. Sci. Sports Exerc.* **2011**, *43*, 1334–1359. [CrossRef] [PubMed]
19. He, F.; Li, J.; Liu, Z.; Chuang, C.C.; Yang, W.; Zuo, L. Redox Mechanism of Reactive Oxygen Species in Exercise. *Front. Physiol.* **2016**, *7*, 486. [CrossRef]
20. Suzuki, K.; Tominaga, T.; Ruhee, R.T.; Ma, S. Characterization and Modulation of Systemic Inflammatory Response to Exhaustive Exercise in Relation to Oxidative Stress. *Antioxidants* **2020**, *9*, 401. [CrossRef]
21. He, Z.; Xu, Q.; Newland, B.; Foley, R.; Lara-Sáez, I.; Curtin, J.F.; Wang, W. Reactive oxygen species (ROS): Utilizing injectable antioxidative hydrogels and ROS-producing therapies to manage the double-edged sword. *J. Mater. Chem. B* **2021**, *9*, 6326–6346. [CrossRef]
22. Schippinger, G.; Wonisch, W.; Abuja, P.M.; Fankhauser, F.; Winklhofer-Roob, B.M.; Halwachs, G. Lipid peroxidation and antioxidant status in professional American football players during competition. *Eur. J. Clin. Invest.* **2002**, *32*, 686–692. [CrossRef] [PubMed]
23. Vincent, H.K.; Powers, S.K.; Demirel, H.A.; Coombes, J.S.; Naito, H. Exercise training protects against contraction-induced lipid peroxidation in the diaphragm. *Eur. J. Appl. Physiol. Occup. Physiol.* **1999**, *79*, 268–273. [CrossRef] [PubMed]
24. Vincent, H.K.; Powers, S.K.; Stewart, D.J.; Demirel, H.A.; Shanely, R.A.; Naito, H. Short-term exercise training improves diaphragm antioxidant capacity and endurance. *Eur. J. Appl. Physiol.* **2000**, *81*, 67–74. [CrossRef] [PubMed]
25. Rothschild, J.A.; Bishop, D.J. Effects of dietary supplements on adaptations to endurance training. *Sports Med.* **2020**, *50*, 25–53. [CrossRef]
26. Battaglia, A.M.; Chirillo, R.; Aversa, I.; Sacco, A.; Costanzo, F.; Biamonte, F. Ferroptosis and Cancer: Mitochondria Meet the "Iron Maiden" Cell Death. *Cells* **2020**, *9*, 1505. [CrossRef] [PubMed]

27. Zeng, Y.; Zhu, G.; Zhu, M.; Song, J.; Cai, H.; Song, Y.; Wang, J.; Jin, M. Edaravone Attenuated Particulate Matter-Induced Lung Inflammation by Inhibiting ROS-NF-κB Signaling Pathway. *Oxid. Med. Cell Longev.* **2022**, *2022*, 6908884. [CrossRef]
28. Marmett, B.; Pires Dorneles, G.; Böek Carvalho, R.; Peres, A.; Roosevelt Torres Romão, P.; Barcos Nunes, R.; Ramos Rhoden, C. Air pollution concentration and period of the day modulates inhalation of PM(2.5) during moderate- and high-intensity interval exercise. *Environ. Res.* **2021**, *194*, 110528. [CrossRef]
29. Kang, J.; Jia, Z.; Ping, Y.; Liu, Z.; Yan, X.; Xing, G.; Yan, W. Testosterone alleviates mitochondrial ROS accumulation and mitochondria-mediated apoptosis in the gastric mucosa of orchiectomized rats. *Arch. Biochem. Biophys.* **2018**, *649*, 53–59. [CrossRef]
30. Tostes, R.C.; Carneiro, F.S.; Carvalho, M.H.; Reckelhoff, J.F. Reactive oxygen species: Players in the cardiovascular effects of testosterone. *Am. J. Physiol. Regul. Integr. Comp. Physiol.* **2016**, *310*, R1–R14. [CrossRef]
31. Katayama, H.; Kogure, T.; Mizushima, N.; Yoshimori, T.; Miyawaki, A. A sensitive and quantitative technique for detecting autophagic events based on lysosomal delivery. *Chem. Biol.* **2011**, *18*, 1042–1052. [CrossRef]
32. Sun, N.; Yun, J.; Liu, J.; Malide, D.; Liu, C.; Rovira, I.I.; Holmström, K.M.; Fergusson, M.M.; Yoo, Y.H.; Combs, C.A.; et al. Measuring In Vivo Mitophagy. *Mol. Cell* **2015**, *60*, 685–696. [CrossRef] [PubMed]
33. Czyżowska, A.; Brown, J.; Xu, H.; Sataranatarajan, K.; Kinter, M.; Tyrell, V.J.; O'Donnell, V.B.; Van Remmen, H. Elevated phospholipid hydroperoxide glutathione peroxidase (GPX4) expression modulates oxylipin formation and inhibits age-related skeletal muscle atrophy and weakness. *Redox Biol.* **2023**, *64*, 102761. [CrossRef] [PubMed]
34. Cho, C.C.; Hsieh, W.Y.; Tsai, C.H.; Chen, C.Y.; Chang, H.F.; Lin, C.S. In Vitro and In Vivo Experimental Studies of PM(2.5) on Disease Progression. *Int. J. Environ. Res. Public Health* **2018**, *15*, 1380. [CrossRef] [PubMed]
35. Dey, S.K.; Sugur, K.; Venkatareddy, V.G.; Rajeev, P.; Gupta, T.; Thimmulappa, R.K. Lipid peroxidation index of particulate matter: Novel metric for quantifying intrinsic oxidative potential and predicting toxic responses. *Redox Biol.* **2021**, *48*, 102189. [CrossRef]
36. Oberdörster, G.; Sharp, Z.; Atudorei, V.; Elder, A.; Gelein, R.; Lunts, A.; Kreyling, W.; Cox, C. Extrapulmonary translocation of ultrafine carbon particles following whole-body inhalation exposure of rats. *J. Toxicol. Environ. Health. Part A* **2002**, *65*, 1531–1543. [CrossRef]
37. Wong, L.N.; Aung, H.; Lamé, M.; Wegesser, T.; Wilson, D. Fine particulate matter from urban ambient and wildfire sources from California's San Joaquin Valley initiate differential inflammatory, oxidative stress, and xenobiotic responses in human bronchial epithelial cells. *Toxicol. Vitr.* **2011**, *25*, 1895–1905. [CrossRef]
38. Phaniendra, A.; Jestadi, D.B.; Periyasamy, L. Free radicals: Properties, sources, targets, and their implication in various diseases. *Indian. J. Clin. Biochem.* **2015**, *30*, 11–26. [CrossRef]
39. Powers, S.K.; Nelson, W.B.; Hudson, M.B. Exercise-induced oxidative stress in humans: Cause and consequences. *Free Radic. Biol. Med.* **2011**, *51*, 942–950. [CrossRef]
40. Powers, S.K.; Goldstein, E.; Schrager, M.; Ji, L.L. Exercise Training and Skeletal Muscle Antioxidant Enzymes: An Update. *Antioxidants* **2022**, *12*, 39. [CrossRef]
41. Shin, Y.-A.; Lee, J.-H.; Song, W.; Jun, T.-W. Exercise training improves the antioxidant enzyme activity with no changes of telomere length. *Mech. Ageing Dev.* **2008**, *129*, 254–260. [CrossRef]
42. Ghozikali, M.G.; Ansarin, K.; Naddafi, K.; Nabizadeh, R.; Yaghmaeian, K.; Jaafari, J.; Dehghanzadeh, R.; Atafar, Z.; Faraji, M.; Mohammadi, A.; et al. Status of TNF-α and IL-6 as pro-inflammatory cytokines in exhaled breath condensate of late adolescents with asthma and healthy in the dust storm and non-dust storm conditions. *Sci. Total Environ.* **2022**, *838*, 155536. [CrossRef]
43. Tsai, D.H.; Amyai, N.; Marques-Vidal, P.; Wang, J.L.; Riediker, M.; Mooser, V.; Paccaud, F.; Waeber, G.; Vollenweider, P.; Bochud, M. Effects of particulate matter on inflammatory markers in the general adult population. *Part. Fibre Toxicol.* **2012**, *9*, 24. [CrossRef]
44. Campbell, A.; Oldham, M.; Becaria, A.; Bondy, S.C.; Meacher, D.; Sioutas, C.; Misra, C.; Mendez, L.B.; Kleinman, M. Particulate matter in polluted air may increase biomarkers of inflammation in mouse brain. *Neurotoxicology* **2005**, *26*, 133–140. [CrossRef]
45. Wang, J.; Huang, J.; Wang, L.; Chen, C.; Yang, D.; Jin, M.; Bai, C.; Song, Y. Urban particulate matter triggers lung inflammation via the ROS-MAPK-NF-κB signaling pathway. *J. Thorac. Dis.* **2017**, *9*, 4398–4412. [CrossRef]
46. Yuan, C.-S.; Lai, C.-S.; Chang-Chien, G.-P.; Tseng, Y.-L.; Cheng, F.-J. Kidney damage induced by repeated fine particulate matter exposure: Effects of different components. *Sci. Total Environ.* **2022**, *847*, 157528. [CrossRef]
47. Peake, J.M.; Della Gatta, P.; Suzuki, K.; Nieman, D.C. Cytokine expression and secretion by skeletal muscle cells: Regulatory mechanisms and exercise effects. *Exerc. Immunol. Rev.* **2015**, *21*, 8–25.
48. Sharma, B.; Dabur, R. Role of Pro-inflammatory Cytokines in Regulation of Skeletal Muscle Metabolism: A Systematic Review. *Curr. Med. Chem.* **2020**, *27*, 2161–2188. [CrossRef]
49. Guo, L.; Bai, S.; Ding, S.; Zhao, L.; Xu, S.; Wang, X. PM2.5 Exposure Induces Lung Injury and Fibrosis by Regulating Ferroptosis via TGF-β Signaling. *Dis. Markers* **2022**, *2022*, 7098463. [CrossRef]
50. Wang, H.; Shen, X.; Liu, J.; Wu, C.; Gao, J.; Zhang, Z.; Zhang, F.; Ding, W.; Lu, Z. The effect of exposure time and concentration of airborne PM2.5 on lung injury in mice: A transcriptome analysis. *Redox Biol.* **2019**, *26*, 101264. [CrossRef]
51. Damluji, A.A.; Alfaraidhy, M.; AlHajri, N.; Rohant, N.N.; Kumar, M.; Al Malouf, C.; Bahrainy, S.; Ji Kwak, M.; Batchelor, W.B.; Forman, D.E.; et al. Sarcopenia and Cardiovascular Diseases. *Circulation* **2023**, *147*, 1534–1553. [CrossRef] [PubMed]
52. Southern, W.M.; Ryan, T.E.; Kepple, K.; Murrow, J.R.; Nilsson, K.R.; McCully, K.K. Reduced skeletal muscle oxidative capacity and impaired training adaptations in heart failure. *Physiol. Rep.* **2015**, *3*, e12353. [CrossRef]
53. Romanello, V.; Sandri, M. Mitochondria Quality Control and Muscle Mass Maintenance. *Front. Physiol.* **2016**, *6*, 422. [CrossRef]

54. Dantas, W.S.; Zunica, E.R.M.; Heintz, E.C.; Vandanmagsar, B.; Floyd, Z.E.; Yu, Y.; Fujioka, H.; Hoppel, C.L.; Belmont, K.P.; Axelrod, C.L.; et al. Mitochondrial uncoupling attenuates sarcopenic obesity by enhancing skeletal muscle mitophagy and quality control. *J. Cachexia Sarcopenia Muscle* **2022**, *13*, 1821–1836. [CrossRef]
55. Sligar, J.; DeBruin, D.A.; Saner, N.J.; Philp, A.M.; Philp, A. The importance of mitochondrial quality control for maintaining skeletal muscle function across health span. *Am. J. Physiol.-Cell Physiol.* **2022**, *322*, C461–C467. [CrossRef]
56. Wu, N.N.; Tian, H.; Chen, P.; Wang, D.; Ren, J.; Zhang, Y. Physical Exercise and Selective Autophagy: Benefit and Risk on Cardiovascular Health. *Cells* **2019**, *8*, 1436. [CrossRef]
57. Chen, C.C.W.; Erlich, A.T.; Hood, D.A. Role of Parkin and endurance training on mitochondrial turnover in skeletal muscle. *Skelet. Muscle* **2018**, *8*, 10. [CrossRef]
58. Guan, Y.; Drake, J.C.; Yan, Z. Exercise-Induced Mitophagy in Skeletal Muscle and Heart. *Exerc. Sport. Sci. Rev.* **2019**, *47*, 151–156. [CrossRef] [PubMed]

antioxidants

MDPI

Article

Effect of Particulate Matter 2.5 on Fetal Growth in Male and Preterm Infants through Oxidative Stress

Sunwha Park [1,†], Eunjin Kwon [2,†], Gain Lee [3], Young-Ah You [1], Soo Min Kim [3], Young Min Hur [1], Sooyoung Jung [1], Yongho Jee [4], Mi Hye Park [5], Sung Hun Na [6], Young-Han Kim [7], Geum Joon Cho [8], Jin-Gon Bae [9], Soo-Jeong Lee [10], Sun Hwa Lee [11] and Young Ju Kim [1,3,*]

1 Department of Obstetrics and Gynecology, College of Medicine, Ewha Womans University, Seoul 07985, Republic of Korea; clarrissa15@gmail.com (S.P.); yerang02@naver.com (Y.-A.Y.); k0507hym@hanmail.net (Y.M.H.); jsmed9006@naver.com (S.J.)

2 Division of Allergy and Respiratory Disease Research, Department of Chronic Disease Convergence Research, Korea National Institute of Health, Cheongju-si 28159, Republic of Korea; friendkej1004@hanmail.net

3 Graduate Program in System Health Science and Engineering, Ewha Womans University, Seoul 07985, Republic of Korea; loveleee0102@gmail.com (G.L.); soomnium@naver.com (S.M.K.)

4 Advanced Biomedical Research Institute, Ewha Womans University Seoul Hospital, Seoul 07804, Republic of Korea; jyongho@ewha.ac.kr

5 Department of Obstetrics and Gynecology, Ewha Womans University Seoul Hospital, Seoul 07804, Republic of Korea; ewhapmh@ewha.ac.kr

6 Department of Obstetrics and Gynecology, School of Medicine, Kangwon National University, Chuncheon-si 24289, Republic of Korea; lahun@kangwon.ac.kr

7 Department of Obstetrics and Gynecology, College of Medicine, Yonsei University, Seoul 03722, Republic of Korea; yhkim522@yuhs.ac

8 Department of Obstetrics and Gynecology, College of Medicine, Korea University, Seoul 02841, Republic of Korea; geumjoon@korea.ac.kr

9 Department of Obstetrics and Gynecology, School of Medicine, Keimyung University, Dongsan Medical Center, Daegu 42601, Republic of Korea; gonmd@dsmc.or.kr

10 Department of Obstetrics and Gynecology, College of Medicine, Ulsan University, Ulsan 44610, Republic of Korea; exsjlee@uuh.ulsan.kr

11 Seegene Medical Foundation, Seoul 04805, Republic of Korea; lshkim@neolab.co.kr

* Correspondence: kkyj@ewha.ac.kr

† These authors contributed equally to this work.

check for updates

Citation: Park, S.; Kwon, E.; Lee, G.; You, Y.-A.; Kim, S.M.; Hur, Y.M.; Jung, S.; Jee, Y.; Park, M.H.; Na, S.H.; et al. Effect of Particulate Matter 2.5 on Fetal Growth in Male and Preterm Infants through Oxidative Stress. *Antioxidants* **2023**, *12*, 1916. https://doi.org/10.3390/antiox12111916

Academic Editors: Reto Asmis and Yasuhiro Yoshida

Received: 19 September 2023
Revised: 17 October 2023
Accepted: 24 October 2023
Published: 26 October 2023
Corrected: 23 January 2024

Abstract: Particulate matter 2.5 ($PM_{2.5}$) levels are associated with adverse pregnancy outcomes. In this retrospective cohort study, we examined whether the concentration of indoor $PM_{2.5}$ affected pregnancy outcomes. Additionally, we evaluated biomarkers of pregnancy-related complications caused by fine dust. We collected clinical information and data based on residential addresses from the Air Korea database to assess $PM_{2.5}$ exposure levels. As a multicenter prospective cohort study, we measured the indoor $PM_{2.5}$ concentration and inflammatory and oxidative stress markers. The $PM_{2.5}$ concentration of the low-birth-weight (LBW) delivery group was 27.21 μg/m^3, which was significantly higher than that of the normal-birth-weight (NBW) group (26.23 μg/m^3) ($p = 0.02$). When the newborns were divided by sex, the $PM_{2.5}$ concentration of the LBW group was 27.89 μg/m^3 in male infants, which was significantly higher than that of the NBW group (26.26 μg/m^3) ($p = 0.01$). In the prospective study, 8-hydroxy-2-deoxyguanosine significantly increased in the high-concentration group (113.55 ng/mL, compared with 92.20 ng/mL in the low-concentration group); in the high-concentration group, the rates of preterm birth (PTB) and small size for gestational age significantly increased ($p < 0.01$, $p = 0.01$). This study showed an association between $PM_{2.5}$, oxidative stress, and fetal growth, with the PTB group being more vulnerable.

Keywords: 8-hydroxy-2-deoxyguanosine; biomarker; indoor air; low birth weight; particulate matter; preterm birth; sex difference; small for gestational age

1. Introduction

Low birth weight (LBW; birth weight < 2500 g, regardless of gestational age), small for gestational age (SGA; birth weight below the 10th percentile for gestational age), and preterm birth (PTB; delivery at <37 weeks of gestation) are complications of pregnancy that directly affect the prognosis of newborns [1–7]. The incidence of LBW varies by country (range, 6–25%); however, LBW births are associated with short- and long-term complications [8,9]. The prevalence of SGA births is approximately twice that of LBW [10]. Previous studies have investigated sociodemographic and medical risk factors as well as environmental risks (i.e., exposure to toxic substances) for LBW [10].

Particulate matter (PM) 2.5 ($PM_{2.5}$, particles with an aerodynamic diameter of $\leq$2.5 μm), one of the major air pollutants, has been reported to be associated with various adverse pregnancy outcomes [11,12]. In particular, associations between $PM_{2.5}$ and LBW, PTB, and SGA infants have been reported [5–7,11]. Most studies have conducted exposure assessments using outdoor air quality measurements [5,7,11,13]. Recently, it was found that 80–90% of individuals live indoors, and the importance of indoor air quality has been gradually emphasized [14,15]. Therefore, the importance of measuring indoor PM concentrations is increasing, and individual indoor fine dust exposure may reveal better causal relationships between $PM_{2.5}$ and pregnancy outcomes [14,16,17].

There is a lack of understanding of the pathogenesis of $PM_{2.5}$ affecting fetal growth. Various studies have been conducted to understand the cause of fetal growth restriction [18,19]. The analysis of sex differences is useful in understanding biological mechanisms. Although biological mechanisms are influenced by PM regardless of sex, sex-specific effects remain controversial [11,16,20]. It is also controversial whether exposure to PM at any stage of pregnancy has a greater effect on fetal growth [21,22].

Recently, various biomarkers have been developed to evaluate exposure to fine dust [23], and attempts have been made to explain fetal growth based on DNA methylation or telomere length in cord blood [16,24–26]. However, cord blood is not suitable for biomonitoring, because it is difficult to sample during pregnancy. Although there are many studies on oxidative stress, inflammation, DNA damage, and epigenetic modulation as biomarkers of exposure to PM, there are few biomarker studies on pregnant women [23].

Therefore, this study aimed to examine whether the concentration of $PM_{2.5}$ affects fetal growth through indoor PM measurement. We also evaluated biomarkers of pregnancy-related complications caused by fine dust as well as indicators for the biomonitoring of $PM_{2.5}$ exposure.

2. Materials and Methods

2.1. Study Design of Cohort I

The first study was a hospital-based retrospective cohort study of 1880 pregnant women who delivered live babies between 2010 and 2015 at the Ewha Womans University Mokdong Hospital (EUMC 2020-07-043). We collected information on maternal age; body mass index (BMI); gestational age at birth (GAB); neonatal sex; weight and height at birth; appearance, pulse, grimace, activity, and respiration (APGAR) score; and place of residence at the time of delivery. The Air Korea database was used for $PM_{2.5}$ exposure assessment, which includes hourly accumulated air pollution monitoring data for components such as sulfur dioxide, PM_{10}, carbon monoxide, nitrogen dioxide, and ozone from the Ministry of Environment of Korea. The level of $PM_{2.5}$ exposure was measured using the Community Multiscale Air Quality (CMAQ) modeling system. $PM_{2.5}$, from CMAQ modeling data, was estimated using meteorological research and forecasting models comprising three overlapping weather data sources at 3, 9, and 27 km for a specific time period. $PM_{2.5}$ exposure levels during pregnancy were determined based on residential address (city, county, and district, or si, gun, and gu in the Korean language). For each address, daily $PM_{2.5}$ concentrations were matched to each individual according to their delivery date. Each individual's outdoor air quality data were collected for each pregnancy trimester, extending to data from the week of delivery. $PM_{2.5}$ data were collected from 2009 to 2015, depending on the pregnancy period of the study participants.

We stratified the groups by birth weight and height and performed Student's *t*-test (Figure 1a). LBW was defined as a weight at birth of <2500 g, and low birth height (LBH) was defined as a height at birth of <46.3 cm for male and <45.6 cm for female infants.

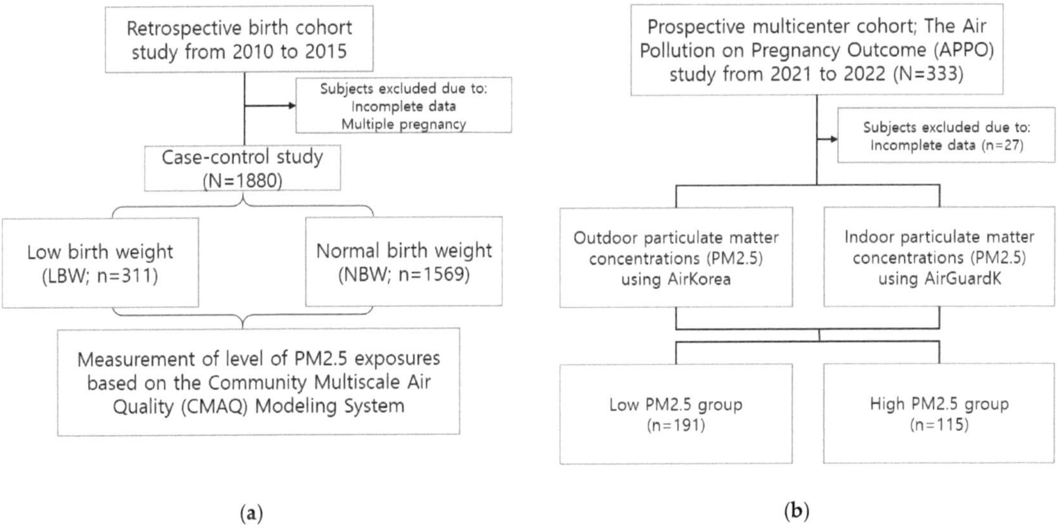

Figure 1. Study flow chart of Cohorts I (**a**) and II (**b**).

2.2. Study Design of Cohort II

A multicenter prospective cohort study of air pollution in pregnant women (APPO) was conducted to investigate the effects of PM on mothers and fetuses by recruiting > 1200 participants between January 2021 and December 2023 in seven university hospitals across Korea. Each hospital was located in a metropolitan area, an industrial complex, or a mountainous area. The participants were singleton pregnant women with no underlying diseases before 28 weeks of gestation.

In the second study, 333 women who had delivered were selected as participants (Figure 1b). We collected data on basic demographic and health-related characteristics including age, BMI, socioeconomic status, and obstetric history. Routine blood tests were conducted to measure the white blood cell (WBC) counts and high-sensitivity C-reactive protein (hs-CRP) levels as inflammatory markers, and urine samples were collected during the second trimester. The pregnancy outcomes were evaluated after delivery.

This study was approved by the Ethical Research Committee of Ewha Womans University Mokdong Hospital (EUMC 2021-04-032), Yonsei University Severance Hospital (4-2021-0414), Kangwon National University Hospital (KNUH-B-2021-04-012-008), Keimyung University Dongsan Medical Center (2021-04-073), Korea University Guro Hospital (2021GR0233), Ewha Womans University Seoul Hospital (2021-04-022), and Ulsan University Hospital (2022-04-020). All participants provided written informed consent.

2.3. PM Exposure Assessment

The indoor $PM_{2.5}$ concentration was measured by placing a fine dust meter at breathing height in the living room of a pregnant woman's house. An AirGuard K (Kweather Co., Seoul, Republic of Korea) instrument was used for measurements using the sensor-based method. The indoor $PM_{2.5}$ was measured online at 1-min intervals. The measured indoor $PM_{2.5}$ data were stored in an indoor air quality monitoring platform (IAQ Station) using Long-Term Evolution. Measurements were performed for at least 1 week, and the measured concentration values were recorded in real time using the Internet of Things and Information and Communication Technology. For indoor $PM_{2.5}$, after removing outliers, the average value of measurements for each trimester of pregnancy was derived, and the average of the two or three trimesters was defined as $PM_{2.5}$. The participants were divided according to the $PM_{2.5}$ exposure level as follows: <10 $\mu g/m^3$ (low $PM_{2.5}$ group) and $\geq$10 $\mu g/m^3$ (high $PM_{2.5}$ group) concentration groups, which are the World Health Organization's annual limit standard concentrations (2005 air quality guidelines).

Outdoor $PM_{2.5}$ concentrations were collected from a nearby urban atmospheric measurement network based on the residential addresses of the study participants. The Urban Air Monitoring Station data used in this study were obtained from the Air Korea database (https://www.airkorea.or.kr/web (accessed on 1 May 2021)) of the Korean Ministry of the Environment.

2.4. Collection of Blood and Urine Samples

We collected 5 mL maternal venous blood and 15 mL urine. Blood and urine samples were collected at regular follow-up visits. Whole blood samples collected in ethylenediaminetetraacetic acid tubes were transferred to cryotubes, and urine was stored in one cryotube. Samples were transferred to the institution (the Seegene Medical Foundation, Seoul, Republic of Korea) on the same day as collection under refrigeration to prevent deterioration (−80 °C).

2.5. Measurement of Oxidative Stress and Inflammatory Markers

The oxidative stress markers 8-hydroxy-2-deoxyguanosine (8-OHdG) and malondialdehyde (MDA) were measured in urine samples collected during the second trimester of pregnancy. The collected urine samples were transported directly to the laboratory and stored at −80 °C before processing. The urine samples were filtered with a 0.2-µm filter (Sartorius, S6534-FMOSK, Göttingen, Germany) for the assay and diluted at 1:20 to measure the concentration of 8-OHdG using enzyme-linked immunosorbent assay (ELISA) kits (Abcam, Ab201734, Waltham, Boston, MA, USA). The concentration of MDA in urine was measured using an MDA ELISA kit (ab118970; Abcam). Among the inflammatory markers, the WBC count was measured using an XN-9000 (Symex, Kobe, Japan) according to the manufacturer's protocol, from the participant's whole blood sample, through an automated complete blood cell count. The hs-CRP level was measured using a particle-enhanced immunoturbidimetric assay according to the manufacturer's protocol using a Cobas 8000 C702 analyzer (Roche, Basel, Switzerland).

2.6. Statistical Analysis

Clinical characteristics were analyzed according to continuous variables (age and BMI) and categorical variables (marital status, education level, occupation, monthly income, gravidity, and pregnancy methods). Pregnancy outcomes were analyzed according to continuous (GAB, birth weight, height, and APGAR score) and categorical variables (delivery mode, neonate sex, and pregnancy complications). Pregnancy complications were defined using the following criteria (LBW: birth weight of <2500 g; LBH: birth height <46.3 cm for male and <45.6 cm for female infants; SGA: birth weight < 10th percentile for gestational age; PTB: delivery at <37 weeks of gestation) and analyzed as categorical variables. Oxidative stress (8-OHdG and MDA) and inflammatory markers (hs-CRP and WBC count) were analyzed as continuous variables. Categorical variables were expressed as frequencies (percentages) and analyzed using chi-square and Fisher's exact tests. Continuous variables were expressed as means ± standard deviations (SDs) and were compared using the t-test or Mann–Whitney U test. Logistic regression model was used to estimate the associations between LBW and $PM_{2.5}$ exposure (1 $\mu g/m^3$) in each trimester with adjustment for covariates (maternal age, parity, maternal pre-pregnancy BMI, preeclampsia, and gestational age). Statistical significance was defined as $p < 0.05$. All statistical analyses were performed using the Statistical Package for the Social Sciences (version 20.0; IBM Corp., Armonk, NY, USA).

3. Results

3.1. Association between $PM_{2.5}$ Exposure and Birth Weight and Height According to Neonatal Sex Using a Retrospective Cohort Study

Among the 1880 neonates, 1569 had a normal birth weight (NBW) and 311 had LBW. The two groups showed significant differences in GAB, birth weight, height, and APGAR scores ($p < 0.01$); however, there were no significant differences in maternal age, pre-pregnancy BMI, and neonatal sex (Table 1). The average outdoor $PM_{2.5}$ concentration was 27.35 μg/m^3 (Table 2 and Supplementary Table S1). The $PM_{2.5}$ concentration in the second trimester for pregnant women who delivered LBW babies was 27.21 μg/m^3, which was significantly higher than the value of 26.23 μg/m^3 in the NBW group ($p = 0.02$, Table 3). When the delivered newborns were divided by sex, the $PM_{2.5}$ concentration in the second trimester for male LBW babies was 27.89 μg/m^3, which was significantly higher than the value of 26.26 μg/m^3 in male infants with NBW ($p = 0.01$), but there was no significant difference in female infants ($p = 0.74$, Table 3).

Table 1. Characteristics of study population of Cohort I.

Characteristics	NBW (*n* = 1569)		LBW (*n* = 311)		*p*-Value
Age (years)	33.20	4.16	32.79	4.47	0.12
Pre-BMI (kg/m^2)	21.46	3.59	21.52	3.65	0.92
GAB (wks)	39.02	1.35	33.30	3.49	<0.01 *
Neonate Sex					
Male	791	82.14%	172	17.86%	0.12
Female	780	84.78%	140	15.22%	
Birth weight (g)	3266.44	394.83	1824.04	497.55	<0.01 *
Birth height (cm)	49.76	1.88	42.08	4.06	<0.01 *
APGAR 1 min	9.30	1.12	6.76	2.54	<0.01 *
APGAR 5 min	9.88	0.55	8.27	2.24	<0.01 *
Pregnancy complications					
PTB	81	24.40%	251	75.60%	<0.01 *

Categorical variables are expressed as frequencies (percentages) and were analyzed using the chi-square test. Continuous variables are expressed as means ± SDs and were compared using *t*-tests. APGAR, appearance, pulse, grimace, activity, respiration; BMI, body mass index; GAB, gestational age at birth; SD, standard deviation. * $p < 0.05$ considered statistically significant.

Table 2. Summary of $PM_{2.5}$ concentration (μg/m^3) by second trimester of pregnancy.

	Mean	SD	Min	25th	50th	75th	Max	*p*-Value
[Cohort I]	27.35	4.48	17.73	23.33	27.89	30.72	41.68	<0.01 *,[1]
[Cohort II]								<0.01 *,[2]
Indoor	10.57	10.47	0.53	4.35	7.8	12.02	69.25	
Outdoor	17.27	7.48	1.69	12.00	16.71	20.14	43.50	

[1] Statistical significance of concentration of $PM_{2.5}$ between Cohorts I and II. [2] Statistical significance between indoor and outdoor $PM_{2.5}$ of Cohort II. $PM_{2.5}$, particulate matter 2.5. * $p < 0.05$ considered statistically significant.

Table 3. Comparison of $PM_{2.5}$ exposure (μg/m^3) by each trimester in NBW, LBW, NBH, and LBH, according to sex.

Exposure Period	All Newborns NBW (*n* = 1567)	All Newborns LBW (*n* = 313)	*p*-Value	Males NBW (*n* = 791)	Males LBW (*n* = 169)	*p*-Value	Females NBW (*n* = 774)	Females LBW (*n* = 141)	*p*-Value
Entire	27.17 ± 4.30	27.40 ± 4.67	0.43	27.24 ± 4.37	27.79 ± 4.82	0.17	27.11 ± 4.24	26.93 ± 4.48	0.65
First trimester	28.75 ± 7.63	28.04 ± 7.70	0.15	28.91 ± 7.87	28.38 ± 7.94	0.45	28.59 ± 7.38	27.71 ± 7.42	0.21
Second trimester	26.23 ± 6.78	27.21 ± 6.53	0.02 *	26.26 ± 6.70	27.89 ± 6.79	0.01 *	26.22 ± 6.86	26.38 ± 6.15	0.78
Third trimester	26.40 ± 6.43	27.04 ± 8.56	0.23	26.39 ± 6.28	26.94 ± 8.63	0.46	26.38 ± 6.57	27.03 ± 8.40	0.40

Continuous variables are expressed as means ± SDs and compared using *t*-tests. LBH, low birth height; LBW, low birth weight; NBH, normal birth height; NBW, normal birth weight. * $p < 0.05$ considered statistically significant.

When the neonates were divided into term and PTB groups, the concentration of $PM_{2.5}$ for the LBW infants in the PTB group was 27.29 ug/m^3, which was significantly higher than the concentration of 25.89 μg/m^3 for NBW babies, respectively (p = 0.02, Table 4). In the PTB group, the $PM_{2.5}$ concentration of the preterm normal birth weight (PNBW) and preterm low birth weight (PLBW) groups showed significant differences between the first and second trimesters, respectively (p = 0.05, p = 0.04), with more significant differences in the male group ($p < 0.01$, p = 0.01, Table 4). In contrast, in female infants in the PTB group, there were no statistically significant differences between the PNBW and PLBW groups in any trimester (p = 0.88, Table 4). In the term birth group, there was no significant difference between the NBW and LBW groups at term in any trimester (p = 0.30; Supplementary Table S2).

Table 4. Subgroup analysis for comparison of $PM_{2.5}$ exposure (μg/m^3) by each trimester in the NBW and LBW at preterm in Cohort I.

Exposure Period	All Newborns			Male Infants			Female Infants		
	PNBW (n = 84)	PLBW (n = 260)	p-Value	PNBW (n = 48)	PLBW (n = 149)	p-Value	PNBW (n = 36)	PLBW (n = 111)	p-Value
Entire	25.89 ± 4.15	27.29 ± 4.78	0.02 *	25.06 ± 4.03	27.61 ± 4.90	<0.01 *	27.00 ± 4.10	26.87 ± 4.64	0.88
First trimester	25.81 ± 6.63	27.58 ± 7.57	0.05	24.35 ± 6.21	27.73 ± 7.65	<0.01 *	27.77 ± 6.76	27.48 ± 7.50	0.84
Second trimester	25.57 ± 6.26	27.25 ± 6.58	0.04 *	24.82 ± 5.87	27.81 ± 6.82	0.01 *	26.58 ± 6.69	26.51 ± 6.22	0.95
Third trimester	26.69 ± 6.15	27.23 ± 8.98	0.55	26.92 ± 5.61	27.27 ± 9.09	0.80	26.39 ± 6.86	27.02 ± 8.77	0.66

Continuous variables were expressed as means ± SDs and compared using t-test. PNBW, preterm normal birth weight; PLBW, preterm low birth weight. * $p < 0.05$ considered statistically significant.

Pregnant women who were exposed to $PM_{2.5}$ had a significantly higher risk of LBW than those who were not exposed, with an adjusted odds ratio (OR) of 1.06 (95% confidence interval [CI]: 1.01–1.10, Table 5). In male infants, the risk of LBW was higher, with an OR of 1.12. The susceptibility period was in the first and second trimester, with ORs of 1.05 and 1.07, respectively (95% CI: 1.01–1.10 and 1.03–1.12, Table 5).

Table 5. Association between $PM_{2.5}$ exposure (per 1 μg/m^3) in each trimester and logistic regression analysis.

	Risk of LBW by $PM_{2.5}$ Exposure (per 1 μg/m^3)								
	All Newborns			Male Infants			Female Infants		
Stage	OR	95% CI	p-Value	OR	95% CI	p-Value	OR	95% CI	p-Value
Entire	1.06	1.01–1.10	0.02 *	1.12	1.04–1.20	<0.01 *	1.00	0.93–1.07	0.94
First trimester	1.02	1.00–1.05	0.08	1.05	1.01–1.10	0.02 *	1.00	0.96–1.04	0.86
Second trimester	1.03	1.00–1.06	0.04 *	1.07	1.03–1.12	<0.01 *	1.00	0.96–1.04	0.90
Third trimester	1.00	0.98–1.03	0.80	1.00	0.96–1.04	0.85	1.01	0.97–1.05	0.73

Models adjusted for maternal age, parity, maternal pre-pregnancy body mass index, preeclampsia, and gestational age. CI, confidence interval; OR, odds ratio. * $p < 0.05$ considered statistically significant.

3.2. *Study Population of APPO and Measurement and Correlation of Indoor/Outdoor $PM_{2.5}$*

An analysis was performed on 306 of 333 delivery participants, excluding those with incomplete data. According to the $PM_{2.5}$ exposure levels of the participants, the numbers in the low and high $PM_{2.5}$ groups were 191 and 115, respectively (Table 6). There were no significant differences in the characteristics of the study population between the high and low $PM_{2.5}$ concentration groups (Table 6).

The average of the indoor $PM_{2.5}$ concentration in the second trimester of pregnancy among the APPO study participants was 10.57 μg/m^3, and that of the outdoor $PM_{2.5}$ was 17.27 μg/m^3. The two measurements showed a statistically significant positive correlation ($p < 0.01$, r^2 = 0.187) (Table 2, Figure 2a). The concentration of outdoor $PM_{2.5}$ in Cohort I was significantly lower than that in Cohort II ($p < 0.01$, Table 2).

Table 6. Characteristics of study population in Cohort II.

Characteristics	Low $PM_{2.5}$ (n = 191)		High $PM_{2.5}$ (n = 115)		p-Value
Age (years)	33.74	±3.62	33.12	±4.50	0.19
Pre-BMI (kg/m^2)	21.86	±3.31	21.56	±3.17	0.43
Married state					
Married	167	100.0%	133	97.8%	0.16
Unmarried	0	0.0%	3	2.2%	
Education level					0.14
High school graduation or below	13	6.8%	13	9.5%	
University graduates	154	92.3%	123	90.4%	
Occupation					
Yes	112	67.1%	94	69.1%	0.70
No	55	32.9%	42	30.9%	
Monthly income					
<4 million won	35	36.1%	35	37.6%	0.95
4–6 million won	27	27.8%	24	25.8%	
>6 million won	35	36.1%	34	36.6%	
Gravidity					
1	80	47.9%	68	50.0%	0.65
2	61	36.5%	50	36.8%	
≥3	26	15.6%	18	13.2%	
Pregnancy methods					
Natural	143	85.6%	116	85.3%	0.91
IUI	2	1.2%	1	0.7%	
IVF-ET	22	13.2%	19	14.0%	

Categorical variables are expressed as frequencies (percentages) and were analyzed using chi-square and Fisher's exact tests. Continuous variables are expressed as means ± SDs and were compared using the t-test or Mann–Whitney U test. IUI, intrauterine insemination; IVF-ET, in vitro fertilization–embryo transfer.

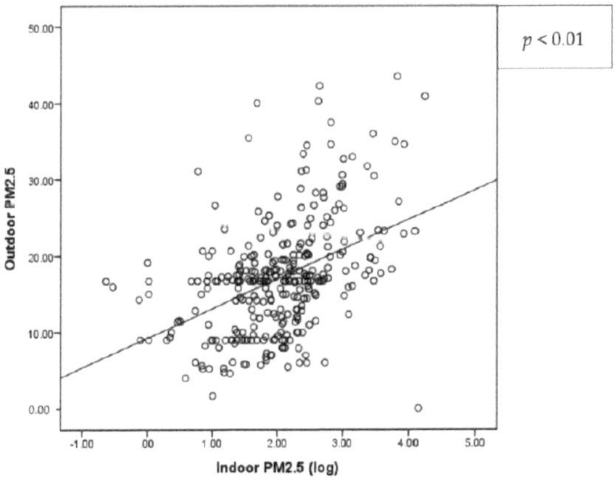

(a)

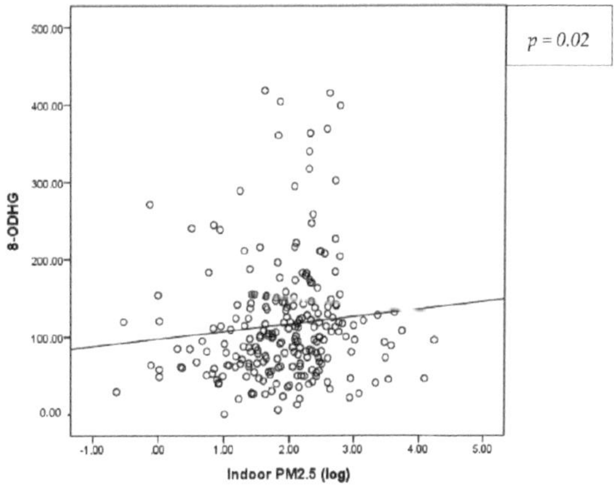

(b)

Figure 2. Correlation analysis between (**a**) indoor and outdoor $PM_{2.5}$, and (**b**) indoor $PM_{2.5}$ and 8-OHdG. 8-OHdG, 8-hydroxy-2-deoxyguanosine; $PM_{2.5}$, particulate matter 2.5.

3.3. Association between $PM_{2.5}$ Exposure and Pregnancy Complications

The oxidative stress marker 8-OHdG was significantly increased in the high-concentration group to 113.55 ng/mL, compared to 92.20 ng/mL in the low-concentration group, and there was a positive correlation (p = 0.02, r^2 = 0.010) (Table 7, Figure 2b). There were no differences in the MDA or inflammatory marker levels between the two groups (Table 7). There were no differences in GAB, delivery mode, neonatal sex, birth weight, or APGAR score (Table 8). The mean birth height was 48.9 cm in the high-concentration group, which was significantly lower than that in the low-concentration group (49.5 cm; $p = 0.04$; Table 8). In the high-concentration group, PTB significantly increased from 4.2% to 13.2%, and SGA significantly increased from 0.6% to 5.1% ($p < 0.01$, $p = 0.01$) (Table 8, Figure 3).

Table 7. Oxidative stress and inflammatory markers between the low and high $PM_{2.5}$ groups.

	Low $PM_{2.5}$ (n = 191)		High $PM_{2.5}$ (n = 115)		p-Value
Oxidative stress marker					
8-OHdG (ng/mL)	92.20	61.16–138.19	113.55	79.69–153.27	0.02 *
MDA (μM)	18.49	8.66–37.18	16.02	9.99–30.34	0.54
Inflammatory marker					
Hs-CRP (mg/L)	2.00	1.00–3.00	1.79	1.00–2.99	0.74
WBC (1 × 10^3/μL)	8.80	7.00–9.72	8.00	7.00–10.00	0.98

Continuous variables are expressed as medians (interquartile ranges) and were compared using the Mann–Whitney U test. 8-OHdG, 8-hydroxy-2-deoxyguanosine; Hs-CRP, high-sensitivity C-reactive protein; MDA, malondialdehyde; WBC, white blood cells. * $p < 0.05$ considered statistically significant.

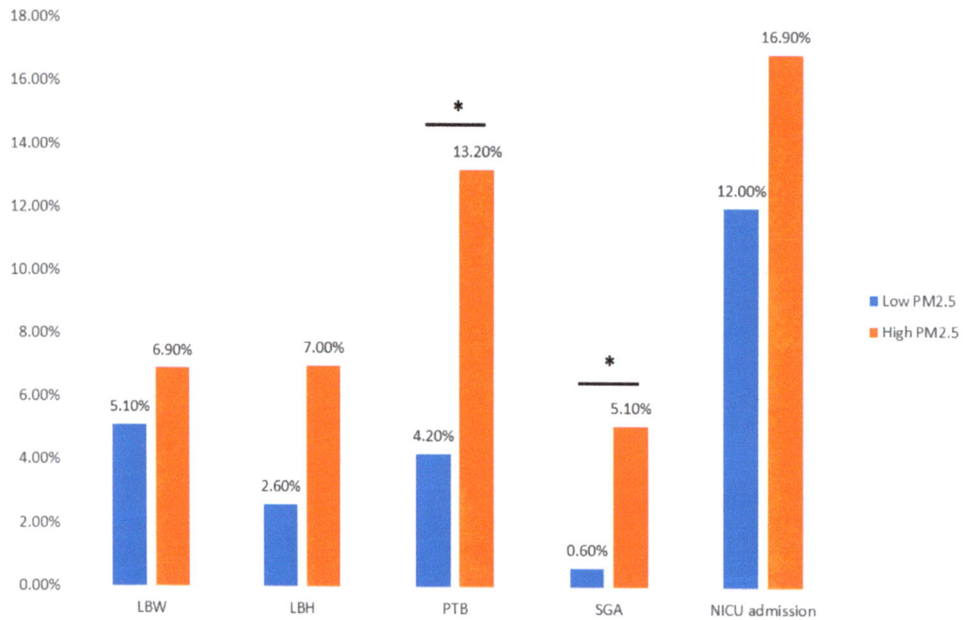

Figure 3. Pregnancy complications in the $PM_{2.5}$ concentration group. LBW, low birth weight; LBH, low birth height; PTB, preterm birth; SGA, small for gestational age; NICU, neonatal intensive care unit. * $p < 0.05$ considered statistically significant.

Table 8. Association between $PM_{2.5}$ exposure and pregnancy complications.

	Low $PM_{2.5}$ (n = 191)		High $PM_{2.5}$ (n = 115)		p-Value
Pregnancy outcome					
GAB (wks)	38.22	±1.59	37.96	±1.95	0.22
Delivery mode					0.86
ND	61	36.5%	51	37.5%	
CS	106	63.5%	85	62.5%	

Table 8. *Cont.*

	Low $PM_{2.5}$ (n = 191)		High $PM_{2.5}$ (n = 115)		p-Value
Neonate Sex					
Male	112	58.9%	56	49.1%	0.10
Female	78	41.1%	58	50.9%	
Birth weight (g)	3150.7	±391.6	3112.0	±522.5	0.50
Birth height (cm)	49.5	±2.2	48.9	±2.8	0.04 *
APGAR 1 min	8.45	±1.22	8.36	±	0.61
APGAR 5 min	9.43	±0.84	9.21	±1.37	0.13
Pregnancy complications					
LBW	4	5.1%	4	6.9%	0.47
LBH	5	2.6%	8	7.0%	0.07
PTB	8	4.2%	15	13.2%	<0.01 *
SGA	1	0.6%	7	5.1%	0.01 *
NICU admission	20	12.0%	23	16.9%	0.32

Categorical variables are expressed as frequencies (percentages) and were analyzed using chi-square and Fisher's exact tests. Continuous variables are expressed as means ± SDs or medians (interquartile range) and were compared using the t-test or Mann–Whitney U test. CS, cesarean section; ND, normal delivery; NICU, neonatal intensive care unit; SGA, small for gestational age. * $p < 0.05$ considered statistically significant.

4. Discussion

In this study, the association between $PM_{2.5}$ and LBW was confirmed through a retrospective cohort study, and its effect was found to be more pronounced in the male and PTB groups. Furthermore, a multicenter prospective cohort study revealed the relationship between $PM_{2.5}$ and birth height, SGA, and PTB by measuring the actual indoor $PM_{2.5}$. The oxidative stress marker 8-OHdG, which was used to explore the pathogenesis of these pregnancy complications and identify biomarkers, showed a positive correlation with $PM_{2.5}$.

Our results are similar to those of previous studies that showed an association between $PM_{2.5}$ and fetal growth and PTB [11,27–30]. Regarding the biological mechanism by which $PM_{2.5}$ affects fetal growth, the effects of oxidative stress, placental inflammation or dysfunction, endothelial dysfunction, and blood coagulation have been reported [18,19]. The pathogenesis of PM-induced PTB involves systemic inflammation caused by the inhalation of toxic particles [31,32]. Many large-scale studies have confirmed these results; however, most have been retrospective [4,11,13], which is a limitation. Although the number of participants was small, pregnancy complications were confirmed in a prospective study using actual indoor $PM_{2.5}$ measurement.

Several studies have shown that male fetuses are vulnerable to intrauterine growth restriction [27,33] and PTB [34]. Although the sex difference in fetuses affected by $PM_{2.5}$ remains controversial [11,16,33], this study showed an association between LBW and $PM_{2.5}$, especially in male infants. To understand the sex differences, it is necessary to understand the biological mechanisms underlying $PM_{2.5}$ [35]. Male fetuses grow faster than female fetuses and require more oxygen; therefore, the possibility that toxic substances inhibit this process has been suggested [33,35]. Another mechanism is that an increase in inflammatory mediators in the blood caused by air pollution increases the blood viscosity [36], which affects placental function. In general, placental dysfunction is more prevalent in male fetuses [37,38]. The cohort of the APPO study could not confirm the sex differences owing to the small number of participants, but it is necessary to confirm whether $PM_{2.5}$ actually affects fetal growth to continuously recruit participants in the future.

Similar to previous studies showing the relationships between $PM_{2.5}$ and environmental pollutants and oxidative stress, this study showed a relationship between $PM_{2.5}$ and 8-OHdG levels [39]. In brief, 8-OHdG is a reactive oxygen species that has been used as a marker of DNA damage because of its mutagenic potential [40]. The results of this study show the possibility of using 8-OHdG as a biomarker of $PM_{2.5}$ exposure. In particular, urine samples have the advantage that they can be collected from patients non-invasively, and since 8-OHdG shows high stability in urine, it can be used as a biomarker [41]. By verifying the effectiveness of 8-OHdG in the future, it can be used as a biomarker for biomonitoring, suggesting the possibility that the use of antioxidants affects the prognosis of newborns [24]. PM not only affects the outcomes of newborns and pregnancy but also long-term diseases in offspring, according to the concept of fetal programming [42,43].

Most existing studies have estimated fine dust exposure concentrations based on differences in air quality by region and location [4,11,28,44]. However, as 80–90% of individuals live indoors, owing to COVID-19, the outdoor activity of pregnant women further decreased, and the importance of indoor air quality has been emphasized [45,46]. In this APPO study, indoor $PM_{2.5}$ measurements were performed in the residences of pregnant women for at least 1 week, and it is thought that this can give causality to the biological mechanism of any relationship between $PM_{2.5}$ and a disease. The results of this study, using correlation analysis between indoor and outdoor $PM_{2.5}$, explain that outdoor conditions significantly affect indoor air quality; however, the correlation can be explained by only 18.7% influence. This indicates that the indoor air quality can be affected by various human activities, including smoking and cooking [14,17], thereby suggesting that education regarding lifestyle changes may also be important for better pregnancy outcomes. Through this survey, we will conduct additional research on the group in which indoor $PM_{2.5}$ increases in preparation for actual outdoor $PM_{2.5}$. We plan to conduct further research on active intervention measures to prevent pregnancy complications.

In this study, there was a difference in $PM_{2.5}$ concentrations between Cohorts I and II, either because of the air pollution reduction effect owing to the COVID-19 lockdown [47] or because the participants in Cohort I were recruited from a metropolitan area, whereas the participants in Cohort II were recruited from the countryside.

To our knowledge, this was the first study to examine the relationship between fine dust and fetal growth in pregnant Korean women with continuously measured indoor $PM_{2.5}$ concentrations. It was also the first study to measure oxidative stress and develop fine dust exposure evaluation indicators in pregnant women with actual indoor $PM_{2.5}$ measurement.

The strength of this study was that it was a prospective multicenter cohort study that investigated the maternal and fetal health effects of PM on pregnancy in patients from various regions of South Korea. Compared with a previous study that measured only outdoor data, it was more reasonable to confirm the causal relationship between fine dust and pregnancy complications through fine dust concentrations measured using individual indoor air quality values.

However, this study had several limitations. Although we measured the indoor fine dust concentrations of the participants for at least 1 week, there was a limitation in that the cumulative concentration during the entire pregnancy could not be calculated. Moreover, other stressful conditions that may affect pregnancy complications, including alcohol abuse and infection, were not considered. Recent, various studies have reported on air pollutants, including $PM_{1.0}$, and their health effects; however, this study focused on $PM_{2.5}$ and did not analyze other air pollutants.

A correlation between $PM_{2.5}$ and 8-OHdG was observed, but no correlation with actual complications was found. Therefore, in the future, we will recruit more participants and make efforts to calculate the cumulative concentrations of fine dust in pregnant women and evaluate the biological mechanisms of 8-OHdG as a biomarker. Moreover, various stressful conditions that can be confounding variables in pregnancy outcomes should be included, and the impact of other air pollutants should be considered during the analysis.

5. Conclusions

This study confirmed the association between $PM_{2.5}$ and LBW through a retrospective cohort study, and its effect was found to be more pronounced in the male and PTB groups. Furthermore, through a multicenter prospective cohort study, we determined the relationships between $PM_{2.5}$ and birth height, SGA, and PTB by measuring the actual indoor

$PM_{2.5}$. The oxidative stress marker showed a positive correlation with $PM_{2.5}$, suggesting its potential as a biomarker for $PM_{2.5}$ exposure. However, further research on the actual mechanisms of action, efficacy, and relevance to pregnancy outcomes is needed. Furthermore, we will continue to identify interventions to lower indoor PM levels.

Supplementary Materials: The following supporting information can be downloaded at: https://www.mdpi.com/article/10.3390/antiox12111916/s1, Table S1: Summary of $PM_{2.5}$ concentration ($\mu g/m^3$) by each trimester of pregnancy of Cohort I; Table S2: Subgroup analysis of comparison of $PM_{2.5}$ exposure ($\mu g/m^3$) by each trimester in the NBW and LBW at full term of Cohort I.

Author Contributions: S.P. enrolled the participants, interpreted the data, and wrote and edited the manuscript. E.K. analyzed the data and wrote the manuscript. G.L. curated the data and performed the experiments. Y.-A.Y. designed the study and edited the manuscript. S.M.K. performed the experiments. Y.M.H. developed the protocol and enrolled the subjects. S.J. performed the data curation and visualization. Y.J. curated and analyzed data. M.H.P., S.H.N., Y.-H.K., G.J.C., J.-G.B. and S.-J.L. enrolled the participants. S.H.L. developed the protocols and performed the experiments. Y.J.K. edited the manuscript, obtained funding, and supervised the study. All authors have read and agreed to the published version of the manuscript.

Funding: This study was supported by a research program funded by the Korea National Institute of Health (grant code# 2021-ER1208-01). This work was partly supported by the Institute of Information Communications Technology Planning Evaluation (grant funded by the Korea Government [MSIT]) (No. RS 2022 00155966 Artificial Intelligence Convergence Innovation Human Resources Development Ewha Womans University) and a Korea Evaluation Institute of Industrial Technology (KEIT) grant funded by the Korea Government (MOTIE). This research was also supported by BK21 FOUR (Fostering Outstanding Universities for Research) funded by the Ministry of Education (MOE, Korea) and National Research Foundation of Korea (NRF-5199990614253).

Institutional Review Board Statement: The first retrospective cohort study was approved by the Ethical Research Committee of Ewha Womans University Mokdong Hospital (EUMC 2020-07-043). As this was a retrospective cohort study, the consent of the subjects was not required. The second prospective cohort study was approved by the Ethical Research Committee of Ewha Womans University Mokdong Hospital (EUMC 2021-04-032), Yonsei University Severance Hospital (4-2021-0414), Kangwon National University Hospital (KNUH-B-2021-04-012-008), Keimyung University Dongsan Medical Center (2021-04-073), Korea University Guro Hospital (2021GR0233), Ewha Womans University Seoul Hospital (2021-04-022), and Ulsan University Hospital (2022-04-020). All the participants provided written informed consent.

Informed Consent Statement: Informed consent was obtained from all subjects involved in the study.

Data Availability Statement: The data supporting the findings of this study are available in the article. Raw data supporting the findings of this study cannot be shared openly to protect personally identifiable information.

Conflicts of Interest: The authors declare no conflict of interest.

References

1. Park, S.; Moon, J.; Kang, N.; Kim, Y.H.; You, Y.A.; Kwon, E.; Ansari, A.; Hur, Y.M.; Park, T.; Kim, Y.J. Predicting preterm birth through vaginal microbiota, cervical length, and WBC using a machine learning model. *Front. Microbiol.* **2022**, *13*, 912853. [CrossRef] [PubMed]
2. Goldenberg, R.L.; Culhane, J.F.; Iams, J.D.; Romero, R. Epidemiology and causes of preterm birth. *Lancet* **2008**, *371*, 75–84. [CrossRef] [PubMed]
3. Yang, T.; Chen, R.; Gu, X.; Xu, J.; Yang, L.; Zhao, J.; Zhang, X.; Bai, C.; Kang, J.; Ran, P.; et al. Association of fine particulate matter air pollution and its constituents with lung function: The China Pulmonary Health study. *Environ. Int.* **2021**, *156*, 106707. [CrossRef] [PubMed]
4. Shen, Y.; Wang, C.; Yu, G.; Meng, X.; Wang, W.; Kan, H.; Zhang, J.; Cai, J. Associations of Ambient Fine Particulate Matter and Its Chemical Constituents with Birth Weight for Gestational Age in China: A Nationwide Survey. *Environ. Sci. Technol.* **2022**, *56*, 8406–8415. [CrossRef]
5. Gray, S.C.; Edwards, S.E.; Miranda, M.L. Assessing exposure metrics for PM and birth weight models. *J. Expo. Sci. Environ. Epidemiol.* **2010**, *20*, 469–477. [CrossRef]

6. Percy, Z.; DeFranco, E.; Xu, F.; Hall, E.S.; Haynes, E.N.; Jones, D.; Muglia, L.J.; Chen, A. Trimester specific $PM_{2.5}$ exposure and fetal growth in Ohio, 2007–2010. *Environ. Res.* **2019**, *171*, 111–118. [CrossRef]
7. Kloog, I.; Melly, S.J.; Ridgway, W.L.; Coull, B.A.; Schwartz, J. Using new satellite based exposure methods to study the association between pregnancy $PM_{2.5}$ exposure, premature birth and birth weight in Massachusetts. *Environ. Health* **2012**, *11*, 40. [CrossRef]
8. Han, Z.; Lutsiv, O.; Mulla, S.; Rosen, A.; Beyene, J.; McDonald, S.D. Low gestational weight gain and the risk of preterm birth and low birthweight: A systematic review and meta-analyses. *Acta Obstet. Gynecol. Scand.* **2011**, *90*, 935–954. [CrossRef]
9. Salihu, H.M.; Garcia, B.Y.; Dongarwar, D.; Maiyegun, S.O.; Yusuf, K.K.; Agili, D.E.A. Maternal pre-pregnancy underweight and the risk of small-for-gestational-age in Asian-American ethnic groups. *Obstet. Gynecol. Sci.* **2021**, *64*, 496–505. [CrossRef]
10. Valero De Bernabé, J.; Soriano, T.; Albaladejo, R.; Juarranz, M.; Calle, M.E.; Martínez, D.; Domínguez-Rojas, V. Risk factors for low birth weight: A review. *Eur. J. Obstet. Gynecol. Reprod. Biol.* **2004**, *116*, 3–15. [CrossRef]
11. Bachwenkizi, J.; Liu, C.; Meng, X.; Zhang, L.; Wang, W.; van Donkelaar, A.; Martin, R.V.; Hammer, M.S.; Chen, R.; Kan, H. Maternal exposure to fine particulate matter and preterm birth and low birth weight in Africa. *Environ. Int.* **2022**, *160*, 107053. [CrossRef]
12. Stieb, D.M.; Chen, L.; Eshoul, M.; Judek, S. Ambient air pollution, birth weight and preterm birth: A systematic review and meta-analysis. *Environ. Res.* **2012**, *117*, 100–111. [CrossRef]
13. Guo, P.; Chen, Y.; Wu, H.; Zeng, J.; Zeng, Z.; Li, W.; Zhang, Q.; Huo, X.; Feng, W.; Lin, J.; et al. Ambient air pollution and markers of fetal growth: A retrospective population-based cohort study of 2.57 million term singleton births in China. *Environ. Int.* **2020**, *135*, 105410. [CrossRef] [PubMed]
14. Zhang, L.; Ou, C.; Magana-Arachchi, D.; Vithanage, M.; Vanka, K.S.; Palanisami, T.; Masakorala, K.; Wijesekara, H.; Yan, Y.; Bolan, N.; et al. Indoor Particulate Matter in Urban Households: Sources, Pathways, Characteristics, Health Effects, and Exposure Mitigation. *Int. J. Environ. Res. Public Health* **2021**, *18*, 1055. [CrossRef]
15. Shezi, B.; Jafta, N.; Naidoo, R.N. Exposure assessment of indoor particulate matter during pregnancy: A narrative review of the literature. *Rev. Environ. Health* **2020**, *35*, 427–442. [CrossRef] [PubMed]
16. Cho, H.J.; Lee, S.H.; Lee, S.Y.; Kim, H.C.; Kim, H.B.; Park, M.J.; Yoon, J.; Jung, S.; Yang, S.I.; Lee, E.; et al. Mid-pregnancy $PM_{2.5}$ exposure affects sex-specific growth trajectories via ARRDC3 methylation. *Environ. Res.* **2021**, *200*, 111640. [CrossRef] [PubMed]
17. Patelarou, E.; Kelly, F.J. Indoor exposure and adverse birth outcomes related to fetal growth, miscarriage and prematurity-a systematic review. *Int. J. Environ. Res. Public Health* **2014**, *11*, 5904–5933. [CrossRef]
18. Kim, J.Y.; Lee, E.Y.; Choi, I.; Kim, J.; Cho, K.H. Effects of the Particulate $Matter_{2.5}$ ($PM_{2.5}$) on Lipoprotein Metabolism, Uptake and Degradation, and Embryo Toxicity. *Mol. Cells* **2015**, *38*, 1096–1104. [CrossRef]
19. Nääv, Å.; Erlandsson, L.; Isaxon, C.; Åsander Frostner, E.; Ehinger, J.; Sporre, M.K.; Krais, A.M.; Strandberg, B.; Lundh, T.; Elmér, E.; et al. Urban PM2.5 Induces Cellular Toxicity, Hormone Dysregulation, Oxidative Damage, Inflammation, and Mitochondrial Interference in the HRT8 Trophoblast Cell Line. *Front. Endocrinol.* **2020**, *11*, 75. [CrossRef]
20. Rosa, M.J.; Hsu, H.L.; Just, A.C.; Brennan, K.J.; Bloomquist, T.; Kloog, I.; Pantic, I.; Mercado García, A.; Wilson, A.; Coull, B.A.; et al. Association between prenatal particulate air pollution exposure and telomere length in cord blood: Effect modification by fetal sex. *Environ. Res.* **2019**, *172*, 495–501. [CrossRef]
21. Liu, X.C.; Strodl, E.; Wu, C.A.; Huang, L.H.; Yin, X.N.; Wen, G.M.; Sun, D.L.; Xian, D.X.; Chen, W.Q. Critical window for the association between prenatal environmental tobacco smoke exposure and preterm birth. *Environ. Res.* **2022**, *212*, 113427. [CrossRef] [PubMed]
22. Chen, W.J.; Rector, A.M.; Guxens, M.; Iniguez, C.; Swartz, M.D.; Symanski, E.; Ibarluzea, J.; Ambros, A.; Estarlich, M.; Lertxundi, A.; et al. Susceptible windows of exposure to fine particulate matter and fetal growth trajectories in the Spanish INMA (INfancia y Medio Ambiente) birth cohort. *Environ. Res.* **2022**, *216*, 114628. [CrossRef] [PubMed]
23. Guo, C.; Lv, S.; Liu, Y.; Li, Y. Biomarkers for the adverse effects on respiratory system health associated with atmospheric particulate matter exposure. *J. Hazard. Mater.* **2022**, *421*, 126760. [CrossRef] [PubMed]
24. Lee, A.G.; Cowell, W.; Kannan, S.; Ganguri, H.B.; Nentin, F.; Wilson, A.; Coull, B.A.; Wright, R.O.; Baccarelli, A.; Bollati, V.; et al. Prenatal particulate air pollution and newborn telomere length: Effect modification by maternal antioxidant intakes and infant sex. *Environ. Res.* **2020**, *187*, 109707. [CrossRef]
25. Kim, Y.J.; Hong, Y.C.; Lee, K.H.; Park, H.J.; Park, E.A.; Moon, H.S.; Ha, E.H. Oxidative stress in pregnant women and birth weight reduction. *Reprod. Toxicol.* **2005**, *19*, 487–492. [CrossRef]
26. Shastri, L.; Pammal, R.S.; Mani, I.; Thomas, T.; Kurpad, A.V. Oxidative stress during early pregnancy and birth outcomes. *Public Health Nutr.* **2016**, *19*, 3210–3215. [CrossRef]
27. Quraishi, S.M.; Hazlehurst, M.F.; Loftus, C.T.; Nguyen, R.H.N.; Barrett, E.S.; Kaufman, J.D.; Bush, N.R.; Karr, C.J.; LeWinn, K.Z.; Sathyanarayana, S.; et al. Association of prenatal exposure to ambient air pollution with adverse birth outcomes and effect modification by socioeconomic factors. *Environ. Res.* **2022**, *212*, 113571. [CrossRef]
28. Liang, Z.; Yang, Y.; Qian, Z.; Ruan, Z.; Chang, J.; Vaughn, M.G.; Zhao, Q.; Lin, H. Ambient $PM_{2.5}$ and birth outcomes: Estimating the association and attributable risk using a birth cohort study in nine Chinese cities. *Environ. Int.* **2019**, *126*, 329–335. [CrossRef]
29. Klepac, P.; Locatelli, I.; Korošec, S.; Künzli, N.; Kukec, A. Ambient air pollution and pregnancy outcomes: A comprehensive review and identification of environmental public health challenges. *Environ. Res.* **2018**, *167*, 144–159. [CrossRef]
30. Yang, S.; Tan, Y.; Mei, H.; Wang, F.; Li, N.; Zhao, J.; Zhang, Y.; Qian, Z.; Chang, J.J.; Syberg, K.M.; et al. Ambient air pollution the risk of stillbirth: A prospective birth cohort study in Wuhan, China. *Int. J. Hyg. Environ. Health* **2018**, *221*, 502–509. [CrossRef]

31. Kannan, S.; Misra, D.P.; Dvonch, J.T.; Krishnakumar, A. Exposures to airborne particulate matter and adverse perinatal outcomes: A biologically plausible mechanistic framework for exploring potential effect modification by nutrition. *Environ. Health Perspect.* **2006**, *114*, 1636–1642. [CrossRef] [PubMed]
32. Bekkar, B.; Pacheco, S.; Basu, R.; DeNicola, N. Association of Air Pollution and Heat Exposure With Preterm Birth, Low Birth Weight, and Stillbirth in the US: A Systematic Review. *JAMA Netw. Open* **2020**, *3*, e208243. [CrossRef]
33. Jedrychowski, W.; Perera, F.; Mrozek-Budzyn, D.; Mroz, E.; Flak, E.; Spengler, J.D.; Edwards, S.; Jacek, R.; Kaim, I.; Skolicki, Z. Gender differences in fetal growth of newborns exposed prenatally to airborne fine particulate matter. *Environ. Res.* **2009**, *109*, 447–456. [CrossRef] [PubMed]
34. Challis, J.; Newnham, J.; Petraglia, F.; Yeganegi, M.; Bocking, A. Fetal sex and preterm birth. *Placenta* **2013**, *34*, 95–99. [CrossRef] [PubMed]
35. Bukowski, R.; Smith, G.C.; Malone, F.D.; Ball, R.H.; Nyberg, D.A.; Comstock, C.H.; Hankins, G.D.; Berkowitz, R.L.; Gross, S.J.; Dugoff, L.; et al. Human sexual size dimorphism in early pregnancy. *Am. J. Epidemiol.* **2007**, *165*, 1216–1218. [CrossRef]
36. Peters, A.; Döring, A.; Wichmann, H.E.; Koenig, W. Increased plasma viscosity during an air pollution episode: A link to mortality? *Lancet* **1997**, *349*, 1582–1587. [CrossRef]
37. Edwards, A.; Megens, A.; Peek, M.; Wallace, E.M. Sexual origins of placental dysfunction. *Lancet* **2000**, *355*, 203–204. [CrossRef]
38. Ghidini, A.; Salafia, C.M. Gender differences of placental dysfunction in severe prematurity. *Bjog* **2005**, *112*, 140–144. [CrossRef]
39. Hu, W.; Wang, Y.; Wang, T.; Ji, Q.; Jia, Q.; Meng, T.; Ma, S.; Zhang, Z.; Li, Y.; Chen, R.; et al. Ambient particulate matter compositions and increased oxidative stress: Exposure-response analysis among high-level exposed population. *Environ. Int.* **2021**, *147*, 106341. [CrossRef]
40. Pilger, A.; Rüdiger, H.W. 8-Hydroxy-2′-deoxyguanosine as a marker of oxidative DNA damage related to occupational and environmental exposures. *Int. Arch. Occup. Environ. Health* **2006**, *80*, 1–15. [CrossRef]
41. Poulsen, H.E.; Loft, S.; Prieme, H.; Vistisen, K.; Lykkesfeldt, J.; Nyyssonen, K.; Salonen, J.T. Oxidative DNA damage in vivo: Relationship to age, plasma antioxidants, drug metabolism, glutathione-S-transferase activity and urinary creatinine excretion. *Free Radic. Res.* **1998**, *29*, 565–571. [CrossRef] [PubMed]
42. Li, S.; Liu, Y.; Liu, B.; Hu, Y.Q.; Ding, Y.Q.; Zhang, J.; Feng, L. Maternal urban particulate matter exposure and signaling pathways in fetal brains and neurobehavioral development in offspring. *Toxicology* **2022**, *474*, 153225. [CrossRef] [PubMed]
43. Öztürk, H.N.O.; Türker, P.F. Fetal programming: Could intrauterin life affect health status in adulthood? *Obstet. Gynecol. Sci.* **2021**, *64*, 473–483. [CrossRef] [PubMed]
44. Yuan, L.; Zhang, Y.; Wang, W.; Chen, R.; Liu, Y.; Liu, C.; Kan, H.; Gao, Y.; Tian, Y. Critical windows for maternal fine particulate matter exposure and adverse birth outcomes: The Shanghai birth cohort study. *Chemosphere* **2020**, *240*, 124904. [CrossRef]
45. Li, Z.; Wen, Q.; Zhang, R. Sources, health effects and control strategies of indoor fine particulate matter ($PM_{2.5}$): A review. *Sci. Total Environ.* **2017**, *586*, 610–622. [CrossRef]
46. Park, S.; Marcotte, R.T.; Staudenmayer, J.W.; Strath, S.J.; Freedson, P.S.; Chasan-Taber, L. The impact of the COVID-19 pandemic on physical activity and sedentary behavior during pregnancy: A prospective study. *BMC Pregnancy Childbirth* **2022**, *22*, 899. [CrossRef]
47. Ju, M.J.; Oh, J.; Choi, Y.H. Changes in air pollution levels after COVID-19 outbreak in Korea. *Sci. Total Environ.* **2021**, *750*, 141521. [CrossRef]

antioxidants

MDPI

Article

Body Mass Index Modulates the Impact of Short-Term Exposure to Air Particulate Matter on High-Density Lipoprotein Function

Alice Ossoli [1,†], Chiara Favero [2,†], Luisella Vigna [3], Angela Cecilia Pesatori [2,3], Valentina Bollati [2,3,*] and Monica Gomaraschi [1,*]

1 Center E. Grossi Paoletti, Department of Pharmacological and Biomolecular Sciences, Università degli Studi di Milano, 20133 Milan, Italy
2 EPIGET Lab, Department of Clinical Sciences and Community Health, Università degli Studi di Milano, 20122 Milan, Italy
3 Fondazione IRCCS Ca' Granda Ospedale Maggiore Policlinico, 20122 Milan, Italy
* Correspondence: valentina.bollati@unimi.it (V.B.); monica.gomaraschi@unimi.it (M.G.)
† These authors contributed equally to this work.

Abstract: Air particulate matter (PM) exposure has been associated with increased cardiovascular risk, especially in obesity. By triggering inflammation and oxidative stress, PM could impact atheroprotection by high-density lipoproteins (HDL). The aim of the study was to assess the relationship between short-term exposure to PM and HDL function, and the modifying effect of body mass index (BMI). Daily exposures to PM_{10} and $PM_{2.5}$ of 50 subjects with overweight/obesity and 41 healthy volunteers with BMI < 30 kg/m^2 were obtained from fixed monitoring stations. HDL function was assessed as promotion of nitric oxide (NO) release by endothelial cells and reduction in cholesterol in macrophages. HDL-induced NO release progressively declined with the increase in BMI. No association was found between HDL function and PM exposure, but a modifying effect of BMI was observed. The positive association between PM_{10} exposure at day −1 and NO production found at normal BMI values was lost in participants with higher BMI. Similar results were obtained for the reduction in macrophage cholesterol. The loss of the compensatory response of HDL function to PM exposure at increasing BMI levels could contribute to the endothelial dysfunction induced by PM and help to explain the susceptibility of subjects with obesity to air pollution.

Keywords: high-density lipoproteins; air pollution; endothelium; nitric oxide; body mass index

check for updates

Citation: Ossoli, A.; Favero, C.; Vigna, L.; Pesatori, A.C.; Bollati, V.; Gomaraschi, M. Body Mass Index Modulates the Impact of Short-Term Exposure to Air Particulate Matter on High-Density Lipoprotein Function. *Antioxidants* **2022**, *11*, 1938. https://doi.org/10.3390/antiox11101938

Academic Editor: Yasuhiro Yoshida

Received: 30 August 2022
Accepted: 24 September 2022
Published: 28 September 2022

Publisher's Note: MDPI stays neutral with regard to jurisdictional claims in published maps and institutional affiliations.

1. Introduction

Ischemic heart disease (IHD) and stroke are the main causes of morbidity and mortality worldwide [1], with hypertension, diabetes, obesity, and dyslipidemia representing the most important modifiable risk factors. Regarding dyslipidemia, higher plasma levels of total and non-high-density lipoprotein cholesterol and reduced high-density lipoprotein cholesterol (HDL-C) are associated with an increased risk of IHD and stroke [2]. Nowadays, both short- and long-term exposures to environmental pollution are also recognized as relevant risk factors for cardio- and cerebrovascular diseases [1]. The detrimental effect of pollution is possibly related to the ability of air pollutants to trigger inflammation and oxidative stress, which can directly affect the cardiovascular system or indirectly impair other key players such as blood pressure and lipids [3]. There is evidence for increased susceptibility to the effects of pollution for elderly subjects, those with obesity, or in secondary prevention [4].

Exposure to air pollutants has been variably associated with changes in plasma lipid profile, including a reduction in HDL-C [5]. HDL can exert a series of protective activities against the development and progression of atherosclerosis [6]. The key role of HDL in the reverse cholesterol transport (RCT) from peripheral cells, especially macrophages in the arterial wall, to the liver has been widely considered the most relevant atheroprotective function of HDL. In particular, HDL are the main extracellular acceptors of cholesterol from cells, which represents the first and rate-limiting step of RCT. Moreover, HDL can exert antioxidant and anti-inflammatory activities and can preserve endothelial homeostasis. Since endothelial dysfunction is deeply involved in atherosclerosis development and progression, the ability of HDL to preserve the endothelial barrier as a continuous and anti-adhesive layer and to contribute to the regulation of vascular tone has gained attention. In particular, the capacity of HDL to promote the release of nitric oxide (NO) by endothelial cells was tested in the context of several clinical conditions [7]. The development of cell-free and cell-based assays to measure the various atheroprotective activities of HDL allowed understanding that many pathologic conditions, as well as therapeutic interventions, can affect HDL function independently from changes in HDL-C, suggesting that HDL function is more relevant than their concentrations (at least when expressed as their cholesterol content) to estimate HDL-mediated cardiovascular protection. In particular, pathologic conditions associated with acute or chronic inflammatory states, such as obesity, diabetes, myocardial infarction, infections, and autoimmune diseases, were associated with a general impairment of HDL function [7]. In this context, since air pollutants trigger inflammation, exposure to environmental pollution could also impact HDL function. To date, the ability of HDL to promote cholesterol efflux and their antioxidant/anti-inflammatory index through a cell-free assay have been investigated in the context of short- or long-term exposure to air pollutants with inconsistent results [8]. On the contrary, the capacity of HDL to promote NO release by endothelial cells has not been investigated yet. Interestingly, it has been shown that acute exposure to diesel exhaust can impair vascular function for the following 24 h, likely due to the reduced availability of NO [9].

Thus, the aim of the present study was (i) to assess the relationship between short-term exposure to air pollutants and HDL function, measured as their ability to reduce cell cholesterol content in macrophages and to promote NO release by endothelial cells, and (ii) to investigate whether body mass index (BMI) could modulate such a relationship.

2. Materials and Methods

2.1. Subjects

Fifty subjects with overweight/obesity were enrolled at the Center for Obesity and Work (COW; Department of Preventive Medicine, IRCCS Fondazione Ca' Granda–Ospedale Maggiore Policlinico, Milan, Italy) as part of the cross-sectional study SPHERE [10]. Forty-one healthy volunteers with BMI $\leq$ 30 kg/m^2 were enrolled by answering an ad hoc developed announcement posted on the SPHERE Project website (http://users.unimi.it/sphere, accessed on 29 August 2022). Subjects were enrolled between January 2013 and March 2015. Each participant signed an informed consent form, which had been approved by the Ethics Committee of Fondazione IRCCS Cà Granda Ospedale Maggiore Policlinico (approval number 1425), in accordance with the principles of the Declaration of Helsinki [11].

2.2. Collection of Personal Data and Biological Samples

Epidemiological and clinical data were collected as described [10]. Blood samples were collected after an overnight fast. Serum was prepared by low-speed centrifugation at 4 °C and immediately frozen at −80 °C. Apolipoprotein B-depleted (apoB-D) serum was obtained by the precipitation of apoB-containing lipoproteins with PEG 6000 in 10 mM HEPES at 4 °C for 30 min, followed by centrifugation at 2200 g for 30 min.

2.3. Exposure Assessment

Short-term particulate matter (PM) with an aerodynamic diameter of 10 μm or less (PM_{10}) and PM with an aerodynamic diameter of 2.5 μm or less ($PM_{2.5}$) exposure was evaluated as one-week lag exposure time window.

Daily PM concentrations were collected from fixed monitoring stations of the Regional Environmental Protection Agency (ARPA Lombardy). Using ArcGIS® software (Desktop:

Release 10.1, Environmental Systems Research Institute, Esri, Redlands, CA, USA), we assigned each subject the daily PM concentrations from the nearest monitor to that subject's home address for the six days preceding recruitment and the nearest monitor to the hospital (Ospedale Maggiore Policlinico, Milan, Italy) for the day of recruitment. Meteorological data were obtained from the ARPA monitoring stations, measuring temperature and relative humidity. The apparent temperature was calculated as previously reported [12]. Detailed information on the exposure assessment method has been previously described [13,14].

2.4. Biochemical Analysis

The concentrations of total cholesterol, triglycerides, and HDL-C levels in serum were assessed by standard enzymatic or turbidimetric techniques on a Cobas c311 autoanalyzer (Roche Diagnostics, Milan, Italy). Low-density lipoprotein cholesterol (LDL-C) was calculated by the Friedewald formula [15]. Inflammatory markers and mitochondrial DNA copy number (mtDNAcn) were measured as described [16].

2.5. Nitric Oxide Production

Primary cultures of human umbilical vein endothelial cells (HUVECs) were purchased from PromoCell (Heidelberg, Germany) and subcultured for 1–3 passages according to manufacturer instructions. Experiments were performed in M199 with 0.75% bovine serum albumin (BSA) and 1% fetal calf serum (FCS) (Euroclone, Milan, Italy). To evaluate NO production, HUVECs were incubated with apoB-D sera for 30 min. The generated NO was measured by fluorescence using a diacetate derivative of 4,5-diaminofluorescein (DAF-2 DA, Sigma-Aldrich Chemie, Steinheim, Germany) [17]. For each sample, fluorescence was normalized by the protein concentration of the total cell lysate, assessed by the microBCA method (Thermo Fischer Scientific, Waltham, MA, USA).

2.6. Macrophage Cholesterol Mass

The human monocytic leukaemia cell line THP-1 was purchased from the American Type Culture Collection (ATCC, Manassas, VA, USA) and cultured in Roswell Park Memorial Institute (RPMI) 1640 medium supplemented with 10% (*v*/*v*) fetal bovine serum (FBS), 1 mM sodium pyruvate, and 50 μM 2-mercaptoethanol. To induce the differentiation into macrophage-like cells, THP-1 cells were cultured in the presence of 100 ng/mL phorbol 12-myristate 13-acetate (PMA) for 72 h. Macrophages were loaded with acetylated LDL (50 μg of protein/mL) for 24 h in the presence of 2 μg/mL of an acyl-CoA:cholesterol acyltransferase inhibitor and then incubated with 2.5% apoB-D serum for 7 h. At the end of the experiment, cells were washed with PBS and lysed overnight in 1% sodium cholate and 10 U/mL of DNase (Sigma-Aldrich Chemie, Steinheim, Germany) at room temperature. Total cholesterol was measured by fluorescence using the Amplex Red Cholesterol Assay Kit (Sigma-Aldrich Chemie, Steinheim, Germany), according to the manufacturer's instructions [18]. For each sample, cholesterol concentration was normalized by the protein concentration of the total cell lysate, assessed by the microBCA assay (ThermoFisher Scientific, Waltham, MA, USA).

2.7. Statistical Analysis

Descriptive statistics were performed on all variables. Categorical variables were presented as absolute numbers and percentages. Continuous data were expressed as the mean ± standard deviation (SD) or as the median and interquartile range (Q1–Q3), as appropriate. Normality assumption was verified by graphical inspection. Baseline demographic, lifestyle, biochemical, and clinical characteristics were compared by BMI groups (normal weight: 18.5–24.99 kg/m^2; overweight: 25–29.99 kg/m^2; obesity: ≥30 kg/m^2) for continuous variables with one-way analysis of variance (ANOVA) or Kruskal–Wallis test as appropriate. Categorical data were compared with the chi-square test or Fisher exact test as appropriate.

Univariate and multivariable linear regression models were used to test the relationship between NO production or macrophage cholesterol mass in HDL-treated cells and PM exposure as the continuous predictor. Multiple analysis was adjusted for a priori covariates (age, gender, BMI, smoking habits, and HDL cholesterol) and for variables that were significantly related to NO production or cholesterol mass in univariate analysis ($p < 0.05$). To determine the best-performing models, we ran several regression equations separately for evaluating the β coefficients, standard errors (SE), *p*-values, as well as the goodness of fit (R2). Finally, the best models selected to predict the association between NO production or macrophage cholesterol mass and PM exposure were adjusted for: age, gender, BMI, smoking habits, HDL cholesterol, triglycerides, interleukin-8, and apparent temperature on the day of recruitment. MtDNAcn, an established biomarker of oxidative stress, was not associated with NO production and macrophage cholesterol mass.

To examine the potential effect modification of BMI, we added the interaction term PM * BMI to the multivariable selected models. We evaluated whether the effect of PM_{10} and $PM_{2.5}$ exposure on NO production and macrophage cholesterol mass differs depending on BMI levels.

All statistical analyses were performed with SAS software (version 9.4; SAS Institute Inc., Cary, NC, USA). Two-sided *p*-values below 0.05 were considered statistically significant.

3. Results

3.1. Anthropometric, Biochemical, and Clinical Features of Enrolled Subjects

Enrolled subjects were mainly females and had a mean age of 52.1 ± 9.6 years (Table 1). On average, their lipid profile was characterized by normal plasma levels of HDL-C and triglycerides and by borderline levels of total and LDL cholesterol (Table 1). Subjects were stratified according to their BMI into three groups: BMI 18.5–24.9 kg/m^2 (n = 23), BMI 25–29.9 kg/m^2 (n = 26) and BMI ≥ 30 kg/m^2 (n = 42) (Table 1). Subjects with higher BMI levels were older and characterized by lower levels of education and occupation. In addition, higher BMI levels were associated with higher triglycerides and lower HDL-C. Interestingly, no differences were detected in inflammatory and oxidative stress markers. The prevalence of diabetes was similar among the three groups, while the prevalence of hypertension significantly increased with BMI (Table 1).

Table 1. Anthropometric, biochemical, and clinical features of enrolled subjects.

	All (n = 91)	BMI 18.5–24.9 (n = 23)	BMI 25–29.9 (n = 26)	BMI ≥ 30 (n = 42)	*p*-Value
Age, years	52.1 ± 9.6	45.0 ± 5.6	52.3 ± 8.8	55.6 ± 9.8	<0.0001
Gender					
Males	27 (29.67%)	4 (17.39%)	13 (50.00%)	10 (23.81%)	0.0306
Females	64 (70.33%)	19 (82.61%)	13 (50.00%)	32 (76.19%)	
BMI, kg/m^2	28.9 ± 5.0	21.8 ± 1.9	28.0 ± 1.6	33.3 ± 1.6	-
Smoking status					
Never smoker	49 (53.85%)	17 (73.91%)	11 (42.31%)	21 (50.00%)	0.0739
Former smoker	26 (28.57%)	3 (13.04%)	12 (46.15%)	11 (26.19%)	
Current smoker	16 (17.58%)	3 (13.04%)	3 (11.54%)	10 (23.81%)	
Education					
Primary school	3 (3.30%)	0 (0.0%)	1 (3.85%)	2 (4.76%)	<0.0001
Secondary school	16 (17.58%)	0 (0.0%)	2 (7.69%)	14 (33.33%)	
High school	35 (38.46%)	5 (21.74%)	9 (34.62%)	21 (50.00%)	
University or more	37 (40.66%)	18 (78.26%)	14 (53.85%)	5 (11.9%)	
Occupation					
Employee	72 (79.12%)	23 (100.0%)	21 (80.77%)	28 (66.67%)	0.0189
Unemployed	0 (0.0%)	0 (0.0%)	0 (0.0%)	0 (0.0%)	
Pensioner	14 (15.38%)	0 (0.0%)	4 (15.38%)	10 (23.81%)	
Housewife	5 (5.49%)	0 (0.0%)	1 (3.85%)	4 (9.52%)	

Table 1. *Cont.*

	All (n = 91)	BMI 18.5–24.9 (n = 23)	BMI 25–29.9 (n = 26)	BMI ≥ 30 (n = 42)	*p*-Value
Total cholesterol, mg/dL	212.6 ± 43.2	213.8 ± 33.1	211.7 ± 44.4	213.0 ± 47.6	0.9368
HDL cholesterol, mg/dL	56.9 ± 17.6	68.0 ± 16.5	59.0 ± 16.3	49.6 ± 15.6	0.0001
LDL cholesterol, mg/dL	133.3 ± 36.9	126.3 ± 27	129.7 ± 34.8	139.9 ± 41.7	0.2664
Triglycerides, mg/dL	113 [77;162]	83.2 [69.7;116.1]	93.4 [73;142.7]	145.5 [109;215]	0.0001
Interleukin-8, pg/mL	6.7 ± 6.83	7.6 ± 6.1	8.4 ± 11.9	5.1 ± 4.8	0.2500
N. of subjects with IL-8 < LLOQ	25 (27.47%)	0 (0.0%)	5 (19.23%)	20 (48.78%)	-
Mitochondrial DNA, cn	1.08 ± 0.31	1.00 ± 0.23	1.13 ± 0.39	1.08 ± 0.29	0.3358
Diabetes					
Yes	9 (9.89%)	0 (0.0%)	2 (7.69%)	7 (16.67%)	0.0930
No	82 (90.11%)	23 (100.0%)	24 (92.31%)	35 (83.33%)	
Hypertension					
Yes	40 (43.96%)	1 (4.35%)	9 (34.62%)	30 (71.43%)	<0.0001
No	51 (56.04%)	22 (95.65%)	17 (65.38%)	12 (28.57%)	

Enrolled subjects (n = 91) were divided into three groups according to BMI categories. For normal distribution, values are expressed as mean ± SD and one-way ANOVA was applied. When not normally distributed, values are expressed as median [Q1;Q3] and Kruskal–Wallis test was used. For categorical variables, values are reported as frequencies and percentages and chi-square test or Fisher exact test was applied, as appropriate. LLOQ, lower limit of quantification; cn, copy number.

The estimated levels of exposure to PM_{10} and $PM_{2.5}$ in the last week before recruitment are reported in Table 2. A similarity in pollutant concentrations across the days of exposure was evident. Mean PM_{10} concentrations were above the annual regional air quality standards of 40 μg/m^3. Mean $PM_{2.5}$ concentrations were higher than the annual limits set at 25 μg/m^3.

Table 2. Mean concentrations of PM_{10} and $PM_{2.5}$ exposure evaluated from one week before the enrollment of participants.

		Mean ± SD
PM_{10} exposure (μg/m^3)		
	Day 0	50.61 ± 22.49
	Day -1	48.35 ± 22.01
	Day -2	47.36 ± 22.02
	Day -3	51.61 ± 27.68
	Day -4	50.67 ± 24.00
	Day -5	53.09 ± 22.97
	Day -6	49.72 ± 24.41
$PM_{2.5}$ exposure (μg/m^3)		
	Day 0	38.19 ± 20.82
	Day -1	36.23 ± 18.47
	Day -2	37.97 ± 19.49
	Day -3	34.37 ± 17.93
	Day -4	37.51 ± 17.38
	Day -5	41.22 ± 18.59
	Day -6	38.92 ± 20.58
Apparent temperature (°C)	Day 0	7.37 ± 4.81

PM, particulate matter.

3.2. *Production of NO by Endothelial Cells*

An average 1.36 ± 0.32-fold increase in NO production was observed in endothelial cells after treatment with HDL. When subjects were divided according to their BMI, HDL's ability to promote NO release progressively declined with the increase in BMI ($p < 0.0001$, Figure 1A). Indeed, NO production was 1.59 ± 0.32-fold in the group with BMI

18.5–24.9 kg/m^2 and decreased to 1.21 ± 0.20 fold in the group with BMI ≥ 30 kg/m^2. No association was found between NO production and PM_{10} or $PM_{2.5}$ exposures (Table 3). Nevertheless, when the interaction term between BMI and PM concentrations was taken into account, a significant modifying effect of BMI on PM_{10} was observed on the day before the recruitment. In subjects with a normal BMI, we found a positive association between day -1 PM_{10} exposure and NO production (Figure 2A): HDL's ability to induce NO release increased by 0.072 fold every 10 μg/m^3 increase in PM_{10} concentration measured at day -1 (p = 0.0061). In overweight subjects, NO production still increased by 0.032 fold every 10 μg/m^3 increase in PM_{10} concentration measured at day -1 (p = 0.0331). On the contrary, HDL compensatory response to PM_{10} exposure was completely lost in participants with BMI ≥ 30 kg/m^2, with a tendency towards a negative association (Figure 2A). Although not statistically significant, a similar interaction between BMI and $PM_{2.5}$ was observed (Figure 2B).

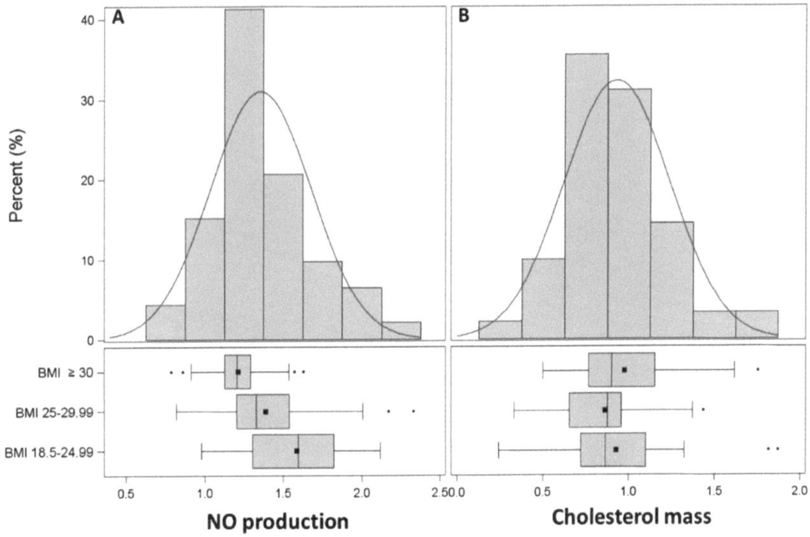

Figure 1. Frequency distribution of HDL function and impact of BMI. (**A**), Frequency distribution of NO production by endothelial cells exposed to apoB-D sera in the whole cohort and by BMI groups. Data are expressed as fold increase in untreated cells. (**B**), Frequency distribution of cholesterol mass of macrophages exposed to apoB-D sera in the whole cohort and by BMI groups. Results are expressed as fold reduction in untreated cells. All data are presented as histograms and boxplots; squares indicate the mean values and dots the outliers.

Table 3. Association between HDL-mediated NO production, HDL-mediated reduction in cholesterol mass, and PM exposures.

	NO Production			Cholesterol Mass		
	β	SE	p-Value	β	SE	p-Value
PM_{10} exposure						
Day 0	0.016	0.014	0.2757	−0.014	0.016	0.3909
Day -1	0.021	0.014	0.1411	−0.016	0.016	0.3320
Day -2	0.003	0.014	0.8424	0.004	0.016	0.8103
Day -3	−0.002	0.011	0.8590	0.011	0.012	0.3721
Day -4	−0.004	0.013	0.7667	0.021	0.014	0.1421
Day -5	0.004	0.013	0.7536	0.009	0.015	0.5560
Day -6	0.006	0.013	0.6399	0.007	0.014	0.6001

Table 3. *Cont.*

	NO Production			Cholesterol Mass		
	β	SE	*p*-Value	β	SE	*p*-Value
$PM_{2.5}$ exposure						
Day 0	0.007	0.015	0.6322	−0.027	0.017	0.1082
Day -1	0.025	0.017	0.1468	−0.034	0.020	0.0916
Day -2	0.020	0.016	0.2119	−0.008	0.018	0.6843
Day -3	0.008	0.017	0.6329	−0.005	0.019	0.8068
Day -4	0.008	0.018	0.6726	0.002	0.020	0.9254
Day -5	0.008	0.018	0.6672	−0.011	0.020	0.5759
Day -6	0.030	0.017	0.0815	−0.002	0.020	0.9037

β regression coefficients represent the increase in NO production/cell cholesterol mass for 10 μg/m^3 increase in PM_{10} or $PM_{2.5}$ concentration. Linear regression models were adjusted for age, gender, BMI, smoking habits, HDL-C, triglycerides, interleukin-8 and apparent temperature on the day of recruitment.

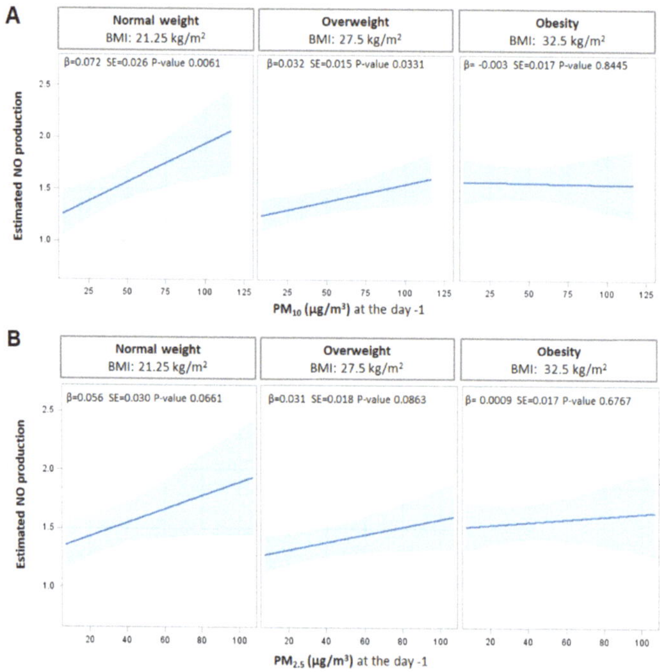

Figure 2. Interaction effect of particulate matter on NO production. Interaction effect of PM_{10} (**A**) and $PM_{2.5}$ (**B**) and BMI on NO production by endothelial cells exposed to apoB-D sera. Strength of association between PM the day before the enrollment and NO production at three selected levels of BMI (normal weight 21.75 kg/m^2, overweight 27.5 kg/m^2, and obesity 32.5 kg/m^2). Adjusted β regression coefficients were reported for 10 μg/m^3 increases in PM concentration, at each level of BMI. *p*-values of interaction terms PM*BMI were 0.0203 and 0.2163, respectively, for PM_{10} and $PM_{2.5}$. Linear regression models in both panels A and panel B were adjusted for age, gender, smoking habit, HDL-C, triglycerides, IL-8, and apparent temperature on the day of recruitment.

3.3. *Macrophage Cholesterol Mass*

The exposure to HDL reduced cholesterol mass in THP-1-derived macrophages by 0.93 ± 0.31 fold. HDL's ability to promote cell cholesterol removal was independent of BMI (p = 0.34, Figure 1B). No association was found between macrophage cholesterol mass after HDL and PM_{10} or $PM_{2.5}$ exposures (Table 3).

BMI interacted with the association between cholesterol mass after HDL and PM_{10} exposure measured the day -1 (Table 4). Indeed, in subjects with normal BMI, a significant negative relationship between cell cholesterol mass after HDL and PM_{10} exposure was observed, with a 0.065-fold decrease every 10 $\mu g/m^3$ increase in PM_{10} (p = 0.034). This association was markedly attenuated in overweight subjects and completely lost in subjects with BMI $\geq$ 30 kg/m^2, with a tendency towards a positive association.

Table 4. Modifying effect of BMI on the association between HDL-mediated reduction in macrophage cholesterol mass and PM exposure the day before the enrollment.

Independent Variable	BMI	β	SE	*p*-Value	*p*-Value for Interaction
PM_{10} Day -1	**Normal weight: 21.25 kg/m²**	−0.0650	0.0301	0.0342	0.0583
	Overweight: 27.5 kg/m²	−0.0270	0.0170	0.1191	
	Obesity: 32.5 kg/m²	0.0067	0.0197	0.7365	
$PM_{2.5}$ Day -1	Normal weight: 21.25 kg/m²	−0.0560	0.0368	0.1309	0.4724
	Overweight: 27.5 kg/m²	−0.0390	0.0211	0.0687	
	Obesity: 32.5 kg/m²	−0.0240	0.0241	0.3211	

Strength of association between PM the day before the enrollment and cholesterol mass at three selected values of BMI. Adjusted β regression coefficients were reported for 10 $\mu g/m^3$ increases in PM concentration, at each level of BMI. Linear regression models were adjusted for age, gender, smoking habit, HDL-C, triglycerides, IL-8, and apparent temperature on the day of recruitment.

The modifying effect of BMI on the relationship between cell cholesterol mass after HDL and $PM_{2.5}$ exposure was not significant, with the three groups for BMI categories displaying similar associations (Table 4).

4. Discussion

This retrospective analysis showed that HDL function, especially HDL's endothelial protective activity, was affected by short-term exposure to air pollutants. More interestingly, the effect of PM exposure on HDL function was modulated by BMI. Indeed, in subjects with normal weight, a compensatory response was observed, i.e., HDL's ability to induce NO production increased with exposure to higher concentrations of air pollutants the day before sample collection. However, this compensatory response was progressively lost at increasing BMI levels. Similarly, the ability of HDL to reduce cholesterol mass in macrophage-like cells increased with exposure to higher levels of PM only in subjects with normal weight. In addition, our results confirmed an impaired HDL ability to preserve endothelial homeostasis in obesity [7] and integrate previous findings showing impaired antioxidant and anti-inflammatory activities of HDL after PM exposure [19]. Exposure to environmental pollutants is known to trigger inflammation and oxidative stress and HDL should be able to limit the extent of such events. However, the loss of the compensatory response observed in subjects with elevated BMI likely indicates that in this condition, which is a chronic inflammatory status per se, HDL lose their ability to manage additional detrimental stimuli. Since inflammation and oxidative stress could directly impact HDL composition and function, related markers, such as interleukin 8 and mitochondrial DNA copy number, and smoking habits were included in the model. Similarly, the outdoor temperature was included, since it could influence the effect of PM on HDL function, as previously shown [20]. Finally, obesity alters plasma lipid profile, resulting in higher triglyceride and lower HDL-C levels in subjects with elevated BMI; thus, these variables were also included in the model.

Nitric oxide is a key player in the regulation of vascular tone and in preserving the endothelium as an anti-adhesive surface for circulating cells [21]. Consequently, a reduction in its availability is associated with impaired vasodilation, as previously observed in many clinical conditions. Rapid alterations in systemic microvascular and macrovascular tone were also observed after short-term exposure to coarse, fine, and ultrafine PM [9,22,23]. With HDL being the main activator of endothelial NO synthase (eNOS) in serum [17], an impairment of its function could explain the reduced vasodilation after PM exposure, together with sympathetic nervous system activation [24]. Interestingly, the significant modifying effect of BMI on the interaction between HDL function and PM exposure is consistent with the higher impairment of vasodilation observed in subjects with obesity [25]. Thus, our results provide additional evidence supporting endothelial dysfunction as a key mechanism leading to ischemic events in subjects exposed to environmental pollutants, especially in those with obesity [26].

Similar results were obtained when another HDL atheroprotective function was assessed, i.e., the ability to promote the removal of cholesterol from peripheral cells, including macrophages in the arterial wall. In this case, HDL's ability to reduce cholesterol mass was not affected by BMI and the modifying effect of BMI on the interaction between HDL functionality and PM exposure was attenuated if compared to that observed for NO production. Previous studies assessed the impact of environmental pollutants on cholesterol efflux to HDL with variable results [8]. These inconsistencies could be partly explained by the different cell types and experimental settings used, which can be specific for one or some efflux pathways. Thus, in the present study, we decided to evaluate the net changes in cholesterol mass in LDL-loaded macrophages after HDL exposure in order to provide an overall estimation of HDL effect, irrespective of the specific underlying efflux mechanisms.

In the literature, there is evidence of an association of both short- and long-term exposures to environmental pollutants with cardiovascular morbidity and mortality [27]. Long-term exposure could favor the development and progression of atherosclerosis by establishing a chronic low-grade inflammatory state and by upregulating the sympathetic nervous system [3]. Regarding the clinical relevance of short-term exposure, the transient loss of a compensatory response by HDL that we observed in subjects with elevated BMI supports the hypothesis that a susceptible individual, who is at high risk for IHD due to multiple risk factors, may experience an event or face an unfavorable prognosis when acutely exposed to high levels of PM.

The present study has some limitations. PM exposure was estimated according to air-quality-monitoring-station data, whose measurements might inaccurately capture the real exposure at residential addresses. The exposure assessment did not include personal monitoring; thus, we were not able to take into account the indoor activities of the subjects. Obesity is a clinical condition characterized by several chronic comorbidities. Thus, although we evaluated the possible contribution of these diseases in modulating HDL function, we cannot completely exclude they are part of the "obesity" effect rather than the BMI only. In addition, we cannot rule out the effect of eNOS gene polymorphisms on the relationship between HDL function and PM exposure, since these DNA variants were previously shown to modify the effect of PM on oxidative stress biomarkers [28]. Even if blood samples were collected, processed, and stored according to standardized procedures, we cannot completely exclude that storage time could have partially affected HDL function. Lastly, the limited sample size and the observational design of the study highlight an association between HDL function and PM exposure but foreclose any causative link.

5. Conclusions

Our results support the hypothesis of a transient impairment of NO-mediated vasodilation after PM exposure [9] and suggest a role for HDL dysfunction among the underlying mechanisms. Moreover, the loss of the compensatory response of HDL function to PM exposure at increasing BMI levels could help to explain why subjects with obesity seem more susceptible to the detrimental effects of air pollution.

Author Contributions: Conceptualization, V.B. and M.G.; Methodology and Formal Analysis A.O. and C.F.; Investigation, A.O., L.V. and A.C.P.; Data Curation, V.B. and M.G.; Writing—Original Draft Preparation, M.G.; Writing—Review and Editing, A.O., C.F., L.V., A.C.P. and V.B.; Visualization, A.O. and C.F.; Supervision, M.G.; Funding Acquisition, V.B. and M.G. All authors have read and agreed to the published version of the manuscript.

Funding: This research was funded by an intramural grant of Università degli Studi di Milano to MG (PSR2019 Linea 2 Azione A) and by the EU Programme "Ideas", European Research Council, to VB (ERC-2011-StG 282413).

Institutional Review Board Statement: The study was conducted in accordance with the Declaration of Helsinki, and the protocol was approved by the Ethics Committee of Fondazione IRCCS Cà Granda Ospedale Maggiore Policlinico (approval number 1425).

Informed Consent Statement: All subjects gave their informed consent for inclusion before they participated in the study.

Data Availability Statement: The data presented in this study are available on request from the corresponding authors. The data are not publicly available due to ethical reasons.

Conflicts of Interest: The authors declare no conflict of interest.

References

1. GBD 2019 Diseases and Injuries Collaborators. Global burden of 369 diseases and injuries in 204 countries and territories, 1990–2019: A systematic analysis for the Global Burden of Disease Study 2019. *Lancet* **2020**, *396*, 1204–1222. [CrossRef]
2. Mach, F.; Baigent, C.; Catapano, A.L.; Koskinas, K.C.; Casula, M.; Badimon, L.; Chapman, M.J.; De Backer, G.G.; Delgado, V.; Ference, B.A.; et al. 2019 ESC/EAS Guidelines for the management of dyslipidaemias: Lipid modification to reduce cardiovascular risk. *Eur. Heart J.* **2020**, *41*, 111–188. [CrossRef] [PubMed]
3. Joshi, S.S.; Miller, M.R.; Newby, D.E. Air pollution and cardiovascular disease: The Paul Wood Lecture, British Cardiovascular Society 2021. *Heart* **2022**, *108*, 1267–1273. [CrossRef] [PubMed]
4. Brook, R.D.; Franklin, B.; Cascio, W.; Hong, Y.; Howard, G.; Lipsett, M.; Luepker, R.; Mittleman, M.; Samet, J.; Smith, S.C.; et al. Air pollution and cardiovascular disease: A statement for healthcare professionals from the Expert Panel on Population and Prevention Science of the American Heart Association. *Circulation* **2004**, *109*, 2655–2671. [CrossRef] [PubMed]
5. Zhang, W.; Liu, J.; Hu, D.; Li, L.; Cui, L.; Xu, J.; Wang, W.; Deng, F.; Guo, X. Joint effect of multiple air pollutants on lipid profiles in obese and normal-weight young adults: The key role of ozone. *Environ. Pollut.* **2022**, *292 Pt A*, 118247. [CrossRef]
6. Calabresi, L.; Gomaraschi, M.; Rossoni, G.; Franceschini, G. Synthetic high density lipoproteins for the treatment of myocardial ischemia/reperfusion injury. *Pharmacol. Ther.* **2006**, *111*, 836–854. [CrossRef] [PubMed]
7. Ossoli, A.; Pavanello, C.; Giorgio, E.; Calabresi, L.; Gomaraschi, M. Dysfunctional HDL as a Therapeutic Target for Atherosclerosis Prevention. *Curr. Med. Chem.* **2019**, *26*, 1610–1630. [CrossRef]
8. Holme, S.A.N.; Sigsgaard, T.; Holme, J.A.; Holst, G.J. Effects of particulate matter on atherosclerosis: A link via high-density lipoprotein (HDL) functionality? *Part. Fibre Toxicol.* **2020**, *17*, 36. [CrossRef]
9. Mills, N.L.; Törnqvist, H.; Robinson, S.D.; Gonzalez, M.; Darnley, K.; MacNee, W.; Boon, N.A.; Donaldson, K.; Blomberg, A.; Sandstrom, T.; et al. Diesel exhaust inhalation causes vascular dysfunction and impaired endogenous fibrinolysis. *Circulation* **2005**, *112*, 3930–3936. [CrossRef]
10. Bollati, V.; Iodice, S.; Favero, C.; Angelici, L.; Albetti, B.; Cacace, R.; Cantone, L.; Carugno, M.; Cavalleri, T.; De Giorgio, B.; et al. Susceptibility to particle health effects, miRNA and exosomes: Rationale and study protocol of the SPHERE study. *BMC Public Health* **2014**, *14*, 1137. [CrossRef]
11. World Medical Association. World Medical Association Declaration of Helsinki: Ethical principles for medical research involving human subjects. *JAMA* **2013**, *310*, 2191–2194. [CrossRef] [PubMed]
12. Analitis, A.; Katsouyanni, K.; Biggeri, A.; Baccini, M.; Forsberg, B.; Bisanti, L.; Kirchmayer, U.; Ballester, F.; Cadum, E., Goodman, P.; et al. Effects of cold weather on mortality: Results from 15 European cities within the PHEWE project. *Am. J. Epidemiol.* **2008**, *168*, 1397–1408. [CrossRef] [PubMed]
13. Pergoli, L.; Cantone, L.; Favero, C.; Angelici, L.; Iodice, S.; Pinatel, E.; Hoxha, M.; Dioni, L.; Letizia, M.; Albetti, B.; et al. Extracellular vesicle-packaged miRNA release after short-term exposure to particulate matter is associated with increased coagulation. *Part. Fibre Toxicol.* **2017**, *14*, 32. [CrossRef] [PubMed]
14. Bonzini, M.; Pergoli, L.; Cantone, L.; Hoxha, M.; Spinazzè, A.; Del Buono, L.; Favero, C.; Carugno, M.; Angelici, L.; Broggi, L.; et al. Short-term particulate matter exposure induces extracellular vesicle release in overweight subjects. *Environ. Res.* **2017**, *155*, 228–234. [CrossRef]
15. Friedewald, W.T.; Levy, R.I.; Fredrickson, D.S. Estimation of the concentration of low-density lipoprotein cholesterol in plasma, without use of the preparative ultracentrifuge. *Clin. Chem.* **1972**, *18*, 499–502. [CrossRef] [PubMed]
16. Pavanello, S.; Dioni, L.; Hoxha, M.; Fedeli, U.; Mielzynska-Svach, D.; Baccarelli, A.A. Mitochondrial DNA copy number and exposure to polycyclic aromatic hydrocarbons. *Cancer Epidemiol. Biomark. Prev.* **2013**, *22*, 1722–1729. [CrossRef] [PubMed]
17. Arnaboldi, L.; Ossoli, A.; Giorgio, E.; Pisciotta, L.; Lucchi, T.; Grigore, L.; Pavanello, C.; Granata, A.; Pasta, A.; Arosio, B.; et al. LIPA gene mutations affect the composition of lipoproteins: Enrichment in ACAT-derived cholesteryl esters. *Atherosclerosis* **2020**, *297*, 8–15. [CrossRef]

18. Calabresi, L.; Nilsson, P.; Pinotti, E.; Gomaraschi, M.; Favari, E.; Adorni, M.P.; Bernini, F.; Sirtori, C.R.; Calandra, S.; Franceschini, G.; et al. A novel homozygous mutation in CETP gene as a cause of CETP deficiency in a Caucasian kindred. *Atherosclerosis* **2009**, *205*, 506–511. [CrossRef]
19. Li, J.; Zhou, C.; Xu, H.; Brook, R.D.; Liu, S.; Yi, T.; Wang, Y.; Feng, B.; Zhao, M.; Wang, X.; et al. Ambient Air Pollution Is Associated With HDL (High-Density Lipoprotein) Dysfunction in Healthy Adults. *Arter. Thromb. Vasc. Biol.* **2019**, *39*, 513–522. [CrossRef]
20. Mathew, A.V.; Yu, J.; Guo, Y.; Byun, J.; Chen, Y.E.; Wang, L.; Liu, M.; Bard, R.L.; Morishita, M.; Huang, W.; et al. Effect of Ambient Fine Particulate Matter Air Pollution and Colder Outdoor Temperatures on High-Density Lipoprotein Function. *Am. J. Cardiol.* **2018**, *122*, 565–570. [CrossRef]
21. Calabresi, L.; Gomaraschi, M.; Franceschini, G. Endothelial protection by high-density lipoproteins: From bench to bedside. *Arterioscler. Thromb. Vasc. Biol.* **2003**, *23*, 1724–1731. [CrossRef] [PubMed]
22. Wauters, A.; Dreyfuss, C.; Pochet, S.; Hendrick, P.; Berkenboom, G.; van de Borne, P.; Argacha, J.F. Acute exposure to diesel exhaust impairs nitric oxide-mediated endothelial vasomotor function by increasing endothelial oxidative stress. *Hypertension* **2013**, *62*, 352–358. [CrossRef] [PubMed]
23. Lucking, A.J.; Lundbäck, M.; Barath, S.L.; Mills, N.L.; Sidhu, M.K.; Langrish, J.P.; Boon, N.A.; Pourazar, J.; Badimon, J.J.; Gerlofs-Nijland, M.E.; et al. Particle traps prevent adverse vascular and prothrombotic effects of diesel engine exhaust inhalation in men. *Circulation* **2011**, *123*, 1721–1728. [CrossRef]
24. Rankin, G.D.; Kabéle, M.; Brown, R.; Macefield, V.G.; Sandström, T.; Bosson, J.A. Acute Exposure to Diesel Exhaust Increases Muscle Sympathetic Nerve Activity in Humans. *J. Am. Heart Assoc.* **2021**, *10*, e018448. [CrossRef] [PubMed]
25. Hemmingsen, J.G.; Rissler, J.; Lykkesfeldt, J.; Sallsten, G.; Kristiansen, J.; Møller, P.P.; Loft, S. Controlled exposure to particulate matter from urban street air is associated with decreased vasodilation and heart rate variability in overweight and older adults. *Part. Fibre Toxicol.* **2015**, *12*, 6. [CrossRef] [PubMed]
26. Münzel, T.; Gori, T.; Al-Kindi, S.; Deanfield, J.; Lelieveld, J.; Daiber, A.; Rajagopalan, S. Effects of gaseous and solid constituents of air pollution on endothelial function. *Eur. Heart J.* **2018**, *39*, 3543–3550. [CrossRef]
27. de Bont, J.; Jaganathan, S.; Dahlquist, M.; Persson, Å.; Stafoggia, M.; Ljungman, P. Ambient air pollution and cardiovascular diseases: An umbrella review of systematic reviews and meta-analyses. *J. Intern. Med.* **2022**, *291*, 779–800. [CrossRef]
28. Kim, J.H.; Choi, Y.H.; Bae, S.; Park, H.Y.; Hong, Y.C. eNOS gene polymorphisms modify the association of PM_{10} with oxidative stress. *Toxicol. Lett.* **2012**, *214*, 263–267. [CrossRef]

MDPI AG
Grosspeteranlage 5
4052 Basel
Switzerland
Tel.: +41 61 683 77 34

Antioxidants Editorial Office
E-mail: antioxidants@mdpi.com
www.mdpi.com/journal/antioxidants

www.ingramcontent.com/pod-product-compliance
Lightning Source LLC
Chambersburg PA
CBHW041343210926
51850CB00013B/2972
9783725828289